MATHEMATICS FOR DATA PROCESSING

MATHEMATICS FOR DATA PROCESSING

ROBERT N. McCULLOUGH
FERRIS STATE UNIVERSITY

wcb

WM. C. BROWN PUBLISHERS
DUBUQUE, IOWA

To Nancy Ann and Jason

Copyright © 1988 by Wm. C. Brown Publishers. All rights reserved

Library of Congress Card Number: 87–21170

ISBN 0–697–06766–1

No part of this publication may be reproduced, stored in a retrieval system, or transmitted, in any form or by any means, electronic, mechanical, photocopying, recording, or otherwise, without the prior written permission of the copyright owner.

Printed in the United States of America
10 9 8 7 6 5 4 3 2

CONTENTS

Preface vii

CHAPTER 1
Basic Algebra 1

1.1 Sets of Numbers 2
1.2 Properties of Real Numbers 6
1.3 Exponents 10
1.4 Order of Operations and Polynomials 15
1.5 Equations and Inequalities 20
1.6 Radicals and Fractional Exponents 25
Chapter Summary 31

CHAPTER 2
Functions and Linear Equations 35

2.1 Relations 36
2.2 Functions 40
2.3 Graphing Linear Equations 44
2.4 Slope of a Line 48
2.5 Forms of Linear Equations 53
Chapter Summary 58

CHAPTER 3
Systems of Linear Equations 61

3.1 Solution by Graphing 62
3.2 Solution by Substitution 66
3.3 Solution by Multiplication–Addition 69
3.4 Dependent and Inconsistent Systems 74
3.5 Applications 78
Chapter Summary 86

CHAPTER 4
Nonlinear Functions 89

4.1 Completing the Square 90
4.2 Quadratic Formula 94
4.3 Graphing Quadratic Functions 98
4.4 Exponential Functions 103
4.5 Logarithmic Functions 107
4.6 Sequences 111
Chapter Summary 116

CHAPTER 5
Number Systems 119

5.1 The Decimal Number System 120
5.2 The Binary Number System 122
5.3 The Octal Number System 124
5.4 The Hexadecimal Number System 126
5.5 Converting from Any Base to Decimal 128
5.6 Converting from Decimal to Any Base 131
5.7 Converting from Binary to Octal or Hexadecimal 137
5.8 Converting from Octal or Hexadecimal to Binary 140
Chapter Summary 142

CHAPTER 6
Computer Arithmetic 145

6.1 Binary Addition 146
6.2 Octal and Hexadecimal Addition 150
6.3 Decimal Subtraction by Complement Addition 154
6.4 Binary Subtraction 158
6.5 Octal and Hexadecimal Subtraction 161
6.6 Multiplication 164
6.7 Division 167
Chapter Summary 170

CHAPTER 7
Computer Considerations 173

7.1 Significant Digits, Accuracy, and Precision 174
7.2 Scientific Notation 176
7.3 Integers and Reals 179
7.4 Truncation, Rounding, and Conversion Error 184
7.5 Order of Operations 188
7.6 Assignment Statements 191
7.7 Coding Nonnumeric Data 193
Chapter Summary 196

CHAPTER 8
Sets   199
- 8.1   Introduction   200
- 8.2   Subsets   203
- 8.3   Operations on Sets   206
- 8.4   Venn Diagrams   209
- 8.5   Basic Properties of Sets   213
        Chapter Summary   217

CHAPTER 9
Logic   219
- 9.1   Simple and Compound Statements   220
- 9.2   Truth Tables: AND, OR, and NOT   223
- 9.3   Other Truth Tables: NAND, NOR, and EOR   227
- 9.4   Conditional and Biconditional Statements   230
- 9.5   Properties of Logic   233
- 9.6   Arguments   237
        Chapter Summary   242

CHAPTER 10
Boolean Algebra   245
- 10.1   Introduction   246
- 10.2   Circuits   249
- 10.3   Combinations of Switches   252
- 10.4   Properties of Networks   258
- 10.5   Simplification of Networks   261
- 10.6   Logic Circuits   264
         Chapter Summary   268

CHAPTER 11
Computer Logic and Structured Programming   273
- 11.1   Algorithms   274
- 11.2   Flowcharts   277
- 11.3   The Decision Structure   280
- 11.4   The Repetition Structure   287
         Chapter Summary   294

CHAPTER 12
Arrays and Matrices   297
- 12.1   Arrays   298
- 12.2   Fundamental Matrix Operations   302
- 12.3   Matrix Multiplication   306
- 12.4   Identity and Inverse Matrices for Multiplication   310
- 12.5   Determinants   316
- 12.6   Systems of Equations: By Matrices   321
- 12.7   Systems of Equations: By Cramer's Rule   326
         Chapter Summary   330

CHAPTER 13
Linear Programming   333
- 13.1   Introduction to Linear Programming   334
- 13.2   Graphing Linear Inequalities   337
- 13.3   Graphical Linear Programming   344
- 13.4   Simplex Method: Standard Case   349
- 13.5   Simplex Method: General Case   355
         Chapter Summary   360

CHAPTER 14
Probability and Statistics   363
- 14.1   Permutations and Combinations   364
- 14.2   Introduction to Probability   369
- 14.3   Summation Notation   375
- 14.4   Measures of Central Tendency   379
- 14.5   Measures of Dispersion   385
         Chapter Summary   390

Answers to Odd-Numbered Problems   393

Index   438

PREFACE

This book is designed for a one- (or possibly two-) semester course in mathematics for those students who are interested in computer programming. Although there will be numerous references to computers throughout the book, it is designed to be a mathematics text.

We will use several different computer languages to illustrate concepts in this book. However, no prior programming knowledge is necessary. The programming knowledge necessary will be explained whenever needed. BASIC will be the primary language used because of its simplicity and wide use.

The first four chapters of the book are designed to provide the students with the necessary algebra background to complete the book. The instructor may selectively choose the topics to be covered depending on the algebraic background of the student and the goals of the course.

The instructor has great flexibility in choosing the order in which to cover the chapters of the book. The following groups of chapters may be covered in any order.

Computer arithmetic	Chapters 5–7
Sets, logic, and Boolean algebra	Chapters 8–10
Structured programming	Chapter 11
Matrices and linear programming	Chapters 12–13
Statistics	Chapter 14

The answers to all odd-numbered problems appear in the back of the book. An instructor's manual is available which includes worked out solutions to all even-numbered problems and sample quizzes and tests.

Acknowledgments

I would like to thank my wife, Nancy Ann, and my son, Jason, without whose support and understanding this book would have been abandoned long ago!

Many reviewers and accuracy checkers have provided numerous suggestions for improvement. I would like to thank all of these people. Their names appear next.

Richard Burgis

Jan Buzydlowski

Kenneth Chapman
Owens Technical College

Dick Clark
Portland Community College

Elton Fors
Northern State College

Lois McBride
Stark Technical College

Roger McCoach
Country College of Morris

Wayne McPherson
Southeast Missouri State University

L. Michael Majeske
Greater Hartford Community College

Janet Matthes
Southwest Wisconsin Vocational
 Technical Institute

Beverly Mugrage
University of Akron

Neil E. Olsen
Northeast Wisconsin Technical Institute

Pamela Strahine
Columbus Technical Institute

Robert N. McCullough

BASIC ALGEBRA

CHAPTER 1

Most of the people responsible for the development of computers have been mathematicians. Therefore, you must understand some mathematics, particularly algebra, before you study how computers work and what they can do. The skills used in solving mathematical problems are similar to those needed to be a successful programmer. In particular, being able to analyze a problem and devise a method of solution is essential in mathematics and data processing.

The purpose of this chapter is to review some of the basic algebraic concepts that you will need to know to understand computers. You should fully understand this chapter, particularly the properties of real numbers, before proceeding in the book.

1.1 Sets of Numbers

set

The concept of a **set** is left undefined in mathematics, but it is intuitively understood to mean a collection of items. In this book, sets will consist of collections of numbers.

Sets will be studied in detail in Chapter 8, but a few comments about notation are needed now. We will designate sets by capital letters (A, B, etc.) and braces $\{\ \}$. If A is the set consisting of 1, 2, and 3, then it will be written as

$$A = \{1, 2, 3\}$$

Most of the sets that we will consider will have an *infinite* number of numbers. To designate this infinite number, we will write the first three or four numbers of the set and then three dots to indicate that the pattern continues indefinitely. For example, if B is to be the set of all even numbers, then we will write

$$B = \{2, 4, 6, \ldots\}$$

Now, we will consider the four basic sets that we will use throughout the book: the natural numbers, the integers, the rational numbers, and the real numbers.

The Set *N* of Natural Numbers

natural numbers

The first set of numbers used by people was the set N of **natural** (counting) **numbers**:

$$N = \{1, 2, 3, 4, \ldots\}$$

There are no problems with this set when we perform the basic operations of

1.1 Sets of Numbers

addition and multiplication. If any two natural numbers are added or multiplied, the result is always another natural number.

However, subtraction can cause problems if a larger number is subtracted from a smaller number. For example, $4-7$ has no answer in N since -3 is not a natural number.

The Set *I* of Integers

integer

The problem $4-7$ is one that is definitely worth solving, for without negative numbers we couldn't consider temperatures below zero, overdrawn checking accounts, the national debt, or other exciting topics. The set I of **integers** consists of N, zero, and the negatives of N:

$$I = \{\ldots, -3, -2, -1, 0, 1, 2, 3, \ldots\}$$

Subtraction in I is not a problem. The difference of two integers is always an integer.

EXAMPLE 1.1 Perform the subtraction.

a. $4-7$

Solution $4-7 = \underline{-3}$

b. $-2-(-3)$

Solution $-2-(-3) = -2+3 = \underline{1}$

However, division causes problems. The quotient of two integers is not always an integer and, therefore, not in the set I. For example, $\frac{7}{2}$ is $3\frac{1}{2}$, which is not an integer.

The Set *Q* of Rational Numbers

We definitely would like to have fractions. The set Q of rational numbers consists of all quotients of two integers, provided that the divisor (i.e., the number in the denominator) isn't zero. Since we cannot list even a representative number of members of Q, we will use set-builder notation.

set-builder notation

In **set-builder notation**, rather than list the members of the set, we will write an arbitrary member of the set, x for example, and then list the properties that x must have to be in the set. We will use the notation $\in$ to indicate that a certain number is an *element of* a set. For example, $1 \in N$ means that 1 is a natural number. In this notation, the set Q of **rational numbers** is written as

rational numbers

$$Q = \{x \mid x = \frac{a}{b}, \text{ where } a, b \in I, b \neq 0\}$$

This notation can be read as "the set of all numbers x such that x equals the quotient of a and b, where a and b are integers, b is not equal to zero."

Since $n/1 = n$ for all numbers n, every integer is rational. For example, $\frac{7}{1} = 7$ and $-\frac{3}{1} = -3$. Again, we have increased our set of numbers without losing any of the previous numbers.

In this set, division no longer causes a problem (except division by zero, which is left undefined). The quotient of two rational numbers is always rational.

EXAMPLE 1.2 Divide $\frac{2}{3}$ by $\frac{7}{9}$.

Solution
$$\frac{\frac{2}{3}}{\frac{7}{9}} = \left(\frac{2}{3}\right) \cdot \left(\frac{9}{7}\right) = \frac{18}{21} = \frac{6}{7}$$

However, there are some important numbers that don't belong to Q. Recall that the square root of a number x is the number $\sqrt{x}$ whose square equals x. Thus, $\sqrt{9} = 3$ since $3^2 = 9$.

Note: Although $(-3)^2$ also equals 9, we will always mean the positive (*principal*) root of x when we write the radical $\sqrt{}$ sign.

Many numbers have square roots that are *not* rational. For example, $\sqrt{2}$ and $\sqrt{3}$ are not rational. That is, $\sqrt{2} = 1.414\ldots$ and $\sqrt{3} = 1.732\ldots$. These decimal expressions never stop and never repeat. The circumference of any circle divided by its diameter is always the same number, π (Greek letter pi). Pi is also not rational. So we again have a problem.

The Set R of Real Numbers

real numbers

We would like to use square roots in this book. The set of **real numbers** R is defined as follows:

$$R = \{x \mid x \text{ can be written in decimal notation}\}$$

Thus, $\sqrt{2} = 1.414\ldots$ and $\pi = 3.14159\ldots$ can be written in decimal notation, although the decimals never stop or repeat. Similarly, any member of Q can be written in decimal notation by simply dividing the denominator into the numerator.

EXAMPLE 1.3 Write $\frac{14}{20}$ in decimal notation.

Solution
$$\frac{14}{20} = \frac{7}{10} = 0.7$$

real number line

Another way of representing R is with a **real number line**, a line with zero in the middle, the positive numbers to the right, and the negative numbers to the left, as illustrated in Figure 1.1. Every real number is represented by a point on the number line, and every point on the number line represents a real number.

1.1 Sets of Numbers

Figure 1.1
Real number line

Note that there is still a problem with R. The square root of a negative number is not real. For example, $\sqrt{-4}$ does not equal 2 or -2 since $2^2 = 4$ and $(-2)^2 = 4 \neq -4$. Although we will not consider such "imaginary" numbers in this book, you should be aware that they exist.

From now on, when you see N, I, Q, and R in this book, it will refer to these sets of numbers.

Inside a computer, a number is usually considered to be an integer if no decimal point is present and real if a decimal point is present. So 5 is stored as an integer, and 5.0 and 5. are stored as real numbers.

EXAMPLE 1.4 Determine to which of the sets N, I, Q, and R the numbers belong.

Solution
a. 0.3 $0.3 = \frac{3}{10}$ Q and R
b. $\sqrt{9}$ $\sqrt{9} = 3$ N, I, Q and R
c. -1.34 $-1.34 = -\frac{134}{100}$ Q and R
d. $\sqrt{7}$ 7 is not a perfect square R
e. $-\frac{15}{1}$ $-\frac{15}{1} = -15$ I, Q and R
f. $\sqrt{-5}$ "Imaginary"   None

Problem Set 1.1

In Problems 1–26, determine to which of the sets N, I, Q, and R the numbers belong.

1. 7
2. 9
3. -1.3
4. -2.7
5. $\sqrt{2}$
6. $\sqrt{8}$
7. -5
8. -10
9. $\frac{23}{37}$
10. $\frac{14}{16}$
11. π
12. $-\pi$
13. $\sqrt{-1}$
14. $\sqrt{-7}$
15. $-\frac{2}{9}$
16. $-\frac{11}{13}$
17. $-\sqrt{7}$
18. $-\sqrt{5}$
19. $\sqrt{4}$
20. $\sqrt{16}$
21. $-\frac{7}{1}$
22. $-\frac{16}{2}$
23. 0
24. A bank balance
25. Your age
26. A temperature

In Problems 27–36, represent the points on a number line.

27. 3
28. -4
29. -1.5
30. $\frac{7}{2}$
31. $\sqrt{3}$
32. $\frac{\pi}{2}$
33. $\frac{9}{4}$
34. -3.8
35. $\frac{\pi}{3}$
36. $\sqrt{4}$

In Problems 37–48, give an example (if possible).

37. A natural number that isn't an integer
38. An integer that isn't a natural number
39. A rational number that isn't a natural number
40. A natural number that isn't rational
41. A natural number that isn't real
42. A real number that isn't a natural number
43. A rational number that isn't an integer
44. An integer that isn't rational
45. An integer that isn't real
46. A real number that isn't an integer
47. A real number that isn't rational
48. A rational number that isn't real

In Problems 49–60, write the numbers in decimal notation to the nearest hundredth.

49. $\frac{14}{35}$
50. $-\frac{17}{4}$
51. $-\frac{3}{7}$
52. $\frac{2}{9}$
53. 3
54. -13
55. $-\sqrt{2}$
56. $-\sqrt{3}$
57. -6
58. -9
59. 0
60. $\pi/2$

■ 1.2 Properties of Real Numbers

We will encounter the basic properties of R often in this book, particularly when we get to sets, logic, Boolean algebra, and matrices, in addition to the chapters on algebra, so they will be summarized here.

Commutative Property   For all real numbers a and b:

$a + b = b + a$

$a \cdot b = b \cdot a$

Order doesn't matter in addition or multiplication of real numbers. Thus, $2 + 3$ and $3 + 2$ are both 5, and $2 \cdot 3$ and $3 \cdot 2$ are both 6. You've probably known this property for a long time, whether you knew it's name or not, and it makes sense.

Subtraction, on the other hand, is *not* commutative. For example,

$5 - 3 = 2$ and $3 - 5 = -2$ so $5 - 3 \neq 3 - 5$

Associative Property   For all real numbers a and b:

$(a + b) + c = a + (b + c)$

$(a \cdot b) \cdot c = a \cdot (b \cdot c)$

When you perform addition or multiplication involving three numbers, it doesn't matter whether the operation is performed on the first two numbers first or the last two numbers first. That is, when you add the items on your shopping list, the sum is the same whether you add from the top down or the bottom up.

1.2 Properties of Real Numbers

EXAMPLE 1.5 Perform the addition or multiplication.

a. $3+5+7$

Solution
$(3+5)+7 = 8+7 = 15$
$3+(5+7) = 3+12 = 15$

b. $3 \cdot 5 \cdot 7$

Solution
$(3 \cdot 5) \cdot 7 = 15 \cdot 7 = 105$
$3 \cdot (5 \cdot 7) = 3 \cdot 35 = 105$

Closure Property For all real numbers a and b:

$a+b$ and $a \cdot b$ are real numbers

In other words, addition and multiplication of real numbers result in real numbers. This property may seem obvious, but our problem with N in Section 1.1 was that subtraction in N was not *closed* ($4-7$ was not in N). Similarly, division in I is not closed ($\frac{7}{2}$ is not an integer).

Identity Property For all real numbers a:

$a+0 = a$
$a \cdot 1 = a$

identity for addition
identity for multiplication

Zero is called the **identity for addition**, and one is the **identity for multiplication**. These numbers leave the number a unchanged when the operation is performed.

Inverse Property For every real number a, there is a number $-a$ and, if $a \neq 0$, there is a number $1/a$ such that

$a+(-a) = 0$
$a \cdot \left(\dfrac{1}{a}\right) = 1$

inverse
reciprocal

The number $-a$ is called the **inverse** of a for addition, and $1/a$ is the *inverse* of a for multiplication, also called the **reciprocal**. These numbers take the number a back to the identity (0 or 1) when the operation is performed. For example,

$7+(-7) = 0$
$7 \cdot \left(\dfrac{1}{7}\right) = 1$

All of these properties are in two parts, one for addition and one for multiplication. The last basic property is the only one that combines the two operations of addition and multiplication into one property.

Distributive Property For all real numbers a, b and c:

$a(b+c) = ab + ac$

This property is the distributive property of multiplication over addition. A few examples should convince you of its validity.

EXAMPLE 1.6 Show the validity of the distributive property.

a. $3(5+7)$

Solution
$3(5+7) = 3 \cdot 12 = 36$
$(3 \cdot 5) + (3 \cdot 7) = 15 + 21 = 36$

b. $-2[3+(-6)]$

Solution
$-2[3+(-6)] = -2 \cdot (-3) = 6$
$(-2 \cdot 3) + (-2 \cdot -6) = -6 + 12 = 6$

The other possible distributive property (addition over multiplication) rarely works. That is,

$a + bc \neq (a+b)(a+c)$

EXAMPLE 1.7 Show that the distributive property of addition over multiplication is *not* valid.

a. $1 + (2 \cdot 3)$

Solution
$1 + (2 \cdot 3) = 1 + 6 = 7$
$(1+2)(1+3) = 3 \cdot 4 = 12$
$7 \neq 12$

b. $-2 + [3 \cdot (-5)]$

Solution
$-2 + [3 \cdot (-5)] = -2 + (-15) = -17$
$(-2+3)(-2+(-5)) = 1 \cdot (-7) = -7$
$-17 \neq -7$

Although there is only one distributive property for real numbers, this "second" property turns out to be very significant when you get to computer logic. You must remember that this "second" property is *not* a property of real numbers.

We have been writing expressions like $a(b+c)$, where the multiplication symbol is assumed between a and $(b+c)$. In most programming languages, the

1.2 Properties of Real Numbers

multiplication sign must be explicitly written in the formula. One language that will allow assumed multiplication is the simulations language DYNAMO. The asterisk (*) is commonly used to represent multiplication in computers. Addition is still represented by the plus sign. We will be referring to the BASIC programming language throughout this book, and these symbols are used in BASIC.

EXAMPLE 1.8 Write these expressions in BASIC notation.

a. $(a+b)(a+c)$ b. $2ab+bc$

Solution $(a+b)*(a+c)$ $(2*a*b)+(b*c)$

Problem Set 1.2

1. List the six basic properties of real numbers.

In Problems 2–18, show that the properties discussed in this section work, and name each property.

2. $6+(-2+3) = [6+(-2)]+3$
3. $6 \cdot (-2 \cdot 3) = (6 \cdot (-2)) \cdot 3$
4. $7+0 = 7$
5. $5 \cdot (\frac{1}{5}) = 1$
6. $2(7+4) = (2 \cdot 7)+(2 \cdot 4)$
7. $-3(6+2) = (-3 \cdot 6)+(-3 \cdot 2)$
8. $5+6 = 6+5$
9. $5 \cdot 1 = 5$
10. $9+(-9) = 0$
11. $4 \cdot 3 = 3 \cdot 4$
12. $7 \cdot 4$ is a real number
13. $ax+ay = a(x+y)$
14. $(a+b)+c = c+(a+b)$
15. $5+7$ is a real number
16. $(x-y)b = b(x-y)$
17. $xy+(-xy) = 0$
18. $(\frac{4}{7}) \cdot (\frac{7}{4}) = 1$

In Problems 19–29, fill in the blank to complete each property of real numbers, and name it.

19. _____ $+0 = 10$
20. $a+c =$ _____ $+a$
21. $7+(8+3) = ($ _____ $)+3$
22. $4 \cdot 5$ is a _____ number
23. $5(4+$ _____ $) = (5 \cdot 4)+(5 \cdot 6)$
24. $2 \cdot$ _____ $= 1$
25. $6 \cdot$ _____ $= 7 \cdot 6$
26. $23 \cdot$ _____ $= 23$
27. $3+2$ is a _____ number
28. $3 \cdot (4 \cdot -2) = (3 \cdot 4) \cdot$ _____
29. $5+$ _____ $= 0$

In Problems 30–39, give an example supporting each answer.

30. Is division commutative? (Does $a/b = b/a$?)
31. Is subtraction associative? [Does $(a-b)-c = a-(b-c)$?]
32. Is division associative?
33. Is subtraction closed in R?
34. Is division closed in R?
35. Is division closed in N?
36. Is there an "identity" for subtraction (a number that can be subtracted from the number a leaving a unchanged)?
37. Is there an "identity" for division?
38. Is there an "inverse" for subtraction (a number that can be subtracted from the number a giving the identity from Problem 36)?

39. Is there an "inverse" for division?

40. Show that the "other" distributive property is not valid for $7 + (2 \cdot 4) \stackrel{?}{=} (7 + 2)(7 + 4)$.

41. Show that the "other" distributive property is not valid for $1 + (-3 \cdot 7) \stackrel{?}{=} (1 - 3)(1 + 7)$.

42. Can you find *any* values of a, b, and c that make $a + (bc) = (a + b)(a + c)$?

In Problems 43–48, write the expressions in BASIC notation.

43. $a(b + c)$
44. $a(bc)$
45. $(ab)c$
46. $3ab + 5c$
47. $a + bc$
48. $ab + ac + abc$

1.3 Exponents

base
power, exponent
exponentiation

Multiplication is repeated addition. What is repeated multiplication, $2 \cdot 2 \cdot 2$, for example? The number being multiplied is called the **base**, and the number of times the number appears is the **power** or the **exponent**. The power is also the number of multiplications plus one. The process is called **exponentiation**. An exponent of 2 is a *square*, and an exponent of three is a *cube*.

Definition 1.1 If b is a real number and n is a natural number, then $b^n = b \cdot b \cdot b \cdot \ldots \cdot b$, where b occurs n times.

EXAMPLE 1.9 Name the base and the exponent.

Solution

a. $2^3 = 2 \cdot 2 \cdot 2 = 8$
   base = 2   exponent = 3

b. $4^2 = 4 \cdot 4 = 16$
   base = 4   exponent = 2

c. $(-3)^4 = -3 \cdot -3 \cdot -3 \cdot -3 = 81$
   base = -3 exponent = 4

d. $0^7 = 0 \cdot 0 \cdot 0 \cdot 0 \cdot 0 \cdot 0 \cdot 0 = 0$
   base = 0   exponent = 7

e. $\left(\frac{1}{2}\right)^3 = \left(\frac{1}{2}\right) \cdot \left(\frac{1}{2}\right) \cdot \left(\frac{1}{2}\right) = \frac{1}{8}$
   base = $\frac{1}{2}$ exponent = 3

f. $-(-3)^3 = -(-27) = 27$
   base = -3 exponent = 3

Exponentiation is one of the basic operations with computers because extremely large numbers and extremely small numbers are often encountered and are easily represented by using exponents.

Rules of Exponentiation For real numbers a and b and natural numbers m and n:

1. $a^m \cdot a^n = a^{m+n}$

2. $\dfrac{a^m}{a^n} = a^{m-n}$

3. $(a^m)^n = a^{mn}$

4. $(ab)^n = a^n b^n$

5. $\left(\dfrac{a}{b}\right)^n = \dfrac{a^n}{b^n}$

1.3 Exponents

These rules are very important for you to know, understand, and be able to use. A few examples should convince you that these rules make sense.

EXAMPLE 1.10 Use the rules of exponentiation.

a. $2^3 \cdot 2^4$

Solution
$$2^3 \cdot 2^4 = (2 \cdot 2 \cdot 2) \cdot (2 \cdot 2 \cdot 2 \cdot 2)$$
$$= 2 \cdot 2 \cdot 2 \cdot 2 \cdot 2 \cdot 2 \cdot 2$$
$$= 2^7 \qquad \text{(Rule 1)}$$

b. $\dfrac{2^5}{2^3}$

Solution
$$\frac{2^5}{2^3} = \frac{2 \cdot 2 \cdot 2 \cdot 2 \cdot 2}{2 \cdot 2 \cdot 2} = 2 \cdot 2 = 2^2 \qquad \text{(Rule 2)}$$

c. $(x^3)^2$

Solution
$$(x^3)^2 = x^3 \cdot x^3 = x^{3+3} = x^6 \qquad \text{(Rule 3)}$$

d. $(2y)^3$

Solution
$$(2y)^3 = (2y) \cdot (2y) \cdot (2y)$$
$$= 2 \cdot 2 \cdot 2 \cdot y \cdot y \cdot y = 2^3 y^3 \qquad \text{(Rule 4)}$$

e. $\left(\dfrac{x}{a}\right)^4$

Solution
$$\left(\frac{x}{a}\right)^4 = \left(\frac{x}{a}\right) \cdot \left(\frac{x}{a}\right) \cdot \left(\frac{x}{a}\right) \cdot \left(\frac{x}{a}\right)$$
$$= \frac{x \cdot x \cdot x \cdot x}{a \cdot a \cdot a \cdot a} = \frac{x^4}{a^4} \qquad \text{(Rule 5)}$$

How should a^0 be defined? Whatever a^0 is to be, the five rules of exponents should still hold. By applying the first rule of exponents, we have

$$a^3 \cdot a^0 = a^{3+0} = a^3$$

So a^0 must be a number that multiplies a^3 and leaves it unchanged. Hence, a^0 *must* be defined to be 1. Similarly, $a^3/a^3 = 1$, and a^3/a^3 also equals a^{3-3}, or a^0. So, again, a^0 must equal 1.

Definition 1.2 $a^0 = 1$ for all real numbers $a \neq 0$.

What about 0^0? If $0^0 = 1$, then $2^0/0^0 = 1/1 = 1$ by Definition 1.2. However, by Rule 5, $2^0/0^0 = (2/0)^0$, and $2/0$ is undefined. So 0^0 is left undefined in mathematics. Note that 0^0 is considered to be 1 in many computer languages (BASIC, for one) and in calculators. You should remember that mathematically this assumption is incorrect.

How should a^{-n} be defined? Again, the five rules of exponents should still hold:

$$a^n \cdot a^{-n} = a^{n+(-n)} = a^0 = 1$$

So a^{-n} must be the inverse for multiplication, $1/a^n$, also called the *reciprocal*, of a^n.

Definition 1.3

$$a^{-n} = \frac{1}{a^n}, \quad a \neq 0$$

Zero and negative exponents must be defined in these ways to preserve the rules of exponents. There is no other way to define them.

EXAMPLE 1.11

Simplify the expressions.

a. 2^{-3} **b.** 7^{-2}

Solution

$2^{-3} = \frac{1}{2^3} = \underline{\frac{1}{8}}$ $7^{-2} = \frac{1}{7^2} = \underline{\frac{1}{49}}$

c. $\frac{3^4}{3^5}$ **d.** $\frac{3^0}{7^0}$

Solution

$\frac{3^4}{3^5} = 3^{4-5} = 3^{-1} = \underline{\frac{1}{3}}$ $\frac{3^0}{7^0} = \left(\frac{3}{7}\right)^0 = \underline{1}$

e. $7^{-2} \cdot 7^2$ **f.** $(-3)^{-2} \cdot 2^{-2}$

Solution

$7^{-2} \cdot 7^2 = 7^0 = \underline{1}$ $(-3)^{-2} \cdot 2^{-2} = (-3 \cdot 2)^{-2} = (-6)^{-2}$

$= \frac{1}{(-6)^2} = \underline{\frac{1}{36}}$

Many times in algebra we will encounter *variables* raised to numerous powers. The rules remain the same when you simplify these problems. Final answers should always be written *without negative exponents*.

EXAMPLE 1.12

Simplify the expressions.

a. $(x^{-2})^{-4}$

Solution

$(x^{-2})^{-4} = x^{(-2) \cdot (-4)} = \underline{x^8}$

1.3 Exponents

b. $\dfrac{x^2 y^{-1}}{x^{-1} y^2}$

Solution
$$\dfrac{x^2 y^{-1}}{x^{-1} y^2} = x^{2-(-1)} y^{-1-2}$$
$$= x^3 y^{-3} = \underline{\dfrac{x^3}{y^3}}$$

c. $(x^2 y^{-1} z^{-3})^3$

Solution
$$(x^2 y^{-1} z^{-3})^3 = (x^2)^3 (y^{-1})^3 (z^{-3})^3$$
$$= x^6 y^{-3} z^{-9} = \underline{\dfrac{x^6}{y^3 z^9}}$$

d. $\dfrac{(xy^{-2} z^0)^2}{(x^{-4} y^2 z^{-3})^{-2}}$

Solution
$$\dfrac{(xy^{-2} z^0)^2}{(x^{-4} y^2 z^{-3})^{-2}} = \dfrac{x^2 y^{-4} z^0}{x^8 y^{-4} z^6}$$
$$= x^{2-8} y^{-4-(-4)} z^{0-6} = x^{-6} y^0 z^{-6}$$
$$= \underline{\dfrac{1}{x^6 z^6}}$$

You should know the five rules of exponents and how to use them, but you should also know what you *can't* do.

■ **EXAMPLE 1.13** Simplify $(2+3)^2$.

Solution
$$(2+3)^2 \stackrel{?}{=} 2^2 + 3^2$$
$$5^2 \stackrel{?}{=} 4 + 9$$
$$25 \ne 13$$

The rules of exponents never mention adding, only multiplying and dividing. Thus, when you are adding or subtracting, shortcuts like Example 1.13 won't work. Example 1.14 shows how to simplify an expression involving a sum.

■ **EXAMPLE 1.14** Simplify $\dfrac{(x+y)^{-1}}{(x+y)^2}$.

Solution
$$\dfrac{(x+y)^{-1}}{(x+y)^2} = (x+y)^{-1-2} = (x+y)^{-3} = \dfrac{1}{(x+y)^3}$$

Note that there is no consensus on the symbol that should be used to represent exponentiation in computers. The standard notation x^n cannot be used because each expression must be written as *one* line, with no number at a different level than another number. Some of the symbols that are used in different versions of BASIC include ↑, **, and ^. So 2^3 can be written as 2↑3, 2**3. or 2^3. We will use the notation 2^3 when writing BASIC expressions in this book. Subtraction in BASIC still uses the minus sign, but division uses the slash (/) so that fractions can be written on one line as exponents are.

EXAMPLE 1.15 Write the expressions in BASIC.

a. x^5 b. $2x^5 - 4x + 7$ c. x^4yz^{-6}

Solution x^5   (2*(x^5)·)−(4*x)+7   (x^4)*y*(z^(−6))

Problem Set 1.3

In Problems 1–10, write the numbers in simplest exponential form.

1. $2 \cdot 2 \cdot 2 \cdot 2 \cdot 2$
2. $3 \cdot 3$
3. $x \cdot x \cdot x \cdot x$
4. $-y \cdot -y \cdot -y \cdot -y$
5. $a \cdot a \cdot a \cdot a \cdot b \cdot b$
6. $(\frac{1}{3}) \cdot (\frac{1}{3}) \cdot (\frac{1}{3}) \cdot x \cdot y \cdot y$
7. $(\frac{1}{4}) \cdot (\frac{1}{4}) \cdot (\frac{1}{4}) \cdot a \cdot a \cdot a \cdot a$
8. $5 \cdot (\frac{1}{2}) \cdot 5 \cdot (\frac{1}{2}) \cdot (\frac{1}{2})$
9. $3 \cdot x \cdot y \cdot y \cdot 3 \cdot x \cdot x \cdot x$
10. $(-\frac{2}{3}) \cdot a \cdot (-\frac{2}{3})$

In Problems 11–24, simplify the expressions.

11. $2 \cdot 7^2$
12. 5^{-3}
13. $\frac{8^3}{8^2}$
14. $(-3)^4$
15. $(-2)^{-5}$
16. $\frac{5^7}{5^9}$
17. $7^2 \cdot 7$
18. $(\frac{4}{3})^{-2}$
19. $\frac{(-2)^2}{(-2)^{-4}}$
20. $(2 \cdot 4)^2$
21. $(\frac{1}{5})^{-4}$
22. $\frac{3^{-4}}{3^2}$
23. $(\frac{2}{5})^4$
24. $(\frac{17}{19})^0$

In Problems 25–44, simplify the expressions.

25. $(2x)^3$
26. $(-3y)^2$
27. $-(2z)^2$
28. $(-2z)^2$
29. $(x^2)^3$
30. $(x^2y)^3$
31. $(2xy)^2$
32. $\frac{-4x^2}{2xy}$
33. $\frac{x^2y^3z}{x^3y^5z^{-3}}$
34. $\left(\frac{4x}{3y}\right)^3$
35. $(xy^{-1}z^3)^4$
36. $(x^3y^2)^{-1}$
37. $(x^4y^{-1}z)^{-3}$
38. $(x^{-1}y^{-2}z^{-4})^0$
39. $(2^{-1}x^2y^{-2})^{-3}$
40. $(3^2x^{-2}y^4z)^{-2}$
41. $(x+y)^{-1}$
42. $(a-b)^{-2}$
43. $\frac{(x^7y^{-2})^3}{(x^9y^2z^3)^2}$
44. $\frac{(x^{-2}y^{-3}z)^2}{(x^3y^{-3}z^0)^3}$

45. Find the first five powers of 8 ($8^1 = ?$, $8^2 = ?$, etc.).

46. Find the first five powers of 16.
47. Find the first ten powers of 2.
48. Compare the answers to Problems 45, 46, and 47. What conclusion can be reached about powers of 2, 8, and 16?
49. If you started with two rabbits on the first day of the month and the number of rabbits doubled every day (four rabbits on the second day, etc.), how many rabbits would there be on the 30th day?
50. Suppose you started with 16 pieces of paper and had to give back one-half of what you had every day. After ten days, how much paper would you have left?

In Problems 51–58, write the expressions in BASIC.

51. x^4
52. x^{-6}
53. $2x^{-3}$
54. $-5x^5$
55. $x^3 - 3x^2 + 2x - 1$
56. $3x^2 - 3x + 8$
57. $x^{-6} y^6 z^2$
58. $a^2 b^{-7} c d^{-3}$

1.4 Order of Operations and Polynomials

constants
variables

When dealing with real numbers, we will consider **constants** (terms that don't change value, like 14, -3.14, and π) and **variables** (terms that can take on different values, like your age and the day of the week). We will use small letters, usually x and y, to represent variables.

expressions

We can then put constants, variables, exponents, and operations together to make **expressions**. For example, $3x^2 y + 2x$ and $14x\sqrt{y} - (2/x)$ are expressions. Evaluation of expressions requires knowledge of the order of operations.

Order of Operations

In evaluating expressions, we must understand the order of operations that is normally followed. Consider the expression $3 + 4 \cdot 2$. If the addition is performed first, the result is $7 \cdot 2$, or $\underline{14}$. If the multiplication is performed first, the result is $3 + 8$, or $\underline{11}$. Both answers can't be right. Here is the order of operations that is followed in mathematics:

1. Parentheses (from the inside out)
2. Exponents
3. Multiplication and division (from left to right)
4. Addition and subtraction (from left to right)

So the correct answer for $3 + 4 \cdot 2$ is $\underline{11}$ since the multiplication is performed first. Since parentheses override all operations, you can force one operation to be performed before another operation by putting the first operation in parentheses. So $(3+4)\cdot 2$ is $\underline{14}$, since the addition must be performed first because of the parentheses.

EXAMPLE 1.16 Evaluate the expressions.

a. $5\cdot 2 + 5\cdot 6$

Solution $5\cdot 2 + 5\cdot 6 = 10 + 30 = \underline{40}$

b. $4[5-(3-7)\cdot 6] + 3$

Solution
$$\begin{aligned}4[5-(3-7)\cdot 6] + 3 &= 4[5-(-4)\cdot 6] + 3 \\ &= 4[5-(-24)] + 3 \\ &= 4(5+24) + 3 = 4\cdot 29 + 3 \\ &= 116 + 3 = \underline{119}\end{aligned}$$

c. $2\cdot 3^3 \cdot 4^2 - 5$

Solution
$$\begin{aligned}2\cdot 3^3 \cdot 4^2 - 5 &= 2\cdot 27 \cdot 16 - 5 \\ &= 864 - 5 = \underline{859}\end{aligned}$$

d. $\dfrac{4\cdot 3 + 2^4}{2 - 4^2 + 5\cdot 2}$

Solution
$$\begin{aligned}\frac{4\cdot 3 + 2^4}{2 - 4^2 + 5\cdot 2} &= \frac{4\cdot 3 + 16}{2 - 16 + 5\cdot 2} \\ &= \frac{12 + 16}{2 - 16 + 10} \\ &= \frac{28}{-4} = \underline{-7}\end{aligned}$$

In Section 7.5, we will again consider order of operations as it applies to computers. Not surprisingly, the order is the same. Remember that mathematicians were very influential in developing computers, so it is natural that the order should be the same.

Polynomials

We are now ready to introduce the concept of a polynomial.

polynomial

Definition 1.4 A **polynomial** is an expression involving constants, variables with natural number exponents, the four basic operations of $+$, $-$, $\cdot$, and $/$, and no variables in denominators.

1.4 Order of Operations and Polynomials

This definition can be used to determine what is a polynomial and what isn't a polynomial.

EXAMPLE 1.17 Determine whether the expressions are polynomials.

Solution
a. $3x^2y + 2x$ Yes
b. $14x\sqrt{y}$ No (no $\sqrt{\ }$ allowed)
c. $\dfrac{-2}{x}$ No (x can't be in the denominator)
d. $5x^3 - 4x^2 + 3x - 4$ Yes
e. $17x^2y^3z - 5xyz^2$ Yes

Understanding that variables are terms that can take on different values allows us to find the values of polynomials for specific values of the variables.

EXAMPLE 1.18 Find the value of the polynomial for the given conditions.

a. $3x^2y + 2x$ when $x = 2$ and $y = -2$

Solution
$3(2)^2(-2) + 2(2)$
$3 \cdot 4 \cdot -2 \ \ +4$
$-24 \ \ \ \ \ +4$
-20

b. $17xy + 2x^2 - 3y^2$ when $x = 1$ and $y = 0$

Solution
$17 \cdot 1 \cdot 0 + 2 \cdot 1^2 - 3 \cdot 0$
$0 \ \ \ \ \ \ \ +2 \cdot 1 \ -3 \cdot 0$
$0 \ \ \ \ \ \ \ +2 \ \ \ \ -0$
2

c. $6abc + 12a^3c^2$ when $a = -1$, $b = -2$, and $c = 4$

Solution
$6 \cdot -1 \cdot -2 \cdot 4 + 12 \cdot (-1)^3 \cdot 4^2$
$-6 \cdot -8 + 12 \cdot -1 \cdot 16$
$48 + -192$
-144

coefficient

A polynomial consists of sums and/or differences of terms consisting of variables and a constant. This constant is called the **coefficient** of that term. So the polynomial $17xy + 2x^2 - 3y^3$ has coefficients of 17, 2, and -3. The polynomial

BASIC ALGEBRA

$2xy - x^2y + xy^3$ has coefficients of 2, -1, and 1. If the coefficient of a term is 1, it is usually not written.

Addition and subtraction of polynomials follow two rules.

Rules for Addition and Subtraction of Polynomials

similar/like terms

1. Two terms cannot be added (or subtracted) unless they consist of the *same* variables to the *same* exponents. These terms are called **similar** (or **like**) **terms**. In other words, they are identical except (possibly) for the coefficients.
2. To add (or subtract) two similar terms, add (or subtract) the coefficients, leaving the variables unchanged.

EXAMPLE 1.19 Determine whether the pairs of terms are similar.

Solution
a. $2x^2y$ and $3x^2y$ Yes
b. $4x^2y^2$ and $7x^2y^3$ No (y is raised to different powers)
c. $3xyz^4$ and $3x^4yz$ No (x and z are raised to different powers)
d. $17x^3y^4z^{-5}$ and $-8x^3y^4z^{-5}$ Yes

EXAMPLE 1.20 Perform the operations.

a. $2xy + 3xy$

Solution $2xy + 3xy = \underline{5xy}$

b. $17x^3y^4z^{-5} + (-8x^3y^4z^{-5})$

Solution $17x^3y^4z^{-5} + (-8x^3y^4z^{-5}) = \underline{9x^3y^4z^{-5}}$

c. $2x^2y - 7x^2y$

Solution $2x^2y - 7x^2y = \underline{-5x^2y}$

d. $2x^3 + 5x^2$

Solution $\underline{2x^3 + 5x^2 = 2x^3 + 5x^2}$ (not similar terms)

Note: Although the commutative property of multiplication allows the coefficient and variables to be written in any order, we will always write the coefficient first, then the variables in alphabetical order.

EXAMPLE 1.21 Perform the operations and simplify.

a. $(2x^2 + 3x + 1) + (7x^2 + 4x + 3)$

Solution $\underline{9x^2 + 7x + 4}$

1.4 Order of Operations and Polynomials

b. $(3x^2y - 7xy + 4xy^2) - (2xy + 3xy^2 - 7x^2y)$

Solution $3x^2y - 7xy + 4xy^2 - 2xy - 3xy^2 + 7x^2y$

Note that a minus sign in front of a polynomial changes the sign of *every* term in the polynomial.

$\underline{10x^2y + xy^2 - 9xy}$

c. $(2x^2 + 3xy + y^2) - (2x^3 - 4xy + y^2) + (4x^3 + xy - 5x^2)$

Solution $2x^2 + 3xy + y^2 - 2x^3 + 4xy - y^2 + 4x^3 + xy - 5x^2$
$\underline{2x^3 - 3x^2 + 8xy}$

To multiply polynomials, we use the distributive property and the rules of exponents.

EXAMPLE 1.22 Multiply.
a. $2x(x^2 + 3x)$

Solution
$$2x(x^2 + 3x) = 2x \cdot x^2 + 2x \cdot 3x$$
$$= \underline{2x^3 + 6x^2}$$

b. $(2x + 3y)(x + 2y)$

Solution
$$(2x + 3y)(x + 2y) = (2x + 3y)x + (2x + 3y) \cdot 2y$$
$$= 2x^2 + 3xy + 4xy + 6y^2$$
$$= \underline{2x^2 + 7xy + 6y^2}$$

Perhaps an easier way to remember how to multiply polynomials is to multiply every term of the first polynomial by every term of the second polynomial.

EXAMPLE 1.23 Multiply.
a. $(x - y)(2x + y)$

Solution
$$(x - y)(2x + y) = x \cdot 2x + xy - y \cdot 2x - y \cdot y$$
$$= 2x^2 + xy - 2xy - y^2$$
$$= \underline{2x^2 - xy - y^2}$$

b. $(3x^2 - 2x + 4)(x^2 + 3x - 6)$

Solution
$$(3x^2 - 2x + 4)(x^2 + 3x - 6) = 3x^4 + 9x^3 - 18x^2 - 2x^3 - 6x^2 + 12x + 4x^2 + 12x - 24$$
$$= \underline{3x^4 + 7x^3 - 20x^2 + 24x - 24}$$

Problem Set 1.4

In Problems 1–8, determine whether the expressions are polynomials.

1. $7xy^3 - 5xy$
2. $12x^4yz + 6x^3y + 7y^4$
3. $3xy - 7x\sqrt{y}$
4. $12\sqrt{xy} + 5z^4$
5. $4x^3 + \dfrac{12x}{x^2}$
6. $-3xyz^2 + 17x - \dfrac{4x^2}{x^3}$
7. $2x^2 + 4x - 3$
8. $3y^2 - 7y + 6$

In Problems 9–14, state the coefficients of the terms.

9. $6x^2$
10. $5y^7$
11. $-6x^2yz$
12. $-17x^5y^6$
13. $-4.5xy^4$
14. $22.1y^5z^8$

In Problems 15–20, find the value of the polynomial for the given conditions.

15. $2x + y$, when $x = 2$ and $y = 1$
16. $3x + 2y$, when $x = 1$ and $y = 3$
17. $2xy^2 - 4x$, when $x = 1$ and $y = 0$
18. $3x^2y + 3y + 2$, when $x = 2$ and $y = -1$
19. $3a^2 + 2b^2 - 4c$, when $a = 3$, $b = -2$, and $c = 2$
20. $b^2 - 4ac$, when $a = 2$, $b = 3$, and $c = -1$

In Problems 21–38, perform the operations, and simplify.

21. $(2x^2 + 3x + 1) + (7x^2 - 2x + 5)$
22. $(4y^2 - 2y + 9) + (-3y^2 + 6y - 4)$
23. $(3x^2y + 6xy^2 - 7x) + (5xy - 3xy^2 + 5y)$
24. $(4ab^2 + 2ac^3 - 6ab^2c^2) + (7ac^3 - 2b^2 + 4ab^2c^2)$
25. $(7xy^2 + 8x^2y + 9y^2) - (8xy^2 + 4xy - 3y^2)$
26. $(2x^2 - 7y + 5) - (16x^2y + x^2 + 6x - 4)$
27. $(2y^3 + 3y^2 - 7y + 6) + (3y^3 + y^2 + 5y - 4) - (2y^3 - 4y^2 + 6)$
28. $(7x^3 + 2x^2 + 5x - 7) - (2x^3 + 5x^2 - 7x + 4) + (x^3 - x^2 + 3x - 6)$
29. $2x(x^3 - 3x^2 + 4x + 5)$
30. $5y(3y^3 + 2y^2 - 3y + 10)$
31. $(3x + 4y)(2x - 5y)$
32. $(x^2 + 3y)(2x - y^2)$
33. $(2x^2 + 4x - 3)(5x + 6)$
34. $(2x - 3)(4x^2 + 6x - 9)$
35. $(x^2 - 3x + 4)(2x^2 + 7x - 5)$
36. $(3y^2 + 2y - 9)(4y^2 - 8y + 1)$
37. $(4x^2y^3 - 2x^3z - 6y^3z^2)(2x^3y - 7z^6 + 2y^2z^4)$
38. $(3x^3y^2 - 5yz^3 + 4x^2yz)(7xyz - 2x^5y^2 + 8y^4)$
39. By our definition, is 7 a polynomial?
40. By our definition, is 0 a polynomial?

In Problems 41–50, use the order of operations to evaluate the expressions.

41. $2 + 5 \cdot 4$
42. $3 \cdot 5 - 5$
43. $3^2 - 1 \cdot 5$
44. $4 - 2^3 \cdot 4$
45. $7[4 + 3(4 + 2)] - 2$
46. $4 - 3[4 - (4 + 1) \cdot 3]$
47. $2 \cdot 1^3 + (2^3)^2$
48. $-(3^2)^2 + 4 \cdot 3 - 2$
49. $\dfrac{3 + (5 - 7^2 \cdot 2)}{2 \cdot 4 - 3^2 + 4 \cdot 2}$
50. $\dfrac{3 + 2(2 - 4^2 \cdot 5)}{(2 - 4) \cdot 5^2 + 8}$

In Problems 51–54, consider the expression $2 \cdot 3^2 + 5 - 4/2$.

51. Which operation will be performed first?
52. Rewrite the expression so that the subtraction will be performed first.
53. Rewrite the expression so that the multiplication will be performed first.
54. Rewrite the expression so that the exponentiation will be performed last.

1.5 Equations and Inequalities

equation

An **equation** is simply an expression involving an equal sign. A thorough understanding of how to work with equations is essential in working with

1.5 Equations and Inequalities

computers. The equal sign is used often in computer programming. Sometimes, the equal sign represents true mathematical equality, as we will discuss in this section. But on other occasions, it is used in assignment statements (Chapter 7), which are fundamentally different from equations.

The important thing to remember about equations is that both sides represent the same value, so to maintain equality, *we must do the same thing to both sides*.

Rules for Equality For all real numbers a, b, and c, if $a = b$, then the following rules are valid.

1. $a + c = b + c$
2. $a - c = b - c$
3. $a \cdot c = b \cdot c$
4. $\dfrac{a}{c} = \dfrac{b}{c}$ whenever $c \neq 0$

In other words, we can perform the four basic operations and preserve equality as long as we do the same thing to both sides.

We can use these rules to manipulate equations and solve for a particular variable. *To solve* for a variable means to get that variable all by itself on one side of the equation.

EXAMPLE 1.24 Solve for x.

a. $x + 5 = 9$

Solution We would like to change $x + 5$ to x, so we will subtract 5 from *both* sides.

$$x + 5 - 5 = 9 - 5$$
$$\underline{x = 4}$$

b. $2x - 3 = 7$

Solution
$$2x - 3 + 3 = 7 + 3$$
$$2x = 10$$
$$\frac{2x}{2} = \frac{10}{2}$$
$$\underline{x = 5}$$

c. $3x + 2 = 5x + 8$

Solution To solve for x, we must first combine the two x terms into one term either by subtracting $3x$ from both sides or by subtracting $5x$ from both sides.

$$3x + 2 - 3x = 5x + 8 - 3x$$
$$2 = 2x + 8$$
$$2 - 8 = 2x + 8 - 8$$
$$-6 = 2x$$
$$\frac{-6}{2} = \frac{2x}{2}$$
$$\underline{-3 = x}$$

d. $2x - 3y = 4z + 7$

Solution Don't let the other variables bother you. To solve for x, we simply get x all by itself on one side of the equation.

$$2x - 3y + 3y = 4z + 7 + 3y$$
$$2x = 3y + 4z + 7$$
$$\frac{2x}{2} = \frac{3y + 4z + 7}{2}$$
$$x = \underline{\frac{3y + 4z + 7}{2}}$$

When dealing with assignment statements in computers, we are only allowed to have operations on the right side of the equal sign. The left side must be a single variable. So solving equations for specified variables is important for setting up assignment statements. This topic will be covered in detail in Chapter 7.

inequality   Just as an equation relates two expressions which are equal, an **inequality** relates two expressions which are *not* equal. The symbols used in inequalities are as follows:

> Greater than   ≥ Greater than or equal to
< Less than   ≤ Less than or equal to

Maintaining the same inequality is a little trickier than maintaining equality. However, the only major problem comes from multiplying or dividing by a negative number.

■ **EXAMPLE 1.25**   What is the result of multiplying both sides of the inequality $2 < 5$ by -1?

Solution
$$2 < 5$$
$$2 \cdot -1 \; ? \; 5 \cdot -1$$
$$\underline{-2 > -5}$$

The inequality changes direction when both sides are multiplied by a

1.5 Equations and Inequalities

negative number, as Example 1.25 shows. The same thing happens when you divide by a negative number.

EXAMPLE 1.26 What is the result of dividing both sides of the inequality $3 > -2$ by -1?

Solution
$$3 > -2$$
$$\frac{3}{-1} \; ? \; \frac{-2}{-1}$$
$$-3 < 2$$

Rules for Inequality For all real numbers a, b, and c, if $a < b$, then the following rules are valid.
1. $a + c < b + c$
2. $a - c < b - c$
3. $a \cdot c < b \cdot c$, whenever $c > 0$
4. $a \cdot c > b \cdot c$, whenever $c < 0$
5. $\frac{a}{c} < \frac{b}{c}$, whenever $c > 0$
6. $\frac{a}{c} > \frac{b}{c}$, whenever $c < 0$

Knowing these rules should allow you to solve inequalities for a variable just as easily as you solve equations.

EXAMPLE 1.27 Solve for x.

a. $x + 5 > 9$

Solution
$$x + 5 - 5 > 9 - 5$$
$$x > 4$$

b. $2x - 3 \geq 7$

Solution
$$2x - 3 + 3 \geq 7 + 3$$
$$2x \geq 10$$
$$\frac{2x}{2} \geq \frac{10}{2}$$
$$x \geq 5$$

c. $3x + 2 < 5x + 8$

Solution
$$3x + 2 - 5x < 5x + 8 - 5x$$
$$-2x + 2 < 8$$

$$-2x+2-2<8-2$$
$$-2x<6$$
$$\frac{-2x}{-2} > \frac{6}{-2} \quad \text{(notice that } < \text{ became } >)$$
$$x > -3$$

Most computer keyboards do not include $\leq$ and $\geq$ symbols. For this reason, many computer languages use $<=$ and $>=$ to represent these symbols.

One of the main differences between calculators and computers is that computers can make **logical decisions**, decisions that are either true or false. The next example illustrates some logical decisions.

logical decisions

■ **EXAMPLE 1.28** Determine whether these logical decisions from BASIC are true or false when $A = 1$, $B = -2$, and $C = 0$.

a. $A > = B$ **b.** $-A < C$

Solution
$1 > + -2$ $-1 < 0$
true   true

c. $(2*A) > -B$ **d.** $(A-C) = (1-B)$

Solution
$(2*1) > -(-2)$ $(1-0) = [1-(-2)]$
$2 > 2$ $1 = 1+2$
false $1 = 3$
     false

e. $(2*A) + B <= 0$

Solution
$(2*1) + (-2) <= 0$
$2 + (-2) <= 0$
$0 <= 0$
true

Problem Set 1.5

In Problems 1–40, solve for x.

1. $x + 4 = 7$
2. $x + 8 = 13$
3. $x - 7 = 6$
4. $x - 5 = 9$
5. $7x = 9$
6. $-2x = 7$
7. $\dfrac{x}{3} = -2$
8. $\dfrac{x}{4} = 8$
9. $3x + 1 = 2$
10. $7x - 3 = 5$
11. $4 - 2x = 9$
12. $5 - 3x = 11$
13. $2x + 3 = 5 - 3x$
14. $7x + 1 = 6x - 4$
15. $4x - 7 = 9x + 3$
16. $x + 3 = 7 - 6x$
17. $3x - 2 + 4x = 5x - 7$
18. $-x + 4 + 2x = 3 - 5x$

19. $4x - y + 2z =$
    $8x + y - 4$
20. $y = 2x - 4z + 7$
21. $x + 5 < 8$
22. $x + 11 < 13$
23. $x - 4 > 7$
24. $x - 1 \geq 10$
25. $3x \leq 7$
26. $-2x < 8$
27. $\dfrac{x}{-4} \geq 7$
28. $\dfrac{x}{4} > 13$
29. $5x - 7 < 9$
30. $4x + 3 \leq 11$
31. $3 - 9x > 6$
32. $7 - 5x > 17$
33. $2x - 7 \leq 4x + 9$
34. $3x + 6 < 8x - 4$
35. $x - 3 > 4x + 7$
36. $4x - 4 \geq 5x - 3$
37. $-2x - 7 < 3x - 7$
38. $7 + 4x > -5x - 3$
39. $3x - 6 + 2x \geq 4x + 5$
40. $-7x + 4 - 2x \leq 2 - 6x$

In Problems 41–46, solve for the specified variable.

41. $P = 2l + 2w$; solve for w.
42. $A = \pi r^2$; solve for r.
43. $F = 1.8C + 32$; solve for C.
44. $V = lwh$; solve for h.
45. $V = (4\pi r^3)/3$; solve for r.
46. $K = 1.6m$; solve for m.
47. What happens to an inequality if both sides are multiplied by 0?
48. What happens to an inequality if both sides are multiplied by the variable x? (Be careful!)
49. What happens to an equation if both sides are multiplied by the variable x?
50. What happens to an inequality if both sides are multiplied by x^2, where $x \neq 0$?

In Problems 51–60, determine whether the logical decisions from BASIC are true or false when A = −3, B = 2, and C = −1.

51. A < B
52. A = C
53. A = (C − 2)
54. B > = C
55. −A < B
56. −A < = C
57. (A − 1) < (A − 2)
58. (2*A) > = (3*B)
59. (C − A) < = B
60. (A*C) > (A*B)

1.6 Radicals and Fractional Exponents

square root

The **square root** $\sqrt{x}$ of a positive number x, as mentioned earlier, is the positive number whose square equals x. Thus, $\sqrt{4} = 2$ since $2^2 = 4$, and $\sqrt{9} = 3$ since $3^2 = 9$. Similarly, the **cube root** $\sqrt[3]{x}$ of x is the number whose cube equals x. Thus, $(\sqrt[3]{x})^3 = \sqrt[3]{x} \cdot \sqrt[3]{x} \cdot \sqrt[3]{x} = x$, and $\sqrt[3]{8} = 2$ since $2^3 = 8$. More generally, we have the following definition.

cube root

nth root

Definition 1.5 The **nth root** of a, denoted $\sqrt[n]{a}$, is the number b such that

$$b^n = \left(\sqrt[n]{a}\right)^n = a.$$

radical, radical sign
radicand, index

Any expression involving a $\sqrt{}$ sign is called a **radical**. The $\sqrt{}$ sign is called the **radical sign**; a is the **radicand**; and n is the **index**.

EXAMPLE 1.29 Simplify the numbers.

a. $\sqrt{36}$
b. $\sqrt[3]{27}$

Solution $\sqrt{36} = \underline{6}$ $(6^2 = 36)$ $\sqrt[3]{27} = \underline{3}$ $(3^3 = 27)$

c. $\sqrt[5]{32}$

Solution $\sqrt[5]{32} = \underline{2}$ $(2^5 = 2 \cdot 2 \cdot 2 \cdot 2 \cdot 2 = 32)$

d. $\sqrt[3]{-64}$

Solution $\sqrt[3]{-64} = \underline{-4}$ $(-4^3 = -64)$

One method for simplifying radical expressions is to *factor* the number inside the radical. To factor means to write the number as a product of numbers. Usually, we want the numbers in the product to be such that no further factoring is possible. If the same number appears n times inside an nth root, then the number can be taken outside the radical sign once. A radical expression is considered *simplified* when no factor in the radicand appears as many or more times than the index. The next example illustrates this procedure.

EXAMPLE 1.30 Simplify the numbers.

a. $\sqrt{50}$

Solution
$$\sqrt{50} = \sqrt{2 \cdot 5 \cdot 5} \quad \text{(no further factoring possible)}$$
$$= \underline{5\sqrt{2}}$$

b. $\sqrt{300}$

Solution
$$\sqrt{300} = \sqrt{3 \cdot 100}$$
$$= \sqrt{3 \cdot 2 \cdot 50}$$
$$= \sqrt{3 \cdot 2 \cdot 5 \cdot 10}$$
$$= \sqrt{3 \cdot 2 \cdot 5 \cdot 2 \cdot 5} \quad \text{(no further factoring possible)}$$
$$= 2 \cdot 5\sqrt{3}$$
$$= \underline{10\sqrt{3}}$$

c. $\sqrt[3]{125}$

Solution
$$\sqrt[3]{125} = \sqrt[3]{5 \cdot 5 \cdot 5}$$
$$= 5 \cdot \sqrt[3]{1}$$
$$= \underline{5}$$

In cases like Example 1.30c, if *every* factor can be written outside the radical sign, then you are left with 1 inside the radical sign.

By definition, $\sqrt{2} \cdot \sqrt{2} = 2$. Now, consider how $2^{1/2}$ should be defined. By Rule 1 for exponents,

$$2^{1/2} \cdot 2^{1/2} = 2^1 = 2$$

1.6 Radicals and Fractional Exponents

So $(2^{1/2})^2 = 2$. Since the square of $2^{1/2}$ is 2, $2^{1/2}$ *must* be defined to be $\sqrt{2}$. Similarly, $2^{1/3}$ must be $\sqrt[3]{2}$. More generally, we have the following definition.

Definition 1.6 For all real numbers a, $a^{1/n} = \sqrt[n]{a}$, if such a real number exists. [$(-4)^{1/2}$, for example, represents no real number.]

EXAMPLE 1.31 Write as a radical expression.
a. $4^{1/5}$ b. $3^{1/2}$

Solution
$4^{1/5} = \underline{\sqrt[5]{4}}$ $3^{1/2} = \underline{\sqrt{3}}$

c. $x^{1/7}$ d. $a^{-1/4}$

Solution
$x^{1/7} = \underline{\sqrt[7]{x}}$ $a^{-1/4} = \dfrac{1}{a^{1/4}} = \dfrac{1}{\sqrt[4]{a}}$

EXAMPLE 1.32 Write as a fractional exponent.
a. $\sqrt[3]{5}$ b. $\sqrt[4]{2}$

Solution
$\sqrt[3]{5} = \underline{5^{1/3}}$ $\sqrt[4]{2} = \underline{2^{1/4}}$

c. $\sqrt[5]{x}$ d. $\dfrac{1}{\sqrt{3}}$

Solution
$\sqrt[5]{x} = \underline{x^{1/5}}$ $\dfrac{1}{\sqrt{3}} = \dfrac{1}{\underline{3^{1/2}}}$ or $\underline{3^{-1/2}}$

Rules for Radical Expressions

1. $\sqrt[n]{a} \cdot \sqrt[n]{b} = \sqrt[n]{ab}$
2. $\dfrac{\sqrt[n]{a}}{\sqrt[n]{b}} = \sqrt[n]{\dfrac{a}{b}}, b \neq 0$
3. $\left(\sqrt[n]{a}\right)^n = a, a \geq 0$
4. $\left(\sqrt[n]{a}\right)^n = a, a < 0$ and n an odd integer

These rules follow directly from Definition 1.6. For example,
$$\sqrt[n]{a} \cdot \sqrt[n]{b} = a^{1/n} \cdot b^{1/n} = (ab)^{1/n} = \sqrt[n]{ab}$$

These rules allow us to simplify radicals in performing multiplication and division of radicals that are difficult, if not impossible, to simplify separately.

EXAMPLE 1.33 Use the rules for radical expressions to simplify.
a. $\sqrt{2} \cdot \sqrt{8}$

Solution
$\sqrt{2} \cdot \sqrt{8} = \sqrt{2 \cdot 8} = \sqrt{16} = \underline{4}$

b. $\dfrac{\sqrt{105}}{\sqrt{35}}$

Solution $\dfrac{\sqrt{105}}{\sqrt{35}} = \sqrt{\dfrac{105}{35}} = \sqrt{3}$

How should $a^{m/n}$ be defined? Again, we must preserve the rules of exponents:

$$a^{m/n} = a^{(1/n) \cdot m} = (a^{1/n})^m = \left(\sqrt[n]{a}\right)^m$$

Definition 1.7

$$a^{m/n} = \left(\sqrt[n]{a}\right)^m = \sqrt[n]{a^m}$$

EXAMPLE 1.34 Write as a radical expression.

a. $2^{3/5}$

Solution $2^{3/5} = \sqrt[5]{2^3}$ or $\left(\sqrt[5]{2}\right)^3$

b. $4^{2/7}$

Solution $4^{2/7} = \sqrt[7]{4^2}$ or $\left(\sqrt[7]{4}\right)^2$

c. $3^{-3/8}$

Solution $3^{-3/8} = \dfrac{1}{3^{3/8}} = \dfrac{1}{\sqrt[8]{3^3}}$ or $\dfrac{1}{(\sqrt[8]{3})^3}$

d. $(-7)^{6/5}$

Solution $(-7)^{6/5} = \sqrt[5]{(-7)^6}$ or $\left(\sqrt[5]{-7}\right)^6$

EXAMPLE 1.35 Write as a fractional exponent.

a. $\sqrt[3]{2^4}$

Solution $\sqrt[3]{2^4} = 2^{4/3}$

b. $\sqrt[10]{(5.1)^3}$

Solution $\sqrt[10]{(5.1)^3} = (5.1)^{3/10}$

c. $\sqrt[9]{2^{-3}}$

Solution $\sqrt[9]{2^{-3}} = 2^{-3/9} = 2^{-1/3} = \left(\dfrac{1}{2}\right)^{1/3}$

d. $\sqrt[5]{(-3)^{10}}$

Solution $\sqrt[5]{(-3)^{10}} = (-3)^{10/5} = (-3)^2$

1.6 Radicals and Fractional Exponents

Remember that in these conversions, the index is the root and goes in the denominator of the exponent.

When we simplify fractional exponents, using the formula $(\sqrt[n]{a})^m$ is usually easier than using $\sqrt[n]{a^m}$. This method will always result in the smallest number inside the radical.

EXAMPLE 1.36 Simplify the expressions.

a. $8^{5/3}$

Solution $8^{5/3} = (\sqrt[3]{8})^5 = 2^5 = \underline{32}$

b. $36^{3/2}$

Solution $36^{3/2} = (\sqrt{36})^3 = 6^3 = \underline{216}$

c. $(\frac{1}{27})^{4/3}$

Solution $(\frac{1}{27})^{4/3} = (\sqrt[3]{\frac{1}{27}})^4 = (\frac{1}{3})^4 = \underline{\frac{1}{81}}$

d. $16^{-3/2}$

Solution $16^{-3/2} = \dfrac{1}{16^{3/2}} = \dfrac{1}{(\sqrt{16})^3} = \dfrac{1}{4^3} = \underline{\dfrac{1}{64}}$

Although $\sqrt{a}$ is $a^{1/2}$ and can be programmed as A^0.5 in BASIC, many computer languages have built-in expressions to find square roots. For example, in BASIC, the expression SQR(X) returns $\sqrt{X}$. To evaluate expressions containing other fractional exponents, we use the normal exponential notation.

EXAMPLE 1.37 Write the expressions in BASIC.

a. $\sqrt{xy}$

Solution $\underline{\text{SQR(X*Y)}}$

b. $6^{2/11}$

Solution $\underline{6\char`\^(2/11)}$

c. $\sqrt[3]{xy^2z^3}$

Solution $\underline{(X*(Y\char`\^2)*(Z\char`\^3))\char`\^(1/3)}$

d. $(-2y)^{-3/4}$

Solution $\underline{(-2*Y)\char`\^(-3/4)}$

Problem Set 1.6

In Problems 1–12, write as a radical expression.

1. $5^{1/3}$
2. $(1.3)^{1/4}$
3. $(2.1)^{-1/7}$
4. $7^{-1/9}$
5. $5^{3/8}$
6. $6^{2/11}$
7. $x^{4/3}$
8. $y^{7/8}$
9. $(2x)^{1/2}$
10. $(-2y)^{-3/4}$
11. $x^{-3/7}$
12. $y^{-11/13}$

In Problems 13–24, write by using a fractional exponent.

13. $\sqrt{7}$
14. $\sqrt{11}$
15. $\sqrt[3]{2^4}$
16. $\sqrt[7]{3^2}$
17. $\sqrt[10]{x^2}$
18. $\sqrt[8]{y^4}$
19. $\sqrt[6]{x^{-5}}$
20. $\sqrt[9]{y^{-4}}$
21. $\dfrac{1}{\sqrt{b}}$
22. $\dfrac{1}{\sqrt{19}}$
23. $\dfrac{1}{\sqrt[4]{x^3}}$
24. $\dfrac{1}{\sqrt[5]{y^2}}$

In Problems 25–50, simplify as much as possible.

25. $\sqrt{25}$
26. $\sqrt{49}$
27. $\sqrt[3]{64}$
28. $\sqrt[4]{81}$
29. $\sqrt{52}$
30. $\sqrt{60}$
31. $\sqrt{98}$
32. $\sqrt{75}$
33. $\sqrt{160}$
34. $\sqrt[4]{32}$
35. $\sqrt[3]{135}$
36. $\sqrt[4]{64^{-1}}$
37. $\sqrt[3]{\tfrac{1}{8}}$
38. $64^{1/2}$
39. $100^{1/2}$
40. $27^{2/3}$
41. $81^{3/4}$
42. $64^{5/6}$
43. $125^{2/3}$
44. $(-27)^{4/3}$
45. $(-32)^{4/5}$
46. $16^{-3/4}$
47. $25^{-3/2}$
48. $\left(\tfrac{4}{9}\right)^{3/2}$
49. $\left(\dfrac{125}{216}\right)^{-2/3}$
50. $\left(\dfrac{81}{256}\right)^{-3/4}$

In Problems 51–66, use the rules for radical expressions to simplify.

51. $\sqrt{10} \cdot \sqrt{5}$
52. $\sqrt{34} \cdot \sqrt{17}$
53. $\sqrt{30} \cdot \sqrt{6}$
54. $\sqrt{60} \cdot \sqrt{75}$
55. $\sqrt[3]{4} \cdot \sqrt[3]{2}$
56. $\sqrt[3]{25} \cdot \sqrt[3]{25}$
57. $\dfrac{\sqrt{32}}{\sqrt{2}}$
58. $\dfrac{\sqrt[3]{32}}{\sqrt[3]{4}}$
59. $\dfrac{\sqrt{14}}{\sqrt{2}}$
60. $\dfrac{\sqrt[4]{18}}{\sqrt[4]{9}}$
61. $\dfrac{\sqrt{30} \cdot \sqrt{2}}{\sqrt{6}}$
62. $\dfrac{\sqrt{42} \cdot \sqrt{3}}{\sqrt{14}}$
63. $\dfrac{\sqrt[3]{15} \cdot \sqrt[3]{25}}{\sqrt[3]{3}}$
64. $\dfrac{\sqrt[4]{24} \cdot \sqrt[4]{2}}{\sqrt[4]{3}}$
65. $\sqrt[3]{x^3}\ (x \geq 0)$
66. $\sqrt[4]{2^4}$

67. Use the rules of exponents to show that $\sqrt[n]{a}/\sqrt[n]{b} = \sqrt[n]{a/b}$.
68. Use the rules of exponents to show that $(\sqrt[n]{a})^n = a, a \geq 0$.
69. Find values of a and n with $a < 0$ such that $(\sqrt[n]{a})^n = a$.
70. Find values of a and n with $a < 0$ such that $(\sqrt[n]{a})^n \neq a$.

In Problems 71–80, write the expressions in BASIC.

71. $\sqrt{7}$
72. $\sqrt{2A}$
73. $\sqrt{xy^2}$
74. $\sqrt{1+4x}$
75. $3^{4/7}$
76. $5^{-3/8}$
77. $(5x)^{-2/3}$
78. $(1.3)^{1/4}$
79. $\sqrt[4]{5^{2/3}}$
80. $\sqrt[5]{x^3yz^4}$

CHAPTER SUMMARY

Many of the topics in this chapter form the algebraic foundation for the rest of the book. You should thoroughly understand these concepts. The main points of the chapter are summarized here.

Sets of numbers:

Natural numbers $N = \{1, 2, 3, \ldots\}$
Integers $I = \{\ldots, -2, -1, 0, 1, 2, \ldots\}$
Rational numbers $Q = \left\{x \mid x = \dfrac{a}{b}, a, b \in I, b \neq 0\right\}$
Real numbers $R = \{x \mid x \text{ can be written in decimal notation}\}$

Properties of real numbers: For all real numbers a, b, and c:

Commutative property $\quad a + b = b + a$
$\qquad\qquad\qquad\qquad\quad a \cdot b = b \cdot a$
Associative property $\quad a + (b + c) = (a + b) + c$
$\qquad\qquad\qquad\qquad a \cdot (b \cdot c) = (a \cdot b) \cdot c$
Closure property $\quad a + b$ and $a \cdot b$ are real numbers
Identity property $\quad a + 0 = a$ and $a \cdot 1 = a$
Inverse property $\quad a + (-a) = 0$
$\qquad\qquad\qquad\quad a \cdot \left(\dfrac{1}{a}\right) = 1$, if $a \neq 0$
Distributive property $\quad a(b + c) = ab + ac$

Rules of exponentiation: For all real numbers a, b, m, and n:

$a^m \cdot a^n = a^{m+n}$ $\qquad \left(\dfrac{a}{b}\right)^n = \dfrac{a^n}{b^n}, \quad b \neq 0$

$\dfrac{a^m}{a^n} = a^{m-n}$ $\qquad a^0 = 1, \quad a \neq 0$

$(a^m)^n = a^{mn}$ $\qquad a^{-n} = \dfrac{1}{a^n}, \quad a \neq 0$

$(ab)^n = a^n b^n$ $\qquad a^{m/n} = \sqrt[n]{a^m} = (\sqrt[n]{a})^m$

Rules of equality: For all real numbers a, b, and c, if $a = b$, then

$a + c = b + c$ $\qquad a \cdot c = b \cdot c$
$a - c = b - c$ $\qquad \dfrac{a}{c} = \dfrac{b}{c}, \quad c \neq 0$

Rules of inequality: For all real numbers a, b, and c, if $a < b$, then

$$a + c < b + c$$
$$a - c < b - c$$
$$a \cdot c < b \cdot c, \quad c > 0$$
$$a \cdot c > b \cdot c, \quad c < 0$$
$$\frac{a}{c} < \frac{b}{c}, \quad c > 0$$
$$\frac{a}{c} > \frac{b}{c}, \quad c < 0$$

Rules of radical expressions:

$$\sqrt[n]{a} = a^{1/n}$$
$$\frac{\sqrt[n]{a}}{\sqrt[n]{b}} = \sqrt[n]{\frac{a}{b}}$$
$$\sqrt[n]{a} \cdot \sqrt[n]{b} = \sqrt[n]{ab}$$
$$(\sqrt[n]{a})^n = a, \quad a \geq 0$$
$$(\sqrt[n]{a})^n = a, \quad a < 0 \text{ and } n \text{ an odd integer}$$

REVIEW PROBLEMS

In Problems 1–6, determine to which of the sets N, I, Q, and R the numbers belong.

1. -4.5
2. $\sqrt{5}$
3. 4
4. $-\frac{45}{76}$
5. $\sqrt{-6}$
6. $\frac{-\pi}{4}$

7. Give three examples of integers that are not natural numbers.

8. Give three examples of real numbers that are not rational.

In Problems 9–14, name which property is being illustrated, and give an example to illustrate the property.

9. $a + (b + c) = (a + b) + c$
10. $a + (-a) = 0$
11. $a \cdot b$ is a real number
12. $a \cdot 1 = a$
13. $a(b + c) = ab + ac$
14. $a \cdot b = b \cdot a$

In Problems 15–26, simplify the expressions.

15. $(\frac{3}{2})^3$
16. $(3x)^4$
17. $(x^{-2}y^0)^{-3}$
18. $(x^3y^2)^2$
19. $\left(\frac{-3y}{5x}\right)^2$
20. $\frac{(-2x^3y)^3}{(2xyz^{-3})^{-2}}$
21. $\sqrt[3]{40}$
22. $16^{3/4}$
23. $(27)^{-2/3}$
24. $64^{-5/6}$
25. $\sqrt[4]{4} \cdot \sqrt[4]{4}$
26. $\frac{\sqrt{42} \cdot \sqrt{2}}{\sqrt{12}}$

In Problems 27–30, perform the operations, and simplify.

27. $(4x^3 - 16xy^2 + 2x) + (6x^3 + 8xy^2 - 7x)$
28. $(5xy - 2xz^3 + 5x - 7y^3) - (6x - 3xz^3 + 9xy + 4z^2)$
29. $(x^2 - 4x + 7)(x^4 + 2x^2 - 5x + 4)$
30. $(3x^5 - 2xy + 7y^4)(2xy - y + 3x^6 - 8x^4yz)$

Review Problems

In Problems 31–38, solve for x.

31. $2x - 6 = 7$
32. $4 - 5x = 11$
33. $2x - 1 = 4x + 5$
34. $-7x - 7 = 5 - 6x$
35. $4x + 7 > 9$
36. $6 - 3x < -4$
37. $3x + 6 < -2x + 9$
38. $-3x - 5 > 2x + 11$

In Problems 39–40, solve for the specified variable.

39. $V = lwh$; solve for w.
40. $B = (1 + c)d$; solve for c.

In Problems 41–44, write the expressions in BASIC.

41. $Ax^2 + Bx + C$
42. $B^2 - 4AC$
43. $\sqrt{A - B^3 C}$
44. $(2J)^{-4.4}$

FUNCTIONS AND LINEAR EQUATIONS

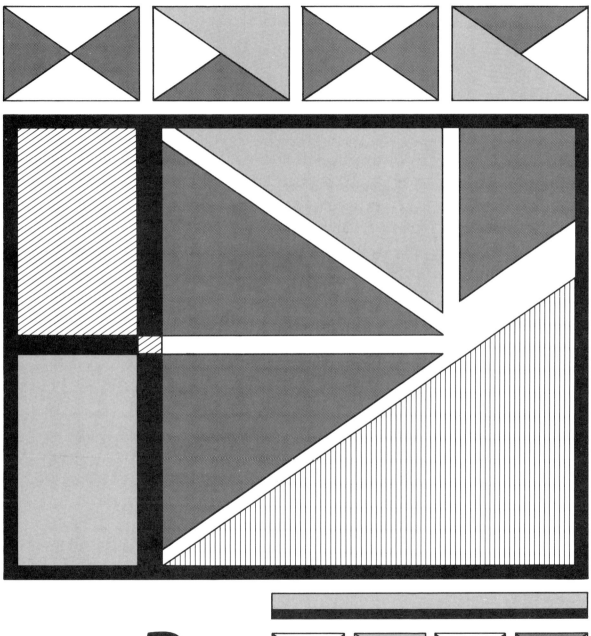

CHAPTER **2**

In this chapter, we will study a certain family of equations called linear equations. This type of equation is used often, as we will see, in real-world problems and is, therefore, an important topic for computer students.

For example, the following situations can be described by linear equations: simple interest, taxes, the speed of an object falling in a gravitational field, commission on a sale, conversion between Fahrenheit and Celsius temperatures, the relationship between circumference and diameter of a circle, and metric conversions. Even Einstein's famous equation $E = mc^2$ is linear equation, since c is a constant.

We will also study the idea of functions, which are directly related to the important concepts of assignment statements and built-in functions in computers.

▬ 2.1 Relations

In Chapter 1, we mentioned polynomials and solved equations in one variable (usually x). Now, we will consider polynomial equations in two variables, x and y. If $x - 2 = 5$, then the *solution* of the equation is $x = 7$. With an equation involving two variables, a solution must include a value for x *and* a value for y which together make the equation true. For example, $x = 2$ and $y = 1$ is a solution to the equation

$$3x = y + 5$$

since

$$3(2) = 1 + 5$$
$$6 = 6$$

ordered pair

We usually write this solution as an **ordered pair** (2, 1), with the x value first and the y value second. The pair (2, 1) is different from (1, 2), which represents $x = 1$ and $y = 2$ and is clearly *not* a solution to the equation

$$3x = y + 5$$

since

$$3(1) \ ? \ 2 + 5$$
$$3 \neq 7$$

2.1 Relations

EXAMPLE 2.1 Show that the specified ordered pair is a solution to the given equation.
  a. $(3, -1)$, $2x - 3y = 9$

Solution
$$2(3) - 3(-1) = 9$$
$$6 - (-3) = 9$$
$$6 + 3 = 9$$
$$9 = 9$$

  b. $(-2, 2)$, $2x^2 + y = 4 + 3y$

Solution
$$2(-2)^2 + 2 = 4 + 3(2)$$
$$2 \cdot 4 + 2 = 4 + 6$$
$$8 + 2 = 10$$
$$10 = 10$$

Consider the equation $x + y = 10$. How many solutions are there to this equation? Three solutions are $(5, 5)$, $(2, 8)$, and $(-2, 12)$. You should be able to convince yourself that there are an *infinite* number of solutions to this equation, which is usually the case with most equations in two variables. The set of all of the solutions to an equation is called the **solution set** for the equation.

solution set

relation
coordinates, domain
range

Definition 2.1 A set of ordered pairs is called a **relation**. The set of all first **coordinates** (x values) is called the **domain**, and the set of all second coordinates (y values) is called the **range**.

EXAMPLE 2.2 Find the domain and range of the relation $\{(1, 2), (3, 4), (3, 5), (-1, 2), (\frac{1}{2}, 0)\}$.

Solution
$$\text{domain} = \{1, 3, 3, -1, \tfrac{1}{2}\} = \{-1, \tfrac{1}{2}, 1, 3\}$$
$$\text{range} = \{2, 4, 5, 2, 0\} = \{0, 2, 4, 5\}$$

The solution set of an equation in two variables is always a relation. When writing out the numbers in a set, we customarily write the numbers in increasing order, and repeating a number is not necessary.

Another way to visualize the relationship between elements in the domain and range is to make a chart consisting of two columns, one for domain and one for range. Then draw arrows showing the element(s) in the range that are matched with each element in the domain. We get the following chart for Example 2.2:

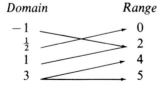

FUNCTIONS AND LINEAR EQUATIONS

We will look at a particular type of relation called a function in the next section.

The solution to an equation or inequality in one variable can be represented on the real number line. To represent the solution set for an equation in two variables, the mathematician René Descartes (1596—1650) had the idea of using *two* number lines, a horizontal line for the x values and a vertical line for the y values, meeting where both lines are zero. This representation is now called the **Cartesian coordinate system**, or simply the **graph**, and the two lines are the **x axis** and the **y axis**. This system is illustrated in Figure 2.1.

Cartesian coordinate system, graph, x axis, y axis

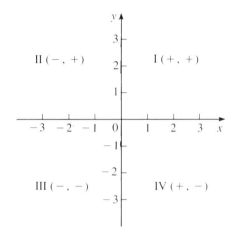

Figure 2.1
Cartesian coordinate system

quadrants

The two lines break the graph into four sections called **quadrants**, numbered counterclockwise, as illustrated in Figure 2.1. To represent a solution (an ordered pair) on the graph, we simply put a point where the x and y values meet.

EXAMPLE 2.3 Plot the points on the Cartesian coordinate system.

a. $A(2, 3)$ d. $D(2, -5)$
b. $B(-3, 1)$ e. $E(4, 0)$
c. $C(0, 6)$ f. $F(-0.5, -2)$

Solution See Figure 2.2.

Look at Figure 2.1 again. Any point in quadrant I has both x and y values (called coordinates) positive: $(+, +)$. Similarly, any point in quadrant II has a negative x value, since it is to the left of 0 on the x axis, and a positive y value, since it is above 0 on the y axis: $(-, +)$. The rules for quadrants III $(-, -)$ and IV $(+, -)$ are determined in a similar fashion.

2.1 Relations

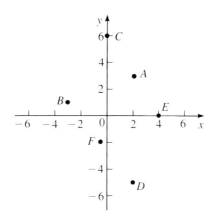

Figure 2.2

If the x coordinate is zero, the point is neither right (+) nor left (−) of the y axis; so it must be on the y axis. Similarly, if the y coordinate is zero, the point is on the x axis.

EXAMPLE 2.4 Determine which quadrant the points are in.
  a. $(-2, 4)$ b. $(0, -2)$

Solution   second quadrant (−, +)   y axis (x = 0)

  c. $(3, -4)$ d. $(-5, 0)$

Solution   fourth quadrant (+, −)   x axis (y = 0)

  e. $(-3, -1)$ f. $(5, 3)$

Solution   third quadrant (−, −)   first quadrant (+, +)

Problem Set 2.1.

In Problems 1–12, determine whether the given point is a solution to the given equation.

1. $(2, 0), 2x - y = 4$
2. $(5, -1), x + 5y = 0$
3. $(0, -3), 2x + y^2 = x + 9$
4. $(-2, 6), 7 = 3x - y$
5. $(2, -2), x^2 - y^2 = 0$
6. $(4, -5), 7x - 4y = 8$
7. $(3, \frac{1}{2}), x + 2y = 3$
8. $(\frac{1}{2}, -\frac{1}{2}), x + y = 0$
9. $(1, \frac{1}{2}), 3x^2 + 2y = 1$
10. $(4, -2), y = \sqrt{x}$
11. $(-3, -2), 2x^2 + y = 16$
12. $(8, 4), y = \sqrt{2x}$

In Problems 13–20, find the domain and range of the relations, and make a chart showing the matchings of domain elements with range elements.

13. $A = \{(0, 1), (2, 1), (-2, 5), (6, -3)\}$
14. $B = \{(1, -1), (3, -1), (-1, 3), (2, 7)\}$
15. $C = \{(2, 2), (3, 2), (4, 2), (5, 2)\}$
16. $D = \{(0, 7), (0, 1), (0, -4), (0, 4)\}$
17. $E = \{(2, -4), (3, -6), (2, 0), (8, 5)\}$
18. $F = \{(3, 1), (-2, -4), (6, -4), (5, 9)\}$
19. $G = \{(x, y) | x, y \in N, x < 4 \text{ and } y \leq 3\}$
20. $H = \{(x, y) | x, y \in I, -2 < x \leq 3 \text{ and } -4 \leq y < 0\}$

In Problems 21–30, plot the points on the Cartesian coordinate system.

21. $A(\frac{3}{2}, 2)$
22. $F(-1, -9)$
23. $B(4, \frac{1}{3})$
24. $G(-3, 4)$
25. $C(0, -3)$
26. $H(-2, 5)$
27. $D(-7, 0)$
28. $I(\frac{1}{2}, -3)$
29. $E(-3, -5)$
30. $J(4, -\frac{2}{3})$

In Problems 31–40, determine which quadrant the points are in.

31. $(7.3, 2.7)$
32. $(-6, -100)$
33. $(0, -2)$
34. $(5, -4)$
35. $(5, 0)$
36. $(-\frac{3}{5}, \frac{6}{7})$
37. $(2.5, -3.25)$
38. $(-\pi, -7)$
39. $(-2.9, 8.3)$
40. $(\sqrt{2}, -\sqrt{3})$

In Problems 41–50, determine which quadrant(s) the relations are in.

41. $\{(x, y) | x > 0, y < 0\}$
42. $\{(x, y) | x < 0, y > 0\}$
43. $\{(x, y) | x > 0\}$
44. $\{(x, y) | x = 0\}$
45. $\{(x, y) | y = 0\}$
46. $\{(x, y) | y < 0\}$
47. $\{(x, y) | x < 0, y < 0\}$
48. $\{(x, y) | x > 0, y > 0\}$
49. $\{(x, y) | y > 0\}$
50. $\{(x, y) | x < 0\}$

51. Ordered pairs $(2, 1)$, $(4, 5)$, and $(0, -3)$ are all solutions to the equation $y = 2x - 3$. Graph these points. Can you guess what the graph of *all* solutions to this equation will look like?
52. Ordered pairs $(-1, 6)$, $(2, 0)$, and $(1, 2)$ are all solutions to the equation $y = 4 - 2x$. Graph these points. Can you guess what the graph of *all* solutions to this equation will look like?
53. Find two solutions to the equation $x + 2y = 3$. (*Hint*: Pick any x value; then find what y must be.)
54. Find two solutions to the equation $3x - y = 6$.

2.2 Functions

We will now consider a special kind of relation called a function.

function

Definition 2.2 A **function** is a relation in which each element in the domain is paired with *exactly one* element in the range.

EXAMPLE 2.5 Determine whether the relations are functions.

a. $A = \{(1, 2), (3, 4), (3, 5), (-1, 2), (\frac{1}{2}, 0)\}$

Solution We get the following chart:

2.2 Functions

```
Domain        Range
  -1            0
  1/2           2
   1            4
   3            5
```

Since the element 3 from the domain is paired with both 4 and 5 [(3, 4) and (3, 5) are both in A], A is <u>not a function</u>.

b. $B = \{(1, 2), (3, 2), (5, -4), (0, 2)\}$

Solution We get the following chart:

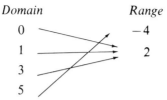

This relation is <u>a function</u>, even though 1, 3, and 0 are all paired with 2. We can reuse a value in the range but not in the domain.

c. The relation that takes any x and triples it.

Solution This relation is <u>a function</u> because if x is a number in the domain, then $3x$ represents a single number. In other words, every number has exactly one triple.

Consider the following nonnumeric example: the function whose domain is the set of students in a class and whose range is the set of final letter grades. Each student gets exactly one grade. It is acceptable (even probable) that two different students will get the same grade, but it would probably *not* be acceptable for the same student to get two different grades.

The functions that we will consider in this book will usually consist of an infinite number of ordered pairs, so listing all of them is clearly impossible. Hence, we will use function notation. In **function notation**, x is the **independent variable**, and y is the **dependent variable**, whose value depends on the chosen value of x. The function takes the value of x and determines the value of y, usually through a mathematical formula. The value of y becomes the value of the function.

*function notation,
independent variable
dependent variable*

If f is the function that takes an x value and doubles it, then we write $f(x) = 2x$ to indicate that f takes x from the domain and pairs it with $2x$ to get the point $(x, 2x)$. The notation $f(x)$ is read "f of x." So y and $f(x)$ are the same: $y = f(x)$.

To find what a function will do with a certain value of x, substitute that value for x into the formula.

FUNCTIONS AND LINEAR EQUATIONS

EXAMPLE 2.6 For $g(x) = x^2 - 2x + 1$, find the indicated value.

  a. $g(0)$

Solution
$$g(x) = x^2 - 2x + 1$$
$$g(0) = 0^2 - 2 \cdot 0 + 1$$
$$= 0 - 0 + 1$$
$$= \underline{1}$$

  b. $g(-2)$

Solution
$$g(-2) = (-2)^2 - 2 \cdot (-2) + 1$$
$$= 4 + 4 + 1$$
$$= \underline{9}$$

  c. $g(1+b)$

Solution
$$g(1+b) = (1+b)^2 - 2(1+b) + 1$$
$$= 1 + 2b + b^2 - 2 - 2b + 1$$
$$= \underline{b^2}$$

built-in functions

Many times, a computer programmer will use **built-in functions**, functions that are used so often that they are written into many computer languages. Finding a square root, for example, is a common occurrence. In the language BASIC, finding a square root is accomplished by the function SQR. When the programmer writes SQR(9), the computer searches its memory for the function SQR and then finds the square root of 9. It returns the *one* value 3. (If we were to get more than one value, things could really get confusing.) Similarly, when you push a button on a calculator, like square root, logarithm, or percent, the calculator uses a built-in function to perform each calculation.

user-defined functions

In some versions of BASIC, the programmer can define new functions in addition to the built-in functions. These functions are called **user-defined functions**, and they are defined by using a DEF FN statement, as follows:

DEF FNfunction name(variable(s)) = expression

The following example makes this notation clear.

EXAMPLE 2.7 Write user-defined BASIC functions for the functions.

  a. $f(x) = 3x$

Solution DEF FNF(x) = 3 * X

  b. $g(x) = x^2 - 2x + 1$

Solution DEF FNG(X) = (X^2) − (2 * X) + 1

2.2 Functions

                c. $h(x) = 7x + 2$

Solution   DEF FNH(X) = (7 * X) + 2

                d. $y = a(b + c)$

Solution   DEF FNF(A, B, C) = A * (B + C)

As Example 2.7d illustrates, we can use this notation with functions that involve more than one variable. Here, the variable y is a function of the three variables a, b, and c. All variables in the expression should be listed in parentheses.

These functions are defined at the beginning of the program and then referred to whenever needed throughout the program, similar to the way that built-in functions are used.

When a computer performs a calculation, we don't want the result to get lost. Usually, the programmer assigns the result of the calculation to a variable in what is called an assignment statement (see Chapter 7). The variable can only take on one value at a time, so assignment statements must also use functions.

▬▬ Problem Set 2.2

In Problems 1–8, determine whether the relations are functions.

1. $A = \{(0, 1), (2, 1), (-2, 5), (6, -3)\}$
2. $B = \{(1, -1), (3, -1), (-1, 3), (2, 7)\}$
3. $C = \{(2, 2), (3, 2), (4, 2), (5, 2)\}$
4. $D = \{(0, 7), (0, 1), (0, -4), (0, 4)\}$
5. $E = \{(2, -4), (3, -6), (2, 0), (8, 5)\}$
6. $F = \{(3, 1), (-2, -4), (6, -4), (5, 9)\}$
7. $G = \{(x, y) | x, y \in N, x < 4, y \leq 3\}$
8. $H = \{(x, y) | x, y \in I, -2 < x \leq 3, -5 \leq y < -1\}$
9. Give the formal definition of a function.
10. Describe the idea of a function in your own words.

In Problems 11–16, $f(x) = 3x + 1$. Find the indicated value.

11. $f(0)$ 12. $f(-1)$ 13. $f(3)$
14. $f(c)$ 15. $f(-x)$ 16. $f(d+1)$

In Problems 17–22, $p(x) = x^2 - 2x + 3$. Find the indicated value.

17. $p(1)$ 18. $p(-2)$ 19. $p(7)$
20. $p(a)$ 21. $p(4a)$ 22. $p(3+x)$

In Problems 23–28, $q(x) = 1/(x-3)$. Find the indicated value.

23. $q(2)$ 24. $q(0)$ 25. $q(-4)$
26. $q(c)$ 27. $q(z+1)$
28. What happens at $q(3)$?

In Problems 29–34, $b(x) = \sqrt{x+3}$. Find the indicated value.

29. $b(1)$ 30. $b(4)$ 31. $b(-2)$
32. $b(h)$ 33. $b(t+1)$
34. What happens at $b(-4)$?

Problems 35–40 list some sample built-in computer functions. Find the indicated value.

35. $s(x) = \sqrt{x}, x = 25$

36. $r(x) = \dfrac{1}{x}$, $x = 7$
37. $s(x) = x^2$, $x = -4$
38. $p(x) = 100x$, $x = 0.34$
39. $c(x) = x^3$, $x = -3$
40. $p(x) = \dfrac{x}{100}$, $x = 47$

Problems 41–46 list some "famous" functions. Find the indicated value.

41. $F(C) = 1.8C + 32$, $C = 100$ (Fahrenheit to Celsius)
42. $C(F) = \tfrac{5}{9}(F - 32)$, $F = 212$ (Celsius to Fahrenheit)
43. $A(r) = \pi r^2$, $r = 3$ (area of a circle)
44. $V(r) = (4\pi r^3)/3$, $r = 6$ (volume of a sphere)
45. $m(y) = 1.09y$, $y = 3$ (yards to meters)
46. $k(m) = 1.61m$, $m = 10$ (miles to kilometers)

In Problems 47–52, decide whether the relations are functions. If they are, write the formula.

47. The relation that takes any x and doubles it.

(*Hint:* Will you always get just one answer for a given x?)

48. The relation that takes any x and adds four to it.
49. The relation that takes any positive x and finds all numbers whose square equals x. (*Hint:* Consider $x = 4$.)
50. The relation that takes any x and finds all numbers whose cube equals x.
51. The relation that matches people with their age.
52. The relation that matches students with their grade point average (GPA).

In Problems 53–62, write user-defined BASIC functions for the functions.

53. $F(C) = 1.8C + 32$
54. $C(F) = \tfrac{5}{9}(F - 32)$
55. $A(r) = \pi r^2$
56. $V(r) = \dfrac{4\pi r^3}{3}$
57. $m(y) = 1.09y$
58. $K(m) = 1.61m$
59. $L(x) = 4x + 8$
60. $E(x) = 5x - 7$
61. $y = ab + ac$
62. $y = a^2 + bcd^2$

2.3 Graphing Linear Equations

For the rest of this chapter, we will consider a certain kind of equation: a linear equation.

linear equation

Definition 2.3 A **linear equation** is a polynomial equation of the form $ax + by = c$, where a, b, and c are real numbers.

Notice that in Definition 2.3, both x and y are raised to the first power. If a or b is zero, then the linear equation contains only one of the variables. Linear equations may not always appear in the form of Definition 2.3, but every term will either be a number or a number times x or y to the first power.

An important skill is being able to recognize what is and what isn't a linear equation.

EXAMPLE 2.8 Determine whether the equation is linear.

  a. $2x + y = 1$ **b.** $x^2 - 2x = y$

Solution linear   nonlinear

2.3 Graphing Linear Equations

	c.	$3 + xy = 2$	d.	$y - 3 = 0$
Solution		nonlinear		linear
	e.	$x = 7 - 2y$	f.	$y = \dfrac{15}{x}$
Solution		linear		nonlinear

As illustrated in Examples 2.8, a linear equation cannot contain an xy term. However, it need not contain both x and y terms.

Consider the linear equation $x + y = 10$. We can find many solutions to this equation by choosing any x value and finding the corresponding y value, or by choosing any y value and finding the corresponding x value. For example, if $x = 2$, then $2 + y = 10$, so $y = 8$, giving the solution $(2, 8)$. If $y = 6$, then $x + 6 = 10$, so $x = 4$, giving the solution $(4, 6)$.

Similarly, we can find other solutions for this equation, like $(3, 7)$, $(9, 1)$, $(-2, 12)$, and $(5, 5)$. If we put all of these points on the same graph, we get Figure 2.3.

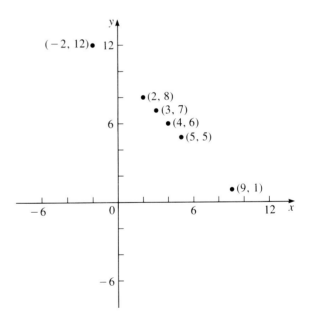

Figure 2.3

All of the points in Figure 2.3 lie on a line. Mathematicians can prove that all of the solutions to this equation will lie on this line, and every point on the line is a solution to the equation. Being able to recognize a linear equation is a help in graphing, because the solution set for a linear equation is *always* a line (hence, the word *linear*). So recognizing a linear equation and finding a few points, two at least and preferably three, is enough to graph the entire solution set! It doesn't matter which two or three points of the line are found since all of the solutions to the equation lie on the same line.

EXAMPLE 2.9 Plot several points and graph the solution set for the lines.

a. $x - 2y = 8$

Solution A good idea is to make a chart of x and y values to keep track of the points:

x	y
8	0
0	−4
4	−2
2	−3

The graph is illustrated in Figure 2.4.

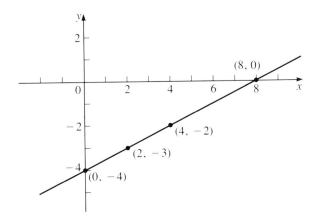

Figure 2.4

b. $y = 3x + 5$

Solution The chart gives a few points.

x	y
0	5
1	8
−1	2
2	11

The graph is illustrated in Figure 2.5.

2.3 Graphing Linear Equations

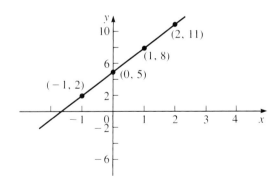

Figure 2.5

y intercept

x intercept

The point where the line crosses the y axis is called the **y intercept** of the line. It must occur when $x = 0$, since any point on the y axis has an x coordinate of zero. Similarly, the **x intercept** of a line is the point where the graph crosses the x axis and must occur when $y = 0$.

EXAMPLE 2.10 Find the intercepts of the line $2x - 4y = -4$.

Solution

x intercept: $y = 0$
$$2x - 4(0) = -4$$
$$2x - 0 = -4$$
$$2x = -4$$
$$x = -2$$

y intercept: $x = 0$
$$2(0) - 4y = -4$$
$$0 - 4y = -4$$
$$-4y = -4$$
$$y = 1$$

Problem Set 2.3

In Problems 1–14, determine whether the equation is linear.

1. $4x - 2y - 7 = 0$
2. $\dfrac{2}{x} + 3y = 5$
3. $2y = x - 4$
4. $7 - xy + y = 2$
5. $x^2 + 3y = 1$
6. $x = 5$
7. $x^3 - y^3 = xy$
8. $y + 3 = 0$
9. $x - y = 9$
10. $\dfrac{x}{2} + \dfrac{y}{3} = 6$
11. $3 + x + 4xy = -9$
12. $\dfrac{y}{7} = 2 - \dfrac{4}{x}$
13. $-4x + \dfrac{7}{y} = 6$
14. $y = mx + b$

In Problems 15–34, plot at least three points, and graph the line.

15. $x + y = 5$
16. $x - y = 10$
17. $y = 2x - 1$
18. $y = 4x + 6$
19. $y = 5x - 8$
20. $y = -4 + 2x$
21. $x = y - 2$
22. $x = 3y + 1$
23. $x = -4y + 7$
24. $x = 6 - 2y$
25. $2x + 3y = 6$
26. $4x - 5y = 20$
27. $2x + 5y = 7$
28. $3x - 6y = 5$
29. $x - 2y + 4 = 0$
30. $3x + y - 7 = -3$
31. $-3x - 3y - 5 = 0$
32. $2x - 4y - 2 = 0$
33. $-4x + 5y - 7 = 0$
34. $-4x + 8y + 4 = 0$

In Problems 35–42, find the intercepts of the line.

35. $y = 4x + 3$
36. $y = -2x - 7$
37. $x = 3y - 2$
38. $x = 7y + 4$
39. $2x - y = 3$
40. $5x + 4y = 10$
41. $-3x + 2y - 5 = 0$
42. $7x - 4y - 3 = 0$

43. Can a linear equation ever have more than one x intercept?
44. Can a linear equation ever have more than one y intercept?
45. Can a linear equation ever have *no* x intercept? If so, what kind of a line would it be?
46. Can a linear equation ever have *no* y intercept? If so, what kind of a line would it be?

In Problems 44–52, graph the line with the given intercepts.

47. y intercept = 4
    x intercept = 3
48. y intercept = 2
    x intercept = 6
49. y intercept = -5
    x intercept = -3
50. y intercept = -5
    x intercept = -8
51. y intercept = 4
    no x intercept
52. no y intercept
    x intercept = -5

53. How many lines have both x and y intercepts of zero?
54. How many lines have both x and y intercepts of 2?

In Problems 55–58, write user-defined BASIC functions for the linear functions.

55. $y = 2x - 7$
56. $y = 3x + 6$
57. $y = 4 - 10x$
58. $y = -9 + 2x$
59. Solve $ax + by = c$ for y.
60. Solve $ax + by = c$ for x.

2.4 Slope of a Line

Perhaps the most important concept to understand about linear equations is the idea of slope The slope of a line is a measure of how "steep" a line is. It is the ratio of how the line changes vertically to how the line changes horizontally. More specifically, given the two points $A(x_1, y_1)$ and $B(x_2, y_2)$, the vertical change between the two points is completely determined by the y values, and the horizontal change is completely determined by the x values.

Consider the triangle in Figure 2.6. Points A and C are at the same vertical level and, therefore, must have the same y values. Similarly, points B and C are at the same position left and right and must have the same x values.

If we move from point A to point B, the horizontal change is the distance from A to C, $(x_2 - x_1)$; and the vertical change is the distance from C to B, $(y_2 - y_1)$. These distances are used in the definition of slope, which follows.

2.4 Slope of a Line

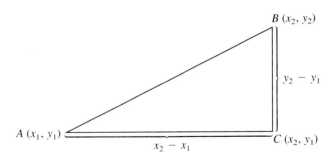

Figure 2.6

slope

Definition 2.4 The **slope** of the line through the points (x_1, y_1) and (x_2, y_2) is $(y_2 - y_1)/(x_2 - x_1)$ and is denoted by the letter m.

The slope is the ratio of vertical change to horizontal change of a line and is the same no matter which two points on the line are chosen.

EXAMPLE 2.11 Find the slope of the lines through the given two points.
  a. $(2, 1)$ and $(0, -3)$ **b.** $(-3, 4)$ and $(2, -5)$

Solution

$$m = \frac{-3-1}{0-2}$$
$$= \frac{-4}{-2}$$
$$= \underline{2}$$

$$m = \frac{-5-4}{2-(-3)}$$
$$= \underline{\frac{-9}{5}}$$

If we were to graph the two lines in Example 2.11 (see Problems 1 and 2 of Problem Set 2.4), we would discover that lines with positive slope slant up from left to right. Lines with negative slope slant down from left to right. See Figure 2.7.

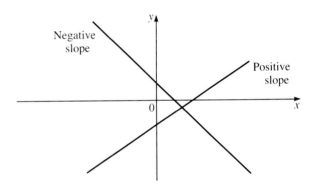

Figure 2.7

equation of a horizontal line

Consider now the line through the points $(3, 2)$ and $(1, 2)$. The slope of this line is

$$m = \frac{2-2}{1-3} = \frac{0}{-2} = 0$$

What does a zero slope mean? Both of these points have the same y value and, therefore, are at the same vertical level. So this line must be a horizontal line, as illustrated in Figure 2.8. Any point on this line has a y value of 2, but x can be any value. Thus, an **equation of a horizontal line** is of the form

$$y = k$$

where k is a constant. In our example, $y = 2$. A horizontal line, then, is the graph of a function which has a domain of all real numbers and a range of one number, k.

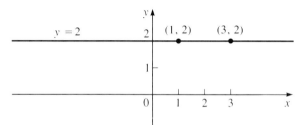

Figure 2.8

equation of a vertical line

Consider the line through the points $(1, 4)$ and $(1, -2)$. The slope of this line is

$$m = \frac{-2-4}{1-1} = \frac{-6}{0}$$

which is undefined. What does an undefined slope mean? Both of the points have the same x value and, therefore, are at the same position left and right. So this line must be a vertical line as illustrated in Figure 2.9. Any point on this line has an x value of 1, but y can be any value. An **equation of a vertical line** is of the form

$$x = k$$

where k is a constant. In our example, $x = 1$.

A vertical line does *not* represent a function since the domain consists of only one number, k, and this number is paired with every real number. All nonvertical linear equations represent functions.

We are now ready to define the concept of parallel lines.

2.4 Slope of a Line

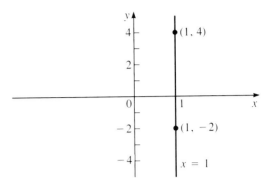

Figure 2.9

parallel lines

Definition 2.5 Two lines are **parallel lines** if they have identical slopes.

Figure 2.10 illustrates two parallel lines. Parallel lines can also be defined as lines that never meet.

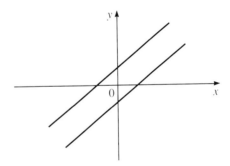

Figure 2.10

Now, consider the idea of perpendicular lines. The usual geometric definition of perpendicular lines is two lines at right angles (90°) to each other. The algebraic definition is in terms of slope.

perpendicular lines

Definition 2.6 Two lines are **perpendicular lines** if they have negative reciprocal slopes.

EXAMPLE 2.12 Find the slope of the line.

a. Any line parallel to $y = 4$

Solution This line must have a slope equal to the slope of $y = 4$. Line $y = 4$ is horizontal, so it has a slope of zero. Thus, any line parallel to $y = 4$ also has a slope of <u>zero</u>.

b. Any line perpendicular to a line with slope $\frac{4}{7}$.

Solution This line must have a slope equal to the negative reciprocal of $\frac{4}{7}$, so the slope must be <u>$-\frac{7}{4}$</u>.

Problem Set 2.4

1. Graph the line of Example 2.11a.
2. Graph the line of Example 2.11b.

In Problems 3–22, find the slope of the line through the pair of points.

3. $(2, 1)$ and $(3, 2)$
4. $(4, 0)$ and $(1, 5)$
5. $(0, 3)$ and $(3, 5)$
6. $(2, 4)$ and $(0, 6)$
7. $(1, 7.6)$ and $(5, 0)$
8. $(-2, 3)$ and $(2, 4)$
9. $(6, 5)$ and $(3, -2)$
10. $(-2, 1)$ and $(-1, 1.4)$
11. $(1, -3)$ and $(\frac{4}{3}, -7)$
12. $(-1, -2)$ and $(2, -5)$
13. $(3, -4.5)$ and $(-2, -1.2)$
14. $(\frac{3}{2}, 7)$ and $(2, 7)$
15. $(4, 6)$ and $(4, -5)$
16. $(3, 0.5)$ and $(2, -2)$
17. $(0.5, -\frac{2}{7})$ and $(4, \frac{3}{7})$
18. $(5, -2.4)$ and $(5.8, 6)$
19. $(-1, -4)$ and $(2, -4)$
20. $(\frac{10}{7}, -9)$ and $(\frac{8}{7}, -3)$
21. $(2.1, 1)$ and $(4, 5.3)$
22. $(-7, -5)$ and $(-7, 7)$

In Problems 23–32, find an equation of the line.

23. The line through $(2, -1)$ and $(2, 3)$ (*Hint:* Find the slope.)
24. The line through $(4, -2)$ and $(1, -2)$
25. The line through $(0, 5)$ and $(-3, 5)$
26. The line through $(-2, 7)$ and $(-2, 0)$
27. The horizontal line through $(2, 4)$
28. The vertical line through $(-1, 3)$
29. The vertical line through $(5, -2)$
30. The horizontal line through $(-2, 6)$
31. The x axis
32. The y axis

In Problems 33–38, find the slope of the line.

33. Any line parallel to the y axis
34. Any line perpendicular to the y axis
35. Any line perpendicular to a line with slope $\frac{2}{3}$
36. Any line parallel to a line with slope $-\frac{5}{8}$
37. Any line perpendicular to the line in Problem 5
38. Any line parallel to the line in Problem 8
39. Find the slopes of the three sides of the triangle ABC with $A(2, -1)$, $B(0, 4)$, and $C(-6, 1)$.
40. Find the slopes of the three sides of the triangle ABC with $A(0, 4)$, $B(2, 1)$, and $C(3, -1)$.
41. The three points $(2, 1)$, $(5, 3)$, and $(-1, -1)$ lie on the same line. Show that the three slopes are the same.
42. The three points $(1, 0)$, $(-1, 2)$, and $(-3, 4)$ lie on the same line. Show that the three slopes are the same.
43. Find *any* three points on the line $2x + y - 3 = 0$. Then find the three slopes between these points, and show that they are equal.
44. Find *any* three points on the line $3x - y + 4 = 0$. Then find the three slopes between these points, and show that they are equal.

In Problems 45–48, show that the points do not lie on the same line by finding the slopes between them.

45. $(0, 2)$, $(4, 7)$, and $(2, 4)$
46. $(0, 1)$, $(3, 4)$, and $(6, 6)$
47. $(0, 0)$, $(9, 0)$, and $(0, 7)$
48. $(1, -2)$, $(3, 3)$, and $(8, 5)$
49. Write a user-defined BASIC function to find the slope of a line in terms of x_1, y_1, x_2, and y_2.
50. Write a user-defined BASIC function to find the equation of a horizontal line.
51. If two lines are both parallel to a given line, what can you conclude about these two lines?
52. If two lines are both perpendicular to a given line, what can you conclude about these two lines?

53. How many lines have a slope of $-\frac{1}{3}$?
54. How many lines have an undefined slope?

2.5 Forms of Linear Equations

In this section, we will consider different forms of linear equations. The first form that we define is the standard form.

standard form

Definition 2.7 The **standard form** of a linear equation is

$$ax + by = c$$

where a, b, and c are real numbers.

The advantage of the standard form is that *every* linear equation, even those with undefined slopes, can be written in this form. Note that some books consider $ax + by + c = 0$ to be standard form. Either form is acceptable.

EXAMPLE 2.13 Write the equation in standard form.

a. $y = 3x - 2$

Solution We must get both variables on one side of the equation:

$$-3x + y = -2$$

Customarily, we like to have the x coefficient positive, so we multiply by -1.

$$\underline{3x - y = 2}$$

b. $2(x - 1) = y + 3$

Solution
$$2x - 2 = y + 3$$
$$2x = y + 5$$
$$\underline{2x - y = 5}$$

When finding the equation of a line, we will usually write the equation in standard form.

Suppose that we are given one point and the slope of a line. There is only one line that goes through this point with this slope. To find an equation of this line, we will use a version of the slope formula, where the given point is (x_1, y_1) and the slope is m. Any other point (x, y) on this line must satisfy the slope formula with (x_1, y_1). So we get

$$m = \frac{y - y_1}{x - x_1}$$

By multiplying both sides of this equation by $x - x_1$, we get the following definition.

FUNCTIONS AND LINEAR EQUATIONS

point-slope form

Definition 2.8 The **point-slope form** of a linear equation with the point (x_1, y_1) and with slope of m is

$$m(x - x_1) = y - y_1$$

EXAMPLE 2.14 Find the equation of the line through the point $(-2, 3)$ with a slope of 7.

Solution

$$\begin{aligned}
m(x - x_1) &= y - y_1 \\
7[x - (-2)] &= y - 3 \quad \text{(point-slope form)} \\
7(x + 2) &= y - 3 \\
7x + 14 &= y - 3 \\
7x &= y - 17 \\
\underline{7x - y = -17} \quad &\text{(standard form)}
\end{aligned}$$

Suppose that we are not given the slope of a line, but we are given *two* points. There is only one line that goes through both of these points. To find an equation of this line, we first find the slope of the line and then use the point-slope form with the calculated slope and either point.

EXAMPLE 2.15 Find the equation of the line through the points $(2, -1)$ and $(4, 3)$.

Solution **Step 1.** Find the slope:

$$m = \frac{3 - (-1)}{4 - 2} = \frac{4}{2} = 2$$

Step 2. Use m and either point to find the equation:

$$\begin{aligned}
2(x - 4) &= y - 3 \quad \text{(point-slope form)} \\
2x - 8 &= y - 3 \\
2x &= y + 5 \\
\underline{2x - y = 5} \quad &\text{(standard form)}
\end{aligned}$$

The result would be the same if the point $(2, -1)$ were chosen instead of $(4, 3)$ (see Problem 61).

Consider the special case where we know the slope m and the y intercept b of a line. There is only one line that crosses the y axis at b with slope m. If the y intercept is b, this point is $(0, b)$. So again, we can use the point-slope form.

EXAMPLE 2.16 Find the equation of the line with a slope of 3 and a y intercept of 7.

Solution

$$\begin{aligned}
m(x - x_1) &= y - y_1 \\
3(x - 0) &= y - 7 \quad \text{(point-slope form)} \\
3x &= y - 7 \\
\underline{3x + 7 = y} &
\end{aligned}$$

2.5 Forms of Linear Equations

The equation in Example 2.16 has been purposely solved for y to illustrate that the slope is the coefficient of x and the y intercept is the number without a variable *if the equation is solved for y*.

slope-intercept form

Definition 2.9 The **slope-intercept form** of a linear equation with a slope of m and a y intercept of b is

$$y = mx + b$$

EXAMPLE 2.17 Find the equation of the line with slope -3 and y intercept 4.

Solution
$$y = mx + b$$
$$y = -3x + 4$$
$$\underline{3x + y = 4}$$

EXAMPLE 2.18 Find the slope and y intercept of the line.

a. $y = 2x - 5$

Solution slope $m = \underline{2}$
y intercept $b = \underline{-5}$

b. $6x + 2y - 4 = 0$

Solution In this case, we can't say that the slope is 6 because the equation is not solved for y. Solve for y first.

$$2y - 4 = -6x$$
$$2y = -6x + 4$$
$$y = -3x + 2$$

Slope $= \underline{-3}$
y intercept $= \underline{2}$

EXAMPLE 2.19 Find the equation of the line that goes through the point $(2, -1)$ and is parallel to the line $3x + y = 4$.

Solution **Step 1.** From the fact that the line we want is parallel to $3x + y = 4$, we know that the two lines have the same slope. So

$$3x + y = 4$$
$$y = -3x + 4$$

The slope of both lines is -3.

Step 2. Now, we can use the point-slope form to find the equation of the line.

$$-3(x-2) = y-(-1)$$
$$-3x+6 = y+1$$
$$-3x = y-5$$
$$-3x-y = -5$$
$$\underline{3x+y = 5}$$

EXAMPLE 2.20 Find the equation of the line that goes through the point $(-3, 1)$ and is perpendicular to the line $2x - 4y - 7 = 0$.

Solution **Step 1.** We find the slope of $2x - 4y - 7 = 0$:

$$2x - 4y - 7 = 0$$
$$2x - 4y = 7$$
$$-4y = -2x + 7$$
$$y = \tfrac{1}{2}x - \tfrac{7}{4}$$

The slope of this line is $\tfrac{1}{2}$. Thus, our line must have slope -2 (negative reciprocal of $\tfrac{1}{2}$).

Step 2. Now, we can use the point-slope form to find the equation of the desired line:

$$-2[x-(-3)] = y-1$$
$$-2(x+3) = y-1$$
$$-2x-6 = y-1$$
$$-2x = y+5$$
$$-2x-y = 5$$
$$\underline{2x+y = -5}$$

Problem Set 2.5

In Problems 1–10, write the equation in standard form.

1. $2x + 3y + 7 = 0$
2. $x - 3y - 8 = 0$
3. $y = 2x + 5$
4. $y = 3x - 4$
5. $x = 7y + 2$
6. $x = -3y - 6$
7. $2(x-3) = y - 5$
8. $4(x+1) = y + 3$
9. $2x - 7 = 5y + 2$
10. $7y - 4 = 2x + 5$

In Problems 11–20, find the equation of the line through the given point with the given slope. Answers should be in standard form.

11. $(2, 1), m = 5$
12. $(5, -2), m = 2$
13. $(-2, 1), m = -1$
14. $(2, 4), m = -2$
15. $(0, 5), m = 3$
16. $(4, 0), m = -4$
17. $(2, -3), m = 0$

2.5 Forms of Linear Equations

18. $(-3, 1)$, m is undefined
19. $(4, 2)$, m is undefined
20. $(-5, 5)$, $m = 0$

In Problems 21–30, find the equation of the line through the two given points. Answers should be in standard form.

21. $(2, 2)$ and $(3, 1)$
22. $(-1, 2)$ and $(0, 4)$
23. $(3, -2)$ and $(4, 2)$
24. $(-2, 0)$ and $(1, 5)$
25. $(4, 5)$ and $(4, -3)$
26. $(2, -3)$ and $(0, -3)$
27. $(5, 1)$ and $(-2, 1)$
28. $(-1, 4)$ and $(-1, 7)$
29. $(-5, -7)$ and $(-7, -5)$
30. $(-2, -3)$ and $(-12, -7)$

In Problems 31–40, find the equation of the line with the given slope and y intercept. Answers should be in standard form.

31. $m = 2, b = 5$
32. $m = 3, b = 6$
33. $m = -3, b = 7$
34. $m = -4, b = 6$
35. $m = -4, b = -7$
36. $m = -8, b = -12$
37. $m = 0, b = -3$
38. m is undefined, no b
39. m is undefined, $b = 0$
40. $m = 0, b = 0$

In Problems 41–50, find the slope and y intercept of the line.

41. $y = 3x + 4$
42. $y = 4x + 9$
43. $y = -5x - 6$
44. $y = -3x + 14$
45. $x = 2y + 7$
46. $x = -5y - 9$
47. $2x + y - 7 = 0$
48. $4x - 5y + 8 = 0$
49. $x = -6$
50. $y = -2$

In Problems 51–60, find the equation of the line through the given point with the given information.

51. $(2, 3)$, parallel to $y = 2x - 3$
52. $(-1, 4)$, parallel to $y = 3x + 4$
53. $(4, -7)$, parallel to $2x - 5y = 4$
54. $(6, 1)$, parallel to $6x + 3y = 7$
55. $(3, 2)$, parallel to $y - 4 = 0$
56. $(3, -3)$, perpendicular to the y axis
57. $(2, -3)$, perpendicular to $y = 3x - 2$
58. $(8, 4)$, perpendicular to $y = 5x + 6$
59. $(2, -7)$, perpendicular to $3x + 7y = 9$
60. $(-3, 4)$, perpendicular to $5x - 4y = 10$

61. Verify that you get the same answer for Example 2.15 by using the point $(2, -1)$ instead of $(4, 3)$.

62. Find the equation of the line in Problem 22 by using the point that you didn't use when first solving that problem.

63. If it costs $4 plus $0.15 per mile to rent a bicycle, then the cost of renting the bicycle is given by the linear equation $C(m) = 0.15m + 4$, where m is the number of miles traveled. How much would it cost to rent a bicycle for a 40-mile trip?

64. If a salesperson makes $400 a month plus 10% of her sales, then her monthly salary is given by the equation $S(x) = 400 + 0.1x$, where x is the amount of sales. What is her salary if she has sales of $2000 this month?

65. The profit from selling x computer monitors is given by the equation $P(x) = mx + b$. The profit from selling 10 monitors is $1200, and the profit from selling 20 monitors is $2200. Find the equation for profit. (*Hint:* You are given two points of the line.)

66. The speed of an object (in feet per second) t seconds after it is thrown from a building is given by $s(t) = mt + b$. Five seconds after being thrown, the object is traveling 200 feet per second. Ten seconds after being thrown, the object is traveling 520 feet per second. Find m and b.

CHAPTER SUMMARY

The main ideas of this chapter were functions and linear equations. These concepts are summarized here.

A relation is a set of ordered pairs. The set of first elements is called the domain, and the set of second elements is called the range. A function is a relation in which each element of the domain is paired with exactly one element of the range.

A linear equation is a polynomial equation of the form $ax + by = c$, where a, b, and c are real numbers. The main concepts concerning linear equations are outlined next.

Slope	$m = \dfrac{y_2 - y_1}{x_2 - x_1}$
Point-slope form	$m(x - x_1) = y - y_1$
Slope-intercept form	$y = mx + b$
Standard form	$ax + by = c$
Horizontal lines	$y = k$
Vertical lines	$x = k$
Parallel lines	Lines with the same slope
Perpendicular lines	Lines with negative reciprocal slopes

REVIEW PROBLEMS

In Problems 1–4, plot the points on the Cartesian coordinate system.

1. $A(2, -3)$
2. $B(4, 1)$
3. $C(-2, -5)$
4. $D(-4, 0)$

5. Find *two* solutions for the equation $2x + 3y = 12$.
6. Find *two* solutions for the equation $-4x = 2y + 8$.

In Problems 7–10, $n(x) = 4x^2 - 3x + 3$. Find the indicated value.

7. $n(-2)$
8. $n(5)$
9. $n(-h)$
10. $n(1 + c)$

11. Find the value of $P(x) = 2(x + 4)$ when $x = -3$.
12. Find the value of $Q(r) = 5(1 - r)$ when $r = -2$.

In Problems 13–18, plot at least three points, and graph the line.

13. $2x - 4y = 6$
14. $x = 4y - 7$
15. $y - 5 = 0$
16. $y = 3x - 2$
17. $5x + 2y - 3 = 0$
18. $x + 3 = 0$

In Problems 19–20, find the intercepts of the line.

19. $2x + 6y = 10$
20. $y = -4x - 9$

In Problems 21–24, find the slope of the line through the pair of points.

21. $(3, 0)$ and $(-2, 2)$
22. $(0, -4)$ and $(3, -5)$
23. $(2, -3)$ and $(-1, -7)$
24. $(4, 7)$ and $(4, -7)$

In Problems 25–38, find the equation, in standard form, of the line.

25. The line through $(2, -1)$ with slope 4

2.5 Review Problems

26. The line through $(-3, 5)$ with slope -5
27. The line through $(2, 4)$ and $(3, -6)$
28. The line through $(0, 6)$ and $(5, -7)$
29. The horizontal line through $(-3, 1)$
30. The vertical line through $(-2, 6)$
31. The line with slope -3 and y intercept 6
32. The line with slope 2 and y intercept -4
33. The line with y intercept 4 and x intercept -3
34. The line with y intercept -5 and x intercept 7
35. The line through $(4, -3)$ parallel to $y = 7x - 4$
36. The line through $(-2, 5)$ parallel to $2x - 3y + 7 = 0$
37. The line through $(0, -9)$ perpendicular to $x - 4y = 5$
38. The line through $(2, -6)$ perpendicular to $2x + y = -3$

In Problems 39–40, find the slope and y intercept of the line.

39. $2x - 3y - 9 = 0$ 40. $x = 2y + 5$

41. If a printing company charges $6 plus $0.10 per page printed, then the cost of printing x pages is given by the linear equation $C(x) = 0.1x + 6$. How much will it cost to have 43 pages printed?

42. The cost of producing x computer disks is given by the linear equation $C(x) = mx + b$. If one disk is produced, then the cost is $8. If five disks are produced, then the cost is $22. Find the equation for the cost of producing x disks.

In Problems 43–44, write user-defined BASIC functions for the functions.

43. $P(x) = 2(x + 4)$ 44. $Q(r) = 5(1 - r)$

SYSTEMS OF LINEAR EQUATIONS

CHAPTER **3**

In the previous chapter, we studied linear equations in detail. Now, we will consider systems of linear equations—problems that involve more than one linear equation. The methods used to solve these problems are relatively easy with two equations, but they get much more cumbersome when the number of equations involved is more than two. These types of problems are encountered often and are easily handled by computers and by using matrices and determinants (Chapter 12). These procedures are also adaptable to the field of linear programming (Chapter 13), which involves systems of linear inequalities.

We will consider applications of these systems to real-world problems and set the stage in this chapter for the development of matrices.

3.1 Solution by Graphing

system of equations

A **system of equations** is simply a set of two or more equations. We will only consider systems of two linear equations in this chapter.

solution to a system

Definition 3.1 A **solution to a system** of equations is a point that is a solution to each equation in the system.

In other words, in order to show that a point is a solution to a system of equations, we must show that the point satisfies each equation in the system.

EXAMPLE 3.1 Show that $(2, -1)$ is a solution to the system

$$2x + 3y = 1$$
$$x - 2y = 4$$

Solution We must show that $(2, -1)$ is a solution for each equation:

$$2(2) + 3(-1) = 1 \qquad 2 - 2(-1) = 4$$
$$4 + (-3) = 1 \qquad 2 + 2 = 4$$
$$1 = 1 \qquad 4 = 4$$

EXAMPLE 3.2 Show that $(-3, 4)$ is *not* a solution to the system

$$2x + y = -2$$
$$2x = y - 2$$

3.1 Solution by Graphing

Solution The point $(-3, 4)$ is a solution to $2x + y = -2$, since $2(-3) + 4 = -6 + 4 = -2$; but it is *not* a solution to $2x = y - 2$, since $2(-3) \neq 4 - 2$ ($-6 \neq 2$). So $(-3, 4)$ is not a solution to the system.

Consider the system of two linear equations in x and y. Any point that is a solution to *both* equations must lie on both lines. Normally, two lines meet in exactly one point, so the solution set consists of one point.

Perhaps the easiest method for finding this point is by graphing both equations on the same graph and visually finding where the two lines meet.

EXAMPLE 3.3 Find the solution set of this system by graphing:

$$2x - y = 5$$
$$3x + 2y = 4$$

Solution We first plot points for each equation.

$2x - y = 5$		$3x + 2y = 4$	
x	y	x	y
0	-5	0	2
1	-3	-2	5
3	1	4	-4

From Figure 3.1, the solution appears to be $(2, -1)$.

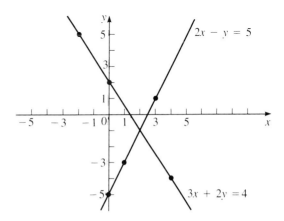

Figure 3.1

By substituting $x = 2$ and $y = -1$ into the two equations of Example 3.3, we can show that $(2, -1)$ is indeed the solution of the system.

This method is easy to use, and it demonstrates what the solution of a system of equations represents. But it has one major drawback. No matter how accurately the graph is drawn, the graphical solution is only an approximation to the real solution. Graphs illustrate, but they don't *prove* anything. Consider the following example.

EXAMPLE 3.4 Approximate the solution to this system by graphing:

$$5x + 10y = 2$$
$$x + 3y = 0$$

Solution From Figure 3.2, the lines appear to meet at an x value between 1 and 2 (closer to 1) and a y value between 0 and -1 (closer to 0). However, it would take a detailed graph and good eyesight to get the true answer, $(1.2, -0.4)$.

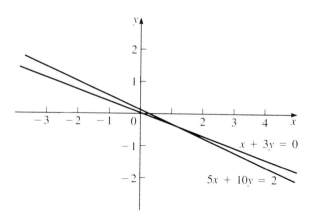

Figure 3.2

We need an algebraic method for solving these systems which will always give the exact answer. Discussing such methods is the purpose of the rest of this chapter.

Problem Set 3.1

In Problems 1–10, determine whether the given point is a solution to the given system of equations.

1. $2x - 4y = 0, (3, 3)$
    $x + y = 6$

2. $x = 2y - 7, (-1, 3)$
    $y = 4x + 7$

3. $3x + y = 9, (3, 0)$
    $2x - y = 6$

3.1 Solution by Graphing

4. $x + y = 7, (4, 3)$
    $3x + 7y = 9$
5. $3x - 5y = 15, (0, -3)$
    $2y + x = -3$
6. $2x - 4y = 6, (1, -2)$
    $3x + 7y = 9$
7. $2x + 2y = 2, (1, 0)$
    $3x + 3y = 3$
8. $x - 3y = 4, (4, 1)$
    $-2x + 6y = -2$
9. $-x - 4y = 0, (-4, 1)$
    $7x - 9y = 9$
10. $6x + 5y = 2, (2, 2)$
    $3x - 4y = -2$

In Problems 11–28, approximate the solution set of the system of equations by graphing.

11. $x = -7$
    $y = 2$
12. $x = 0$
    $y = 3$
13. $x + 3 = 0$
    $y - 4 = 0$
14. $y + 3 = 0$
    $x - 5 = 0$
15. $y = 3x + 2$
    $y = 2x + 4$
16. $y = 2x - 3$
    $y = x + 2$
17. $x = 3y - 1$
    $x = 4 - 2y$
18. $x = -5y + 1$
    $x = -11 + y$
19. $y = 3x - 2$
    $x = -3y + 1$
20. $x = 5y + 10$
    $y = -2x + 2$
21. $x = 2y - 6$
    $2x - 3y = -9$
22. $y = x$
    $2x - y = -1$
23. $2x + 5y = 0$
    $x - 2y = -9$
24. $7x - 5y = 7$
    $3x + 2y = 6$
25. $-3x + 4y = 12$
    $x + 5y = 10$
26. $4y + 2x = 5$
    $3x - y = 7$
27. $4x + y - 6 = -5$
    $2x + 2y - 7 = 1$
28. $5y + 4x - 9 = 0$
    $7x - 6y + 4 = 5$

In Problems 29–32, try to find the solution to the system of equations by graphing. What can you conclude about each solution?

29. $x + 3y = 0$
    $-2x - 6y = 6$
30. $2x + y = 4$
    $4x + 2y = 8$
31. $3x - y = 4$
    $-9x + 3y = -12$
32. $-2x - 4y = 6$
    $x + 2y = 4$

33. Show that $(2, 3)$ and $(0, -1)$ are both solutions to the system

    $2x - y = 1$
    $4x - 2y = 2$

    How can two lines meet in more than one point?

34. Show that $(2, 3)$ and $(0, 1)$ are both solutions to the system

    $x - y = -1$
    $-x + y = 1$

35. Where do the lines $y = 8$ and $y = 4$ meet?
36. Where do the lines $x = -3$ and $x = 4$ meet?
37. One line crosses the x axis at 2 and the y axis at 7. A second line crosses the x axis at -3 and the y axis at 3. Graphically approximate the intersection of these two lines.
38. One line goes through the points $(2, 0)$ and $(-3, 4)$. A second line goes through the points $(1, 6)$ and $(-1, -2)$. Graphically approximate the intersection of these two lines.
39. One line has a slope of 3 and a y intercept of 2. A second line has a slope of -3 and a y intercept of 0. Graphically approximate the intersection of these lines.
40. Graphically find the intersection of the line $2x - 6y = 14$ and the line perpendicular to $2x - 6y = 14$ that goes through the point $(3, 0)$.
41. Find an example of a system of two linear equations in x and y that has $(2, -3)$ as the solution to the system.
42. Find an example of a system of two linear equations in x and y that has $(-1, 5)$ as the solution to the system.

3.2 Solution by Substitution

We are working with two equations in two unknowns (variables). If we could get one equation in one unknown, we could solve for that unknown. We can get one equation in one unknown by the method of substitution.

method of substitution

The **method of substitution** consists of solving for one of the variables in one of the equations and substituting that expression into the other equation. The result is an equation in one unknown. After solving this equation for the unknown, we substitute the result into either of the two original equations to find the other unknown.

EXAMPLE 3.5 Solve this system by substitution:

$$y = 3x + 1$$
$$2x + 2y = 10$$

Solution **Step 1.** Since the first equation is solved for y, we can substitute $3x + 1$ for y in the second equation and solve for x:

$$2x + 2(3x + 1) = 10$$
$$2x + 6x + 2 = 10$$
$$8x + 2 = 10$$
$$8x = 8$$
$$x = 1$$

This result is not yet the solution to the system. We still need to find y.

Step 2. We can find y by substituting 1 for x in either of the original equations:

$$y = 3(1) + 1 = 3 + 1 = 4$$

The solution is the point $(1, 4)$. This answer can be checked by substituting 1 for x and 4 for y into the original equations and showing that it is a solution to both equations.

This method is a good one to use if one equation is solved for one of the variables. However, it can be used on any system by first solving one of the equations for one of the variables. Unlike the graphical method, this method will always give the exact solution.

EXAMPLE 3.6 Solve this system by substitution:

$$-2x + 6y = 18$$
$$4x + 4y = -4$$

3.2 Solution by Substitution

Solution **Step 1.** We can solve the first equation for x (for example):

$$-2x + 6y = 18$$
$$-2x = 18 - 6y$$
$$x = -9 + 3y$$
$$x = 3y - 9$$

Step 2. Now, substitute $3y - 9$ for x in the second equation:

$$4(3y - 9) + 4y = -4$$
$$12y - 36 + 4y = -4$$
$$16y - 36 = -4$$
$$16y = 32$$
$$y = 2$$

Step 3. Substituting 2 for y into the first equation, we get

$$-2x + 6(2) = 18$$
$$-2x + 12 = 18$$
$$-2x = 6$$
$$x = -3$$

The solution is $(-3, 2)$.

Example 3.6 could have been done by first solving the first equation for y or by solving the second equation for either variable. The main goal is to get one equation in one unknown.

EXAMPLE 3.7 Solve this system by substitution:

$$5x - 7y = -12$$
$$4x + y = -3$$

Solution **Step 1.** The easiest method is to solve the second equation for y:

$$y = -4x - 3$$

Step 2. Now, substitute $-4x - 3$ into the first equation for y:

$$5x - 7(-4x - 3) = -12$$
$$5x + 28x + 21 = -12$$
$$33x + 21 = -12$$
$$33x = -33$$
$$x = -1$$

Step 3. To find y, we substitute -1 for x in either of the two original equations. We use the second equation:

$$4(-1) + y = -3$$
$$-4 + y = -3$$
$$y = 1$$

The solution is $(-1, 1)$.

Problem Set 3.2

In Problems 1–24, solve the system of equations by substitution.

1. $x = 4$
   $2x - y = 5$
2. $y = 9$
   $x - 3y = -11$
3. $3x + 5y = -2$
   $y = -1$
4. $2x - 7y = 8$
   $x = -3$
5. $y = 2x - 1$
   $2x + y = 11$
6. $y = 3x - 1$
   $3x - 2y = -4$
7. $3x - 6y = 0$
   $x = 4y - 4$
8. $2x + 9y = 24$
   $x = 6 - 3y$
9. $x + y = 7$
   $x - 2y = 10$
10. $x - y = 3$
    $2x + 3y = 21$
11. $2x + y = 4$
    $3x - y = 16$
12. $x - 3y = 2$
    $4x + 3y = 8$
13. $3x - 5y = 1$
    $2x + 3y = 7$
14. $4x - 3y = -11$
    $-3x + 6y = 0$
15. $5x + 2y = -16$
    $3x - 4y = -6$
16. $4x - 7y = 5$
    $-3x - 2y = -6$
17. $3x - 10y = 2$
    $-3x + 5y = 8$
18. $9x - 8y = -5$
    $-7x + 3y = 5$
19. $3y - 2x = 7$
    $5y - 4x = -1$
20. $5y + 10x = 0$
    $3y + 8x = 4$
21. $2x - 4y + 7 = 0$
    $4x - 5y + 14 = 0$
22. $3x + 6y - 15 = 0$
    $7x - 4y + 10 = 0$
23. $6x - 7y = 2$
    $4x + 6y = 9$
24. $3x - 4y = 22$
    $-5x + 7y = -9$

25. Solve Problem 9 by a different substitution scheme, and show that the result is the same.
26. Solve Problem 10 by a different substitution scheme, and show that the result is the same.
27. Can you use substitution to solve the system $x - 4 = 0$ and $y + 3 = 0$? How will you find the solution?
28. Can you use substitution to solve the system consisting of the x axis and the y axis?
29. The formula for converting Celsius temperature to Fahrenheit temperature is given by $F = 1.8C + 32$. Find the temperature at which Fahrenheit and Celsius temperatures are equal. (*Hint:* The second equation of the system is $F = C$.)
30. The formula $C = \frac{5}{9}(F - 32)$ is used to convert Fahrenheit temperatures to Celsius temperatures. Find the temperature at which Celsius and Fahrenheit temperatures are equal.
31. Find the intersection of two lines if one line has a slope of 2 and a y intercept of 7, and the second line has a slope of -2 and a y intercept of 4.
32. Find the intersection of two lines if one line has a slope of 3 and a y intercept of 6, and the second line has a slope of 1 and a y intercept of 0.
33. Find the intersection of the line $2x + 5y = 3$ and the line through $(2, -3)$ that never crosses the x axis.
34. Find the intersection of the line $x - 5y = 6$ and the line through $(4, 1)$ that never crosses the y axis.
35. Graph the two lines in Problem 33, and approximate their intersection.
36. Graph the two lines in Problem 34, and approximate their intersection.

3.3 Solution by Multiplication-Addition

Substitution is not always the best method to use to solve a system of equations. Whatever method is used, though, the first step in solving a system of equations is to get one equation in one unknown. We can use the rules of equality to get one equation. In particular, we will use the following two rules of equality.

Rules of Equality

1. If both sides of an equation are *multiplied* by the same nonzero number, then the result is still an equation. In other words, if $a = b$, then $ac = bc$ for any $c \neq 0$.
2. If the same expression is *added* to both sides of an equation, then the result is still an equation. In other words, if $a = b$, then $a + c = b + c$.

Consider the following example.

EXAMPLE 3.8 Solve this system:

$$2x + 3y = 1$$
$$x - 3y = 5$$

Solution We must find one equation in one unknown. In this example, we can simply add the left sides and the right sides, and the y terms will cancel, leaving $3x = 6$. We can do addition because we are adding equals to equals, which is one of the rules of equality (Rule 2).

From this point on, solving the system is similar to the method used in the previous section:

$$3x = 6$$
$$x = 2$$
$$2(2) + 3y = 1$$
$$4 + 3y = 1$$
$$3y = -3$$
$$y = -1$$
$$(2, -1)$$

Example 3.8 is a special case where one of the variables just happens to cancel out. This procedure can always be used, even if it isn't set up as nicely as the previous example. The idea is to eliminate either one of the variables. If adding the two equations doesn't eliminate a variable, we can multiply one or both equations by a constant (Rule 1) so that adding the two resulting equations will eliminate a variable.

EXAMPLE 3.9 Solve this system:

$$2x - y = -5$$
$$-3x + 2y = 9$$

Solution **Step 1.** Adding these two equations will do no good, because it will not eliminate one of the variables. However, if we use one of the other rules of equality and multiply both sides of the first equation by 2, we get

$$4x - 2y = -10$$
$$-3x + 2y = 9$$

Step 2. Now, adding these equations leaves one equation in one unknown, and the system can be solved:

$$x = -1$$
$$2(-1) - y = -5$$
$$-2 - y = -5$$
$$-y = -3$$
$$y = 3$$
$$\underline{(-1, 3)}$$

multiplication-addition method

The method illustrated in Example 3.9 is called the **multiplication-addition method**. It involves possibly multiplying one or both equations by a constant and then adding the equations.

EXAMPLE 3.10 Solve this system:
$$6x + 4y = -8$$
$$2x - 3y = 19$$

Solution If we multiply the second equation by -3, we can eliminate the x terms. It doesn't matter which variable we eliminate as long as we end up with one equation in one unknown.

$$6x + 4y = -8$$
$$-3(2x - 3y) = -3 \cdot (19)$$
$$\overline{}$$
$$6x + 4y = -8$$
$$-6x + 9y = -57$$
$$\overline{}$$
$$13y = -65$$
$$y = -5$$

3.3 Solution by Multiplication–Addition

$$6x + 4(-5) = -8$$
$$6x - 20 = -8$$
$$6x = 12$$
$$x = 2$$
$$(2, -5)$$

Many times, we may have to multiply an equation by a fraction or to multiply both equations by different numbers in order to get one variable to drop out. As the next example illustrates, if you don't wish to multiply by fractions, you can simply multiply each equation by an integer.

EXAMPLE 3.11 Solve this system:

$$5x - 2y = -7$$
$$-3x + 7y = -19$$

Solution We could multiply the first equation by $\frac{3}{5}$ and eliminate the x terms, but using fractions can get cumbersome. Instead, we will multiply the first equation by 3 and multiply the second equation by 5. In this way, we can eliminate the x terms without using fractions. We get

$$3(5x - 2y) = 3 \cdot (-7)$$
$$5(-3x + 7y) = 5 \cdot (-19)$$
$$15x - 6y = -21$$
$$-15x + 35y = -95$$
$$29y = -116$$
$$y = -4$$
$$5x - 2(-4) = -7$$
$$5x + 8 = -7$$
$$5x = -15$$
$$x = -3$$
$$(-3, -4)$$

The multiplication-addition method will work on any system of two linear equations, even those systems that may be solved easier by another method.

EXAMPLE 3.12 Solve this system:

$$2x = 3y + 8$$
$$y = -3x + 1$$

Solution The best way to do this problem is by substitution, substituting $-3x+1$ for y in the first equation. However, if we want to use multiplication-addition, we must line up the x's and y's:

$$2x - 3y = 8$$
$$3x + y = 1$$

Now, multiply the second equation by 3 and eliminate the y term:

$$2x - 3y = 8$$
$$\underline{9x + 3y = 3}$$
$$11x = 11$$
$$x = 1$$
$$2(1) - 3y = 8$$
$$2 - 3y = 8$$
$$-3y = 6$$
$$y = -2$$
$$(1, -2)$$

This procedure is the most widely used method and is the one that will be used for most examples from now on.

We can use the multiplication-addition method to solve the general system of two equations in two unknowns (Problems 39 through 44):

$$ax + by = c$$
$$dx + ey = f$$

The solution gives us formulas for x and y in terms of a, b, c, d, e, and f. This method is used by a computer in solving these types of equations. We simply tell the computer the values of the coefficients, and the computer performs the calculations by using these formulas (providing, of course, that the computer has been properly programmed for this task). We will consider these formulas in Chapter 12 when we study a general method of solving systems of equations.

■ Problem Set 3.3

In Problems 1–24, solve the system of equations by multiplication-addition.

1. $2x - 2y = 4$
   $3x + 2y = 6$

2. $6x + 3y = 3$
   $-6x - 4y = 2$

3. $7x + 3y = -1$
   $-7x - 2y = 3$

4. $5x + 5y = 10$
   $4x - 5y = 8$

5. $2x + y = -6$
   $3x - 2y = 12$

6. $3x + 5y = 3$
   $x - 6y = 24$

7. $x - 4y = 9$
   $7x + 5y = 30$

8. $11x + 4y = 21$
   $-5x - y = -12$

9. $3x - 2y = 4$
   $6x + 5y = -31$

10. $9x - 2y = 7$
    $4x - 6y = 44$

3.3 Solution by Multiplication–Addition

11. $4x + 7y = 14$
    $-3x + 5y = -31$
12. $3x + 6y = 3$
    $-5x - 4y = 29$
13. $2x - 8y = 12$
    $-11x + 7y = 45$
14. $10x + 9y = -19$
    $11x - 14y = 3$
15. $y = 2x + 7$
    $2x - 3y = -1$
16. $4y = 6x - 2$
    $x = 3y + 5$
17. $6x - 7y - 5 = 0$
    $5x - 3y - 7 = 0$
18. $10x + 6y = 0$
    $7x + 5y - 8 = 0$
19. $2x + 10y - 9 = 0$
    $7x - 14y + 8 = 0$
20. $3x - 11y + 17 = 0$
    $-4x + 9y + 3 = 0$
21. $y = 7$
    $2x + y = 9$
22. $3x - 4y = 8$
    $x = 17$
23. $x - 9 = 0$
    $4x - 2y = 11$
24. $4x - 2y = 4$
    $y + 6 = 0$

25. Solve Problem 19 by eliminating the variable you didn't eliminate originally.
26. Solve Problem 20 by eliminating the variable you didn't eliminate originally.
27. How would you use multiplication-addition on the system $x = 4$ and $y = 3$?
28. How would you use multiplication-addition on the system $y - 5 = 0$ and $x + 7 = 0$?

In Problems 29–32, describe what happens when multiplication-addition is used to try to solve the system.

29. $2x - 6y = 8$
    $x - 3y = 4$
30. $-5x + 3y = 11$
    $15x - 9y = -33$
31. $4x + y = 5$
    $8x + 2y = 2$
32. $3x - 6y = 3$
    $9x - 18y = 6$

33. The equation of a line is of the form $y = mx + b$. If this line goes through the points $(2, -4)$ and $(5, 1)$, then find m and b by using a system of two equations. (*Hint:* Substitute the values for x and y into the equation. The result is a system of two equations in m and b.)

34. A line has an equation of the form $y = mx + b$. If this lines goes through the points $(5, -5)$ and $(3, 2)$, then find m and b by using a system of two equations.

35. Find the equation of the line that goes through the point $(2, 1)$ and the intersection of the lines $2x - 3y = 6$ and $4x + 5y = 6$.

36. Find the equation of the line that is parallel to the line $2x - 5y = 6$ and goes through the intersection of the lines $2x - 4y = 1$ and $2x + 3y = 7$.

37. Find the intersection of the line $x + y = 4$ and the line perpendicular to $x + y = 4$ that goes through the origin $(0, 0)$.

38. Find the intersection of the line $x + 3y = 6$ and the line perpendicular to $2x + 3y = 6$ that goes through the origin.

For Problems 39–44, consider the following general system:

$$ax + by = c$$
$$dx + ey = f$$

39. Use the multiplication-addition method to solve for y in terms of $a, b, c, d, e,$ and f. (*Hint:* Multiply the first equation by d and the second equation by $-a$. Adding the two equations will then eliminate the x terms.)

40. Use the multiplication-addition method to solve for x in terms of $a, b, c, d, e,$ and f.

41. When will the formula for y in Problem 39 not result in a real number for y?

42. When will the formula for x in Problem 40 not result in a real number for x?

43. Use the formulas from Problems 39 and 40 to solve this system:

    $$3x - 4y = 13$$
    $$-2x + 5y = -11$$

44. Use the formulas from Problems 39 and 40 to solve this system:

    $$2x + 5y = 12$$
    $$10x - 2y = 6$$

3.4 Dependent and Inconsistent Systems

Two lines do not always meet in one point. Consider the following example.

EXAMPLE 3.13 Solve this system:

$$2x + y = 4$$
$$-4x - 2y = 4$$

Solution We can use multiplication-addition and multiply the first equation by 2:

$$4x + 2y = 8$$
$$\underline{-4x - 2y = 4}$$
$$0 = 12$$

What do we do when *both* variables disappear? We are left with the equation $0 = 12$, which is obviously never a true statement. When both variables drop out and you are left with a statement which is not true, you have a system with *no solution*. This system is an inconsistent system. A graph of the system (Figure 3.3) indicates that the two lines are *parallel*.

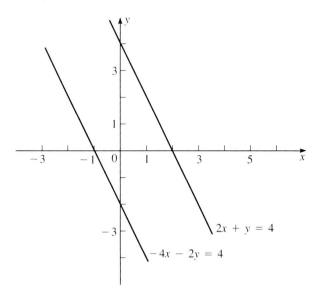

Figure 3.3
Inconsistent System

inconsistent system
consistent system

Definition 3.2. A system of linear equations which has no solution is called an **inconsistent system**. A system of linear equations that has at least one solution is a **consistent system**.

3.4 Dependent and Inconsistent Systems

In the case of a system of two linear equations, an inconsistent system occurs when we have two distinct parallel lines.

There are several procedures for determining whether a given system of linear equations is inconsistent or consistent. One method is to use either substitution or multiplication-addition. When both variables disappear and you are left with a statement which is not true, like $0 = 12$, the system is inconsistent and has no solution.

Another method is to write each equation in slope-intercept form. If the slopes of the lines are equal and the y intercepts are different, then we have an inconsistent system. Otherwise, we have a consistent system. Writing the two equations of Example 3.13 in slope-intercept form, we get

$$y = -2x + 4$$
$$y = -2x - 2$$

Both lines have a slope of -2, so they are parallel lines. Since the y intercepts are different, we have two distinct lines. So this system is inconsistent.

Now, consider the following example, where a similar problem occurs.

EXAMPLE 3.14 Solve this system:

$$-2x + 6y = 12$$
$$x - 3y = -6$$

Solution To solve this system, we multiply the second equation by 2 and add the two equations:

$$-2x + 6y = 12$$
$$\underline{2x - 6y = -12}$$
$$0 = 0$$

Both of the variables in Example 3.14 drop out, just as they did in Example 3.13, but now we are left with $0 = 0$, which is fundamentally different from $0 = 12$. The result $0 = 0$ is a *true* statement. No matter what the values of the variables are, we always get the true statement $0 = 0$.

This result seems to imply that *any* values of x and y will work. Actually, any values of x and y that work in either equation will be a solution to this system. Thus, there are an *infinite* number of solutions to this system. How can two lines meet in an infinite number of points?

A graph of this system (Figure 3.4) shows that we really only have *one* line in the system. The first equation is simply -2 times the second equation. In other words, we have two representations of the *same* line. Such a system is said to be dependent.

SYSTEMS OF LINEAR EQUATIONS

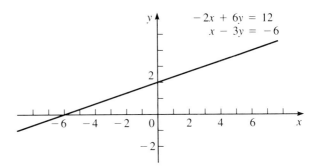

Figure 3.4
Dependent System

Alternatively, solving each equation for y gives

$y = \frac{1}{3}x + 2$
$y = \frac{1}{3}x + 2$

This result again verifies that they are the same line.

dependent system

independent system

Definition 3.3 A system of linear equations consisting of two representations of the same line is a **dependent system** and has an infinite number of solutions. A system of linear equations consisting of distinct lines is an **independent system**.

To determine whether a system of equations is dependent or independent, we can use substitution or multiplication-addition. When both variables drop out and you are left with an equation which is true, like $0 = 0$, the system is dependent and has an infinite number of solutions.

If both equations are solved for y, you should be able to tell which systems are inconsistent (identical slope, different y intercept), which systems are dependent (identical lines), and which systems are neither—that is, they are independent and consistent (different slopes). Any system of two linear equations which has a single point as its solution is an independent and consistent system.

■ **EXAMPLE 3.15** Determine whether the system is dependent or independent, consistent or inconsistent.

    a. $y = 3x - 6$
           $y = 2x - 6$

Solution   different slopes: <u>consistent and independent</u>

    b. $2x + y = 7$
           $6x + 3y = 21$

3.4 Dependent and Inconsistent Systems

Solution   Solve for y:
$$y = -2x + 7$$
$$y = -2x + 7$$
identical line: <u>consistent and dependent</u>

c. $\quad 6x - 3y = 9$
$\quad\quad -2x + y = 11$

Solution   Solve for y:
$$y = 2x - 3$$
$$y = 2x + 11$$
identical slope: <u>inconsistent and independent</u>

After working a few problems, you should be able to recognize which of the three possibilities applies to the particular problem without solving for y.

Before a computer can solve a system of equations, it must check whether the system is inconsistent or dependent. These special cases must be considered first. This point should be remembered when you are programming a computer.

Problem Set 3.4

In Problems 1–18, determine whether the system of equations is dependent or independent, consistent or inconsistent.

1. $y = 2x + 1$
   $y = 2x - 1$
2. $y = 3x - 6$
   $y = 6x - 3$
3. $y = -7x + 4$
   $y = -4x + 7$
4. $y = 6x + 8$
   $y = 6x + 8$
5. $y = x - 8$
   $y = x - 8$
6. $y = 7x - 9$
   $y = 7x + 9$
7. $y = 5x + 2$
   $-y = -5x - 2$
8. $y = 5x - 3$
   $2y = 10x - 6$
9. $y = 4x + 7$
   $3y = 12x + 21$
10. $y = -7x + 3$
    $-y = 7x - 3$
11. $2x + 3y = 7$
    $2x + 3y = 9$
12. $3x - 4y = 2$
    $6x - 8y = 4$
13. $-2x + 5y = 4$
    $4x - 10y = -8$
14. $7x - 6y = 11$
    $7x - 6y = 3$
15. $8x - 4y = 8$
    $-2x + y = 2$
16. $3x + 2y = 6$
    $2x + 3y = 6$
17. $y = 6$
    $y = 8$
18. $x = -7$
    $x = 9$

In Problems 19–30, solve the system, if possible.

19. $y = 5x - 1$
    $y = 5x + 1$
20. $y = 3x + 3$
    $y = 3x + 3$
21. $y = 7x - 7$
    $y = 7x - 7$
22. $y = 4x + 9$
    $y = 4x - 6$
23. $5x - 6y = 11$
    $5x + 7y = 13$
24. $3x - 2y = 3$
    $-6x + 4y = -6$
25. $2x + 3y = 1$
    $-8x \quad 12y - -4$
26. $2x + 5y = 7$
    $4x - 10y = 14$
27. $7x - 2y = 10$
    $14x - 4y = 5$
28. $-5x + 4y = 12$
    $15x - 12y = 1$
29. $x - 9 = 0$
    $x + 4 = 0$
30. $y + 6 = 0$
    $y - 6 = 0$

31. Can a system of two linear equations be both dependent and inconsistent? Why or why not?
32. Does it matter when both variables drop out whether you get $0 = 0$ or $7 = 7$?
33. Does it matter when both variables drop out whether you get $0 = 12$ or $-3 = 9$?
34. If a linear system consists of two perpendicular lines, can this system be dependent? Why or why not?

In Problems 35–38, consider the following system of linear equations:

$$2x + 3y = 6$$
$$-2x - 3y = f$$

35. What value(s) of f will make this system dependent?
36. What value(s) of f will make this system inconsistent?
37. What value(s) of f will make this system consistent?
38. What value(s) of f will make this system independent?
39. Find the intersection of the line $2x - y = 8$ and the line that goes through the point $(4, 0)$ and is perpendicular to the line $-x - 2y = 4$ (if possible).
40. Find the equation of the line that is parallel to the line $3x + y = 8$ and goes through the intersection of the lines $3x - 4y = 6$ and $-6x = 8 + 8y$ (if possible).

3.5 Applications

There are many applications of systems of linear equations, only a few of which will be illustrated here. Many students do not like such "word problems," although these problems show the value behind learning the mathematics. The only new part to these problems is being able to translate words into mathematical formulas. Although each problem is different, there are some general guidelines which can be used in solving these problems:

Step 1 Read the problem, identify the variables, and represent them by letters.

Step 2 Find equations in these variables.

Step 3 Solve the system of equations by using methods discussed in this chapter.

Step 4 Answer the original problem.

Let's look at these four steps one at a time. Perhaps the hardest part of these problems is translating the words into mathematical expressions. You should first read the problem and identify what you are trying to find. Often, this variable (or unknown) appears in the question at the end of the problem. If you are looking for a number, assign a letter to that number, and use this letter in your equations. If you are looking for two numbers, then you will have two variables to be represented by letters.

Once the variables are identified and represented by letters, you must find equations to represent the relationships between these numbers. The following example illustrates some of the techniques used to translate words into mathematical equations.

EXAMPLE 3.16 Identify the variables, and translate the words into mathematical equations.

a. Twice a certain number is 24.

Solution Let $x =$ the number. Twice this number is then $2x$, so we get the equation

$$2x = 24$$

3.5 Applications

b. Ten percent of a number is 16.

Solution If x represents the number, then 10% of the number is $0.10x$. So we get the equation

$$0.10x = 16$$

c. Three more than a certain number is 26.

Solution If x represents the number, then three more than the number is $x + 3$. So we get the equation

$$x + 3 = 26$$

d. The sum of two consecutive integers is 43.

Solution If we let $x =$ the first integer, then the second integer is $x + 1$, since the integers are consecutive. The equation then becomes

$$x + (x + 1) = 43$$

e. The sum of two numbers is 10, and their difference is 5.

Solution We are looking for two numbers, so we must represent these numbers by letters (variables).

Let $x =$ first number
$y =$ second number

Since the sum of these two numbers is 10, we get the equation

$$x + y = 10$$

Since the difference of these numbers is 5, we get the equation

$$x - y = 5$$

Once the problem has been translated into mathematical expressions, the problem may be solved by normal mathematical procedures. In the case of a system of two linear equations, we use substitution or multiplication-addition.

The fourth step may not seem necessary, but the solution to the system of equations may not be exactly what the problem asks for. For example, $x = -3$ may be the algebraic solution, but if x represents an age, -3 doesn't make sense. Another example is finding the length and width of a rectangle by using a system of equations in order to answer the original problem, which may be to find the area of the rectangle.

Here are a few examples which should give you a feel for working with word problems.

EXAMPLE 3.17 Two numbers add to 7. The large number is three more than the smaller number. Find the numbers.

Solution Although you can probably do this problem in your head, we would like to use algebra.

Step 1 x = smaller number
y = larger number

Step 2 $x + y = 7$
$y = x + 3$

Step 3 Now, we can use substitution to solve the system:

$$x + (x + 3) = 7$$
$$2x + 3 = 7$$
$$2x = 4$$
$$x = 2$$
$$y = 2 + 3 = 5$$
$$(2, 5)$$

Step 4 2 and 5

EXAMPLE 3.18 A car leaves town traveling 40 miles per hour (mph). Later, its speed increases to 50 mph for the rest of the trip. The entire trip was 400 miles (mi) and took 9 hours (h). How long did the car travel at each speed?

Solution **Step 1** x = time traveling at 40 mph
y = time traveling at 50 mph

Step 2 To do problems involving distance, we use the familiar formula: distance = speed · time. The distance traveled at 40 mph is $40x$, and the distance traveled at 50 mph is $50y$. We get the following system:

$$40x + 50y = 400$$
$$x + y = 9$$

Step 3 We can use multiplication-addition to solve this system:

$$\begin{array}{r} 40x + 50y = 400 \\ -40x - 40y = -360 \\ \hline 10y = 40 \end{array}$$
$$y = 4$$
$$x + 4 = 9$$
$$x = 5$$
$$(5, 4)$$

3.5 Applications

Step 4 5 h at 40 mph and 4 h at 50 mph

EXAMPLE 3.19 A 30 milliliter (ml) solution of chemicals X and Y is to contain 10% acid. Chemical X contains 14% acid, and chemical Y contains 8% acid. How many milliliters of each chemical should be mixed together to achieve the desired result?

Solution **Step 1** x = amount of chemical X to be mixed
y = amount of chemical Y to be mixed

Step 2 The total solution is 30 ml, so we get

$$x + y = 30$$

The amount of acid in chemical X is 14% of x, or $0.14x$. Similarly, the amount of acid in chemical Y is $0.08y$. The final solution is 10% acid, $0.10 \cdot 30$, or 3. Thus, we obtain the equation

$$0.14x + 0.08y = 0.10 \cdot 30 = 3$$

Step 3 To solve this system, we can first multiply the second equation by 100 to eliminate decimals. Then we solve the system:

$$
\begin{aligned}
x + y &= 30 \\
14x + 8y &= 300 \\
\hline
-8x - 8y &= -240 \\
14x + 8y &= 300 \\
\hline
6x &= 60 \\
x &= 10 \\
10 + y &= 30 \\
y &= 20 \\
(10, &\, 20)
\end{aligned}
$$

Step 4 10 ml of chemical X and 20 ml of chemical Y

Example 3.19 illustrates a simple mixture problem. Consider a more complicated example where a company is trying to mix 30 or more chemicals together to obtain a certain secret formula. We might have 30 equations in 30 unknowns. Solving such a system of equations by hand is very tedious, but computers can solve much more complicated problems very quickly. This chapter lays the foundation for the general method of solving such systems of equations, which will be developed in Chapter 12.

EXAMPLE 3.20 A teacher received a total of $1000 in interest from two accounts. The first account paid 5% interest yearly, and the second account paid 6% interest yearly. If her total investment was $17,800, how much was invested at each rate?

Solution **Step 1** x = amount invested in the first account at 5%
y = amount invested in the second account at 6%

Step 2 The total investment was 17,800, so we get

$$x + y = 17{,}800$$

The amount of interest from the first account is 5% of x, or $0.05x$. Similarly, the amount of interest from the second account is $0.06y$. The second equation then becomes

$$0.05x + 0.06y = 1000$$

Step 3 This system is solved in the same manner that Example 3.18 was solved:

$$\begin{aligned} x + y &= 17{,}800 \\ 5x + 6y &= 100{,}000 \\ \underline{-5x - 5y} &= \underline{-89{,}000} \\ 5x + 6y &= 100{,}000 \\ y &= 11{,}000 \end{aligned}$$

$$x + 11{,}000 = 17{,}800$$
$$x = 6800$$
$$(6800, 17{,}800)$$

Step 4 $6800 is invested in the first account; $11,000 is invested in the second account

■ **EXAMPLE 3.21** A rectangle is three times longer than it is wide. The perimeter is 72 inches (in.). Find the area.

Solution **Step 1** In order to find the area, we need to first find the length and width. Let

$$x = \text{length}$$
$$y = \text{width}$$

Step 2 From the first sentence, we get $x = 3y$. The perimeter is the distance around the rectangle:

$$2x + 2y = 72$$

Step 3 This system is best solved by substitution:

$$\begin{aligned} 2(3y) + 2y &= 72 \\ 6y + 2y &= 72 \\ 8y &= 72 \\ y &= 9 \\ x = 3 \cdot 9 &= 27 \end{aligned}$$

$$(27, 9)$$

3.5 Applications

Step 4 We aren't finished yet. We must find the area, using the formula

$$\text{area} = \text{length} \cdot \text{width}$$

Thus,

$$\text{area} = 27 \cdot 9$$
$$= \underline{243 \text{ in.}^2}$$

EXAMPLE 3.22 A seismometer left on the moon by the Apollo astronauts records a primary and a secondary shock wave from a moonquake. The primary wave of the quake travels 1000 feet per second (ft/s), and the secondary wave travels 800 ft/s. Find the distance from the seismometer to the moonquake if the secondary wave arrives 10 s after the primary wave.

Solution **Step 1** We are looking for the distance between the seismometer and the quake, so we can represent this distance as d. We are given information regarding velocity and time. We will define two more variables:

t_1 = time it takes for the primary wave to reach the seismometer
t_2 = time it takes for the secondary wave to reach the seismometer

Step 2 Since the distance traveled by both waves is the same (d), we get the following equation relating t_1 and t_2:

$$d = \text{speed} \cdot \text{time} = \text{speed} \cdot \text{time} = d$$
$$1000 t_1 = 800 t_2$$

Since the second wave arrives 10 s after the first wave, we get the equation

$$t_2 = t_1 + 10$$

Now, we have a system of two equations in two unknowns.

Step 3 Using substitution, we can solve this system:

$$1000 t_1 = 800(t_1 + 10)$$
$$1000 t_1 = 800 t_1 + 8000$$
$$200 t_1 = 8000$$
$$t_1 = 40 \text{ s}$$

$$t_2 = 40 + 10$$
$$= 50 \text{ s}$$

Step 4 The problem asked for the distance, not the times. The distance is

$$1000 \cdot 40, \text{ or } \underline{40{,}000 \text{ ft}}$$

By coordinating the results from several seismometers left on the moon, one can determine the exact location of the moonquake.

A computer cannot take word problems like those discussed in this section and translate them into equations. All the computer can be used for here is for step 3, solving the system of equations. The setup of the problem and the answer is up to the programmer.

Problem Set 3.5

1. Solve the problem in Example 3.16d.
2. Solve the system in Example 3.16e.

In Problems 3–10, identify the variables, and translate the words into mathematical equations.

3. Five less than a certain number is 12.
4. Seven more than a certain number is 45.
5. Three times a certain number is 24.
6. Two more than three times a number is 35.
7. The sum of two numbers is 45.
8. Thirty-five percent of a number is three more than 12.
9. One number is three more than another number.
10. Twice a larger number is four less than three times a smaller number.
11. Two numbers add to 42. If the larger number is ten more than the smaller number, then find the numbers.
12. The difference between two numbers is 4. If one number is $1\frac{1}{2}$ times the other number, find the numbers.
13. The smaller number plus three times the larger number is 10. The larger number minus twice the smaller number is 1. Find the numbers.
14. Five times the smaller number is one more than the larger number. The difference of the larger number and four times the smaller number is 3. Find the numbers.
15. The length of a rectangle is 5 in. more than its width. The perimeter is 70 in. Find the area of the rectangle.
16. Two times the length of a rectangle is five less than three times the width. The length is one more than the width. Find the length and width of the rectangle.
17. An isosceles triangle is one that has two equal sides. Find the three sides of an isosceles triangle if the two equal sides are four more than the third side and the perimeter is 20.
18. Find the three sides of an isosceles triangle if the two equal sides are 15 more than the third side and the perimeter is 180.
19. Ten pounds (lb) of corn and five lb of beans cost $4.45. Seven pounds of corn and four lb of beans cost $3.27. Find the cost per pound of each.
20. Seven cans of orange juice and three cans of tomato juice cost $9.30. Five cans of orange juice and nine cans of tomato juice cost $12.06. Find the cost per can for both items.
21. Two chemicals X and Y are to be mixed together to achieve a 10-ml solution that is 15% iodine. Chemical X is 10% iodine, and chemical Y is 30% iodine. How much of each chemical should be mixed?
22. Two chemicals X and Y are to be mixed together to achieve a 20-ml solution that is 6.5% water. Chemical X is 6% water, and chemical Y is 8% water. How much of each chemical should be mixed?
23. A restaurant owner wishes to mix orange juice and grapefruit juice together to achieve a 5-gallon (gal) solution that is 60% orange juice. How much of each juice should be added to the mixture?

3.5 Applications

24. A company wants to mix two types of coffee, one containing no caffeine and the other containing 25% caffeine, to achieve a 50-lb blend that has 5% caffeine. How much of each type should be added to the blend?

25. A football team scored twice as many touchdowns (7 points each) as field goals (3 points each). The total points they scored were 34. How many touchdowns and field goals did they make?

26. A basketball team scored nine more field goals (2 points) than they scored free throws (1 point). The total points scored were 66. How many field goals were made?

27. A shopper received 19 coins in change. All of the coins were nickels and dimes. If she got $1.35 back, how many dimes and nickels did she receive?

28. A customer bought 17 stamps. They were all 22¢ stamps or 32¢ stamps. If he spent $4.24 for stamps, how many of each type did he get?

29. A student paid $5000 for tuition and room and board. If tuition was $500 more than twice room and board, how much did the student spend for room and board?

30. An athlete swam at 2 mph and then ran at 10 mph to complete a total distance of 20 mi. If it took her a total of 4 h to complete her activities, how long did she spend at each activity?

31. A worker drove at 50 mph and then walked at 4 mph to cover a total distance of 266 mi. If his trip took him a total of 9 h, how much time did he spend at each activity?

32. A line goes through the points (2, 1) and (4, 5). Find m and b for this line ($y = mx + b$), using systems of equations.

33. A linear equation is of the form $ax + by = 8$ and goes through the points (1, −3) and (2, 9). Find a and b, using a system of equations.

34. The sum of two numbers is 4. Twice the first number plus twice the second number is 7. *Try* to find the numbers.

35. One number is twice the other number. Four times the smaller number minus two times the larger number is 0. *Try* to find the numbers.

36. Two companies, A and B, both produce computers and sell them to a school. If the total number of computers ordered by the school is 170 and company A sells 20 more than twice the number of computers company B sells to the school, then find how many computers each company sells to the school.

37. On a set of test scores, the range (the difference between the highest and the lowest test score) was 84 and the average was 52. If the high and the low test scores are dropped, the average of the remaining test scores is also 52. Find the high and the low test scores. (*Hint:* The average of the high and the low test scores must be 52.)

38. On a set of scores for a diving performance, the difference between the high and the low score is 2.6. The average score is 8.9 whether or not the high and the low scores are dropped. Find the high and the low scores.

39. Two drivers arrive at Halifax, New Hampshire, 2 h apart. If both drivers started from the same town, the first driver averaged 50 mph, and the second driver averaged 40 mph, how far did they drive?

40. A Russian Venera spacecraft on Venus records primary and secondary waves from a quake 22 s apart. If the waves travel at 1200 ft/s and 400 ft/s respectively, find the distance from Venera to the quake.

41. The cost y of producing x units is given by the formula $y = a + bx$, where a is the fixed (setup) cost and b is the variable cost. If machine A has a fixed cost of $500 and a variable cost of $150 and machine B has a fixed cost of $400 and a variable cost of $200, then find where the cost equations for these two machines intersect.

42. The total time for a certain computer program to be performed and printed is 1 h and 17 minutes (min). If the printer takes ten times longer to print the program than the computer takes to run the program, then find how long each machine takes for the program.

CHAPTER SUMMARY

In this chapter, we studied solving systems of linear equations by using the following three methods:

1. By *graphing*, where the two lines are plotted on the same graph and their intersection is visually determined
2. By *substitution*, where one of the equations is solved for either variable and that expression is then substituted into the other equation
3. By *multiplication-addition*, where the equations are multiplied by constants and then added to eliminate one of the variables

The method to be used depends on how the original problem is set up, but the graphical method is not recommended if an exact answer is required.

REVIEW PROBLEMS

In Problems 1–4, solve the system by graphing.

1. $y = 2x - 6$
   $x + y = 6$
2. $4x = y - 2$
   $x - 2y = 11$
3. $2x + 3y = 4$
   $x + 4y = 7$
4. $2x - 2y = -2$
   $5x - 3y = 1$

In Problems 5–8, determine whether the system is consistent or inconsistent, dependent or independent.

5. $3x - 2y - 6 = 0$
   $12x - 8y = -6$
6. $x - 5y = 10$
   $2x - 10y = 20$
7. $4x - 4y = 8$
   $-x + y = 4$
8. $x + 4y = 7$
   $2x - 4y = 7$

In Problems 9–12, solve the system by substitution.

9. $y = 4x - 3$
   $2x + y = 15$
10. $x = 5 - 2y$
    $3x - 7y = 15$
11. $2x + 7y = 4$
    $3x - 2y = 16$
12. $6x - 4y = 20$
    $4x + 5y = -16$

In Problems 13–16, solve by multiplication-addition.

13. $3x + y = 6$
    $-2x + 3y = -4$
14. $4x + 7y = -10$
    $2x - 3y = 8$
15. $7x - 3y = -2$
    $6x + 5y = -32$
16. $7x + 2y = 8$
    $-4x - 5y = 34$

In Problems 17–24, solve by any method.

17. $3x + 2y = 7$
    $2x - y = 0$
18. $4x + 6y = 2$
    $2x - 6y = -8$
19. $y = 6x - 7$
    $2x - 5y = 7$
20. $13x + 14y = -2$
    $-5x - 9y = 8$
21. $x - 8 = 6$
    $3x - 5y = -8$
22. $3x - 6y = 9$
    $x = 4y + 3$
23. $5x - 6y + 7 = 0$
    $-3x + 5y - 9 = 0$
24. $4x + 7y - 11 = 0$
    $-5x + 4y + 8 = 0$

25. Seven times the smaller number is two less than twice the larger number. The larger number is four more than the smaller number. Find the numbers.

26. Five bushels of potatoes and three bushels of beets cost $5.97. Seven bushels of potatoes and six bushels of beets cost $9.33. Find the cost per bushel of potatoes and beets.
27. Two drinks are to be mixed to achieve a 36 ounce (oz) drink that is 20% alcohol. The first drink is 6% alcohol, and the second drink is 24% alcohol. Find how many ounces of each drink should be mixed.
28. The length of a rectangular field is 5 yards (yd) longer than the width. Three times the length is 5 yd less than five times the width. Find the amount of fencing necessary to enclose the field.
29. A line has an x intercept of 7 and a y intercept of -4. Find the equation ($y = mx + b$) for this line, using a system of equations.
30. The cost of operating two computers is given by the formulas $c_1 = 600 + 100x$ and $c_2 = 400 + 125x$, where x is the number of hours of operation for the computers. Find the number of hours of operation for which the cost of operating the two computers is equal.

NONLINEAR FUNCTIONS

CHAPTER 4

In the previous chapters, we have studied functions that are linear in form. In this chapter, we will look at several types of nonlinear functions which are used in computers.

We will consider quadratic functions in detail, including how to graph them and find the values of x which give function values of zero.

We will also look at exponential functions, logarithms, and sequences, and we will see how these functions are handled by computers.

4.1 Completing the Square

quadratic equation

Definition 4.1 A **quadratic equation** in x is an equation of the form $ax^2 + bx + c = 0$, where a, b, and c are real numbers, $a \neq 0$.

In other words, a quadratic equation is a polynomial equation in which the highest power of x is 2.

There are several methods for finding the values of x which make the equation zero. One method that you may have seen before is **factoring**, in which the equation is written as a product of two linear factors. Solving a quadratic equation by factoring is a straightforward procedure. However, many quadratic equations don't factor easily.

factoring

We need a method that will allow us to solve all quadratic equations, not just the ones that factor easily. Consider the equation $x^2 = 4$. There are two values of x whose square is 4, namely 2 and -2. For every positive number p, there are always two numbers, $\sqrt{p}$ and $-\sqrt{p}$, whose square equals p. To solve the equation $x^2 = 4$, we simply take the square root of both sides of the equation to get $x = \pm\sqrt{4}$. We write $\pm$ to indicate that there are two answers: $+\sqrt{4}$ and $-\sqrt{4}$.

extraction of roots

This procedure works whenever the left side of the equation is a perfect square and the right side of the equation is not negative. This method is called the method of **extraction of roots**. If the right side of the equation is negative, we will be taking the square root of a negative number. In this case, there are no real solutions to the equation.

EXAMPLE 4.1   Solve for x by extraction of roots.

   a. $(x-4)^2 = 25$

4.1 Completing the Square

Solution
$$x - 4 = \pm\sqrt{25}$$
$$x - 4 = \pm 5$$
$$x = 4 \pm 5$$

$x = 4 + 5$	or	$x = 4 - 5$
$x = 9$	or	$x = -1$

b. $\quad x^2 + 6x + 9 = 10$

Solution
$$(x+3)^2 = 10$$
$$x + 3 = \pm\sqrt{10}$$
$$x = -3 \pm \sqrt{10}$$

c. $\quad 9x^2 - 6x + 1 = -5$

Solution
$$(3x-1)^2 = -5$$
$$3x - 1 = \pm\sqrt{-5}$$

Since $\sqrt{-5}$ is not real, there are no real solutions.

Most quadratic equations are not perfect squares. However, we can use a procedure called completing the square to make one side of the equation a perfect square.

We will illustrate the method by considering an example, $2x^2 + 4x - 6 = 0$. The left side is not a perfect square. The first step is to add 6 to both sides of the equation:

$$2x^2 + 4x = 6$$

Now, divide both sides by 2:

$$x^2 + 2x = 3$$

This step gets the two x terms all alone on the left side of the equation and makes the coefficient of x^2 equal to 1.

The expression $x^2 + 2x$ is not a perfect square, but $x^2 + 2x + 1$ is a perfect square. In general, adding the square of $\frac{1}{2}$ times the coefficient of x will give a perfect square at this point:

$$(\tfrac{1}{2} \cdot 2)^2 = 1^2 = 1$$

Why does this step always give a perfect square? Consider $x^2 + dx$. Add $(\tfrac{1}{2} \cdot d)^2$ and we get $x^2 + dx + d^2/4$. This quadratic equals $[x + (d/2)]^2$, which is a perfect square.

This number (1 for our example) must be added to both sides of the equation to preserve equality:

$$x^2 + 2x + 1 = 3 + 1 = 4$$

At this point, the method of extraction of roots can be used to finish the problem:

$$x^2 + 2x + 1 = 4$$
$$(x+1)^2 = 4$$
$$x + 1 = \pm\sqrt{4}$$
$$x = -1 \pm 2$$

$x = -1 + 2$ or $x = -1 - 2$
$x = 1$ or $x = -3$

The steps of this procedure are listed in the following rule.

Rule for Completing the Square For the quadratic $ax^2 + bx + c = 0$:

1. Subtract c from both sides of the equation.
2. Divide both sides of the equation by a.
3. Calculate the square of one-half of the coefficient of x, and add this number to both sides of the equation. [At this point, the left side of the equation will be the perfect square $(x+d)^2$, where d is one-half of the coefficient of x.]
4. Take the square root of both sides of the equation.
5. Solve for x.

EXAMPLE 4.2 Solve for x by completing the square.

a. $3x^2 - 12x - 6 = 0$

Solution
$$3x^2 - 12x = 6$$
$$x^2 - 4x = 2$$
$$x^2 - 4x + 4 = 2 + 4$$
$$(x-2)^2 = 6$$
$$x - 2 = \pm\sqrt{6}$$
$$x = 2 \pm \sqrt{6}$$

b. $4x^2 + 16x + 4 = 0$

Solution
$$4x^2 + 16x = -4$$
$$x^2 + 4x = -1$$
$$x^2 + 4x + 4 = -1 + 4$$
$$(x+2)^2 = 3$$
$$x + 2 = \pm\sqrt{3}$$
$$x = -2 \pm \sqrt{3}$$

4.1 Completing the Square

c. $-2x^2 + 6x + 20 = 0$

Solution
$$-2x^2 + 6x = -20$$
$$x^2 - 3x = 10$$
$$x^2 - 3x + \tfrac{9}{4} = 10 + \tfrac{9}{4}$$
$$(x - \tfrac{3}{2})^2 = \tfrac{49}{4}$$
$$x - \tfrac{3}{2} = \pm\sqrt{\tfrac{49}{4}}$$
$$x - \tfrac{3}{2} = \pm\tfrac{7}{2}$$
$$x = \tfrac{3}{2} \pm \tfrac{7}{2}$$

$x = \tfrac{3}{2} + \tfrac{7}{2}$ or $x = \tfrac{3}{2} - \tfrac{7}{2}$
$x = 5$ or $x = -2$

The method of completing the square is useful in many types of mathematical problems besides solving quadratic equations. For example, a variation of completing the square will be used again in Section 4.3 to aid graphing quadratic functions. Other applications will be illustrated in the problem set.

Problem Set 4.1

In Problems 1–20, solve for x by the method of extraction of roots.

1. $x^2 = 16$
2. $x^2 = 49$
3. $x^2 = 50$
4. $x^2 = 44$
5. $(x - 3)^2 = 36$
6. $(x + 4)^2 = 100$
7. $(x + 5)^2 = 60$
8. $(x - 7)^2 = 24$
9. $x^2 + 10x + 25 = 17$
10. $x^2 - 6x + 9 = 14$
11. $x^2 - 8x + 16 = 4$
12. $x^2 + 14x + 49 = 56$
13. $49x^2 - 28x + 4 = 12$
14. $25x^2 + 60x + 36 = 27$
15. $4x^2 - 20x + 25 = 0$
16. $9x^2 + 6x + 1 = -9$
17. $100x^2 - 20x + 1 = -16$
18. $36x^2 - 84x + 49 = 0$
19. $x^2 - \tfrac{1}{2}x + \tfrac{1}{16} = \tfrac{3}{4}$
20. $x^2 + \tfrac{2}{3}x + \tfrac{1}{9} = \tfrac{7}{3}$

In Problems 21–32, determine what must be added to the expression to make it a perfect square?

21. x^2
22. $(x - 1)^2$
23. $x^2 - 10x + 25$
24. $x^2 + 4x + 4$
25. $x^2 + 6x$
26. $x^2 - 8x$
27. $x^2 - 12x$
28. $x^2 + 14x$
29. $x^2 + 7x$
30. $x^2 - 3x$
31. $x^2 + \left(\dfrac{b}{a}\right)x$
32. $x^2 - \left(\dfrac{b}{a}\right)x$

In Problems 33–52, solve for x by the method of completing the square.

33. $x^2 + 4x = 0$
34. $x^2 - 6x = 0$
35. $x^2 + 4x - 6 = 0$
36. $x^2 - 6x + 3 = 0$

37. $x^2 - 2x - 6 = 0$
38. $x^2 + 8x - 4 = 0$
39. $x^2 + 10x - 5 = 0$
40. $x^2 + 12x + 9 = 0$
41. $2x^2 - 4x + 2 = 0$
42. $3x^2 - 12x + 12 = 0$
43. $5x^2 + 20x - 40 = 0$
44. $6x^2 + 12x + 18 = 0$
45. $-4x^2 + 16x + 12 = 0$
46. $-16x^2 + 32x - 32 = 0$
47. $x^2 = 4 - 6x$
48. $8x = 8 - x^2$
49. $2x^2 + 8 = 6x - 4$
50. $3x^2 + 3x + 5 = x^2 - x + 3$
51. $5x^2 - 5x + 7 = 0$
52. $3x^2 - 4x + 9 = 0$
53. Complete the square for the equation $ax^2 + bx + c = 0$ in general, thus getting a formula for x in terms of a, b, and c.
54. Use the formula in Problem 53 to solve Problem 37.
55. What happens in step 2 of completing the square if $a = 0$?
56. Why is $(b/2a)^2$ added to both sides? [*Hint:* Factor $x^2 + (b/a)x + (b/2a)^2$.]

4.2 Quadratic Formula

We can use the method of completing the square to find a general formula for solving quadratic equations. If we follow the steps for completing the square with the equation $ax^2 + bx + c = 0$, we will get a formula for solving quadratic equations in terms of a, b, and c:

$$ax^2 + bx + c = 0$$
$$ax^2 + bx = -c$$
$$x^2 + \frac{b}{a}x = \frac{-c}{a}$$
$$x^2 + \frac{b}{a}x + \frac{b^2}{4a^2} = \frac{-c}{a} + \frac{b^2}{4a^2}$$
$$\left(x + \frac{b}{2a}\right)^2 = \frac{b^2 - 4ac}{4a^2}$$
$$x + \frac{b}{2a} = \pm\sqrt{\frac{b^2 - 4ac}{4a^2}}$$
$$x + \frac{b}{2a} = \pm\frac{\sqrt{b^2 - 4ac}}{2a}$$
$$x = \frac{-b}{2a} \pm \frac{\sqrt{b^2 - 4ac}}{2a}$$
$$= \frac{-b \pm \sqrt{b^2 - 4ac}}{2a}$$

4.2 Quadratic Formula

quadratic formula

Thus, we have the **quadratic formula**:

$$x = \frac{-b \pm \sqrt{b^2 - 4ac}}{2a}$$

Memorizing this formula won't do you any good unless you understand what a, b, and c represent. They are the coefficients of the general quadratic equation $ax^2 + bx + c = 0$. This formula can be used to solve *any* quadratic equation. Simply substitute the values of a, b, and c into the formula and perform the calculations.

EXAMPLE 4.3 Solve the quadratic equations.
  a. $2x^2 + 3x + 1 = 0$

Solution Here, $a = 2$, $b = 3$, and $c = 1$. So

$$x = \frac{-3 \pm \sqrt{3^2 - 4 \cdot 2 \cdot 1}}{2 \cdot 2}$$

$$= \frac{-3 \pm \sqrt{9 - 8}}{4}$$

$$= \frac{-3 \pm \sqrt{1}}{4}$$

$$= \frac{-3 + 1}{4} \quad \text{or} \quad \frac{-3 - 1}{4}$$

$$= -\tfrac{1}{2} \quad \text{or} \quad -1$$

  b. $3x^2 - 7x - 9 = 0$

Solution Here, $a = 3$, $b = -7$, and $c = -9$. So

$$x = \frac{7 \pm \sqrt{(-7)^2 - 4 \cdot 3 \cdot (-9)}}{2 \cdot 3}$$

$$= \frac{7 \pm \sqrt{49 + 108}}{6}$$

$$= \frac{7 \pm \sqrt{157}}{6}$$

  c. $-2x^2 + x - 5 = 0$

Solution Here, $a = -2$, $b = 1$, and $c = -5$. So

$$x = \frac{-1 \pm \sqrt{1^2 - 4 \cdot (-2) \cdot (-5)}}{2 \cdot (-2)}$$

$$= \frac{-1 \pm \sqrt{1 - 40}}{-4}$$

$$= \frac{-1 \pm \sqrt{-39}}{-4}$$

Thus, there are no real solutions (the square root of a negative number is not real).

Note that we cannot use the quadratic formula until we have rewritten the equation in the form $ax^2 + bx + c = 0$. We need to rewrite the equation in this form in order to find the values of a, b, and c.

EXAMPLE 4.4 Solve this quadratic equation:

$$4x^2 + 1 = 3x^2 - 6x + 7$$

Solution We cannot use the quadratic formula until we have rewritten the equation in the form $ax^2 + bx + c = 0$. Thus,

$$x^2 + 1 = -6x + 7$$
$$x^2 + 6x + 1 = 7$$
$$x^2 + 6x - 6 = 0$$

So $a = 1$, $b = 6$, and $c = -6$. Now, we solve for x:

$$x = \frac{-6 \pm \sqrt{6^2 - 4 \cdot 1 \cdot (-6)}}{2 \cdot 1} = \frac{-6 \pm \sqrt{36 + 24}}{2}$$

$$= \frac{-6 \pm \sqrt{60}}{2} = \frac{-6 \pm 2\sqrt{15}}{2}$$

$$= -3 \pm \sqrt{15}$$

Quadratic equations can have two, one, or no real solutions. The number of solutions is determined by the discriminant.

Definition 4.2 For the quadratic equation $ax^2 + bx + c = 0$, $b^2 - 4ac$ is the **discriminant**.

discriminant

If the discriminant is positive, then there are two real solutions, since the square root of a positive number is real. If the discriminant is negative, then there

4.2 Quadratic Formula

are no real solutions, since the square root of a negative number is not real. If the discriminant is zero, then there is one real solution, since we are adding and subtracting zero.

EXAMPLE 4.5 Determine how many real solutions the equation has.

a. $4x^2 - 5x + 7 = 0$

Solution
$$b^2 - 4ac = (-5)^2 - 4 \cdot 4 \cdot 7 = 25 - 112$$
$$= -87$$

no real solutions

b. $5x^2 + 7x + 2 = 0$

Solution
$$b^2 - 4ac = 7^2 - 4 \cdot 5 \cdot 2 = 49 - 40$$
$$= 9$$

two real solutions

c. $x^2 - 4x + 4 = 0$

Solution
$$b^2 - 4ac = (-4)^2 - 4 \cdot 1 \cdot 4 = 16 - 16$$
$$= 0$$

one real solution

EXAMPLE 4.6 The profit from selling x packages of computer paper is given by the formula $p = 3x^2 + 12x - 35$. How many packages must be sold to achieve a profit of $100?

Solution We must solve the formula for x after substituting 100 for p:

$$100 = 3x^2 + 12x - 35$$
$$0 = 3x^2 + 12x - 135$$
$$0 = x^2 + 4x - 45$$

$a = 1, \quad b = 4, \quad c = -45$

$$x = \frac{-4 \pm \sqrt{4^2 - 4 \cdot 1 \cdot (-45)}}{2 \cdot 1}$$

$$= \frac{-4 \pm \sqrt{16 + 180}}{2} = \frac{-4 \pm \sqrt{196}}{2}$$

$$= \frac{-4 \pm 14}{2}$$

$x = \frac{10}{2}$ and $x = -\frac{18}{2}$
$x = 5$ and $x = -9$

In this problem, -9 makes no sense for a value of x (we can't sell -9 packages). So the answer becomes 5 packages.

Problem Set 4.2

In Problems 1–10, use the discriminant to find the number of real solutions the equation has.

1. $x^2 + x + 1 = 0$
2. $x^2 + 3x + 2 = 0$
3. $x^2 - 4x + 7 = 0$
4. $x^2 - 3x + 6 = 0$
5. $x^2 - 9x + 12 = 0$
6. $x^2 + 8x + 16 = 0$
7. $2x^2 - 20x + 50 = 0$
8. $5x^2 - 3x + 7 = 0$
9. $9x^2 - 6x + 8 = 0$
10. $-3x^2 + 12x + 12 = 0$

In Problems 11–34, use the quadratic formula to find all real solutions for the equation.

11. $x^2 + x + 1 = 0$
12. $x^2 + 3x + 2 = 0$
13. $x^2 - 4x + 7 = 0$
14. $x^2 - 3x + 6 = 0$
15. $x^2 - 9x + 12 = 0$
16. $x^2 + 8x + 16 = 0$
17. $2x^2 - 20x + 50 = 0$
18. $5x^2 - 3x + 7 = 0$
19. $9x^2 - 6x + 8 = 0$
20. $-3x^2 + 12x + 12 = 0$
21. $x^2 + 4x - 7 = 9$
22. $6x^2 + 5 - 7x = 0$
23. $x^2 = 9$
24. $x^2 - 4 = 0$
25. $x^2 + 4x - 6 = 2x^2 + 9$
26. $x^2 - 5x = 6x^2 + 10x$
27. $7x^2 + 16 = 0$
28. $3x^2 + 12x = 0$
29. $(x - 3)(x - 4) = 0$
30. $(2x + 1)(x - 4) = 0$
31. $(x + 1)(x + 2) = 5$
32. $(x - 4)^2 = 0$
33. $(2x + 5)^2 = 9$
34. $(x + 1)^2 = (x - 1)^2$

35. If the discriminant is zero, there is only one real solution. Find a formula for this one solution.
36. Find a formula for the sum of the two solutions given by the quadratic formula.
37. Find a formula for the product of the two solutions given by the quadratic formula.

In Problems 38–40, write the quadratic equation whose solutions are given.

38. $x = \dfrac{-4 \pm \sqrt{4^2 - 4 \cdot 3 \cdot (-5)}}{2 \cdot 3}$

39. $x = \dfrac{5 \pm \sqrt{(-5)^2 - 4 \cdot 5 \cdot (-4)}}{2 \cdot 5}$

40. $x = \dfrac{3 \pm \sqrt{(-3)^2 - 4 \cdot (-2) \cdot 3}}{2 \cdot (-2)}$

41. Write user-defined BASIC functions for the two solutions given by the quadratic formula.
42. Write a user-defined BASIC function for the discriminant.
43. The cost of producing x computer desks is given by the formula $C(x) = 1600 + 100x - 2x^2$. How many desks can be produced at a cost of $400?
44. For what value of c will the equation $2x^2 + 16x - c = 0$ have exactly one real root?

4.3 Graphing Quadratic Functions

quadratic function

Definition 4.3 A **quadratic function** is a function of the form

$$f(x) = ax^2 + bx + c$$

4.3 Graphing Quadratic Functions

or simply $y = ax^2 + bx + c$, $a \neq 0$.

Since this function is not linear, the graph will not be a straight line. If we want to investigate what the graph of a quadratic function looks like, we simply plot points until the shape of the graph becomes clear.

EXAMPLE 4.7 Graph $y = 2x^2 - 4x - 16$.

Solution The points are tabulated in the following table. The graph is shown in Figure 4.1.

x	y
5	14
4	0
3	−10
2	−16
1	−18
0	−16
−1	−10
−2	0
−3	14

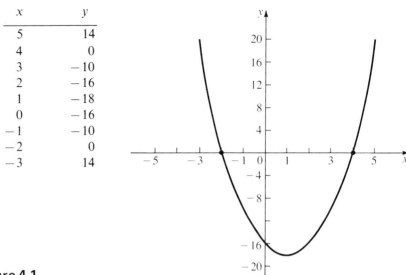

Figure 4.1

parabola

The graph in Example 4.7 is a **parabola**, and every quadratic function has a parabolic graph.

EXAMPLE 4.8 Graph $y = -x^2 + 4x - 4$.

Solution The points are tabulated in the following table. The graph is shown in Figure 4.2.

x	y
4	−4
3	−1
2	0
1	−1
0	−4

Figure 4.2

The point(s) where the graph crosses the x axis ($y = 0$) correspond to the solutions in the previous sections. Notice the difference in the graphs of Examples 4.7 and 4.8. If $a > 0$, the parabola has a low point (*minimum*) and opens up. If $a < 0$, the parabola has a high point (*maximum*) and opens down. This turning point is called the *vertex* of the parabola.

The vertical line through the vertex is called the *axis of symmetry*. If you go the same distance left or right of the axis of symmetry, the graph will always have the same y value. For example, in Example 4.7, the vertex is $(1, -18)$, and the axis of symmetry is $x = 1$. Moving three units to the right, we get $x = 4$ $(4, 0)$, and moving three units to the left we get $x = -2$ $(-2, 0)$. Both of these points have the same y value (0).

Knowing this symmetry makes graphing quadratic functions easier. For example, if the axis of symmetry is $x = 3$ and the graph contains the point $(1, -14)$, we know that it must also contain the point $(5, -14)$ since 1 and 5 are both two units from 3.

There is an easy way to find the vertex and axis of symmetry of a parabola. Rewrite the function in the form $y = a(x-h)^2 + k$ by factoring a from the x terms, using the distributive property. Then complete the square.

The steps of the procedure are listed next.

Procedure for Finding the Vertex and Axis of Symmetry For the expression $y = ax^2 + bx + c$:

1. Factor a from the x terms.
2. Add and subtract $[b/(2a)]^2$ to the x terms.
3. Simplify to the form $y = a(x-h)^2 + k$.

We are now ready for the following definition.

Definition 4.4 For a quadratic function of the form $y = a(x-h)^2 + k$, the **vertex** is the point (h, k), and the **axis of symmetry** is $x = h$.

vertex, axis of symmetry

The following two examples illustrate the procedure for finding the vertex and axis of symmetry of a quadratic equation.

EXAMPLE 4.9 Find the vertex and axis of symmetry of the function $y = 2x^2 + 8x - 4$.

Solution

$y = 2(x^2 + 4x) - 4$ factor 2 from the x terms
$\quad = 2(x^2 + 4x + 4 - 4) - 4$ add and subtract 4 to the x terms
$\quad = 2[(x^2 + 4x + 4) - 4] - 4$
$\quad = 2[(x+2)^2 - 4] - 4$
$\quad = 2(x+2)^2 - 8 - 4$
$\quad = 2(x+2)^2 - 12$
$\quad = 2[x - (-2)]^2 + (-12)$ simplify to the correct form

vertex: $(-2, -12)$
axis of symmetry: $x = -2$

4.3 Graphing Quadratic Functions

EXAMPLE 4.10 Find the vertex and axis of symmetry, and graph the function $y = 3x^2 + 6x + 1$.

Solution
$$y = 3(x^2 + 2x) + 1 = 3(x^2 + 2x + 1 - 1) + 1$$
$$= 3[(x^2 + 2x + 1) - 1] + 1 = 3[(x + 1)^2 - 1] + 1$$
$$= 3(x + 1)^2 - 3 + 1 = 3(x + 1)^2 - 2$$
$$= 3[x - (-1)]^2 + (-2)$$

vertex: $(-1, -2)$
axis of symmetry: $x = -1$

The points are tabulated in the following table. The graph is shown in Figure 4.3.

x	y
2	25
1	10
0	1
-1	-2
-2	
-3	
-4	

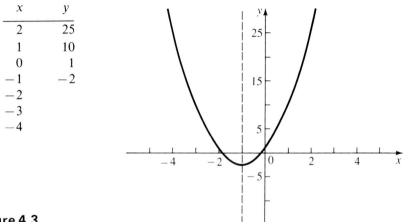

Figure 4.3

Note that the points $(-2, 1)$, $(-3, 10)$, and $(-4, 25)$ can be inferred from symmetry and don't need to be calculated.

Quadratic functions are used to describe many things, including the pull of gravity, profit, and cost. The following example gives one such application.

EXAMPLE 4.11 In an engineering test of its engines, an unmanned Surveyor spacecraft on the moon fired its small engines to "hop" on the moon. If the height above the moon (in feet) t seconds later is given by the formula $h = 60t - 3t^2$, how high will the spacecraft get?

Solution We can use the method of the two previous examples and find the vertex of the parabola, since the high point will occur at the vertex:

$$h = -3t^2 + 60t = -3(t^2 - 20t) = -3(t^2 - 20t + 100 - 100)$$
$$= -3[(t^2 - 20t + 100) - 100] = -3[(t - 10)^2 - 100]$$
$$= -3(t - 10)^2 + 300 \text{ vertex: } (10, 300)$$

The maximum height reached is 300 feet, which occurs 10 seconds (s) after the firing of the engines.

More of these applications will be illustrated in the problem set.

Problem Set 4.3

In Problems 1–18, find the vertex and axis of symmetry of the function.

1. $y = (x-1)^2$
2. $y = (x+4)^2$
3. $y = (x+2)^2 - 3$
4. $y = (x-5)^2 + 7$
5. $y = 2(x-6)^2 + 4$
6. $y = 3(x+8)^2 - 5$
7. $y = -4(x-4)^2 + 3$
8. $y = -7(x-1)^2 - 11$
9. $y = 17 - 5(x+3)^2$
10. $y = -45 + 7(x+5)^2$
11. $y = x^2 + 4x + 2$
12. $y = x^2 - 6x + 11$
13. $y = x^2 + 5x - 9$
14. $y = x^2 + 9x + 9$
15. $y = 2x^2 - 12x + 4$
16. $y = 3x^2 + 18x - 14$
17. $y = -4x^2 + 6x - 7$
18. $y = -5x^2 + 13x - 6$

In Problems 19–34, plot at least five points, and graph the function.

19. $y = x^2 + 2x + 1$
20. $y = x^2 - 4x + 4$
21. $y = x^2 + 3x - 10$
22. $y = x^2 - 5x - 14$
23. $y = -x^2$
24. $y = -x^2 - 9$
25. $y = (x-5)^2$
26. $y = (x+4)^2$
27. $y = -(x+1)^2 + 4$
28. $y = -(x-7)^2 - 10$
29. $y = 3(x-4)^2 - 3$
30. $y = 2(x+6)^2 + 4$
31. $y = 3x^2 + 5x - 3$
32. $y = 2x^2 + 5x - 9$
33. $y = -2x^2 + 5x - 4$
34. $y = -x^2 + 3x + 2$

35. What happens to a quadratic function if $a = 0$?
36. What can you say about a quadratic function in which $c = 0$?
37. If the axis of symmetry of a quadratic equation is $x = 0$, what do you know about the graph of the equation?
38. If $a < 0$ and the vertex is $(-1, -3)$, how many points of this quadratic equation will be on the x axis?
39. Find the equations of two different quadratic functions that both have a vertex of $(3, 4)$.
40. Find the equations of two different quadratic functions that both have an axis of symmetry of $x = -7$.
41. The profit P for selling x calculators is given by the formula $P(x) = -14x^2 + 84x - 70$. How many calculators should be sold to *maximize* profit?
42. The profit $P(x)$ in dollars for selling x computer books is given by the formula $P(x) = -10x^2 + 100x - 160$. What is the maximum profit?

4.4 Exponential Functions

43. The production cost $C(x)$ for x hours of work is given by the formula $C(x) = 7x^2 - 21x + 28$. What is the minimum production cost?
44. The production cost $C(x)$ for x hours of work is given by the formula $C(x) = 11x^2 - 22x - 880$. How many hours of work will *minimize* cost?

For Problems 45–46, consider the Surveyor spacecraft of Example 4.11. Suppose it had performed the same experiment on Mimas (a moon of Saturn). In this case, its height is given by the formula $h = -0.5t^2 + 120t$.

45. How high will the spacecraft get?
46. When will the spacecraft land?
47. Transform the general equation $y = ax^2 + bx + c$ into $y = a(x-h)^2 + k$, and find general formulas for h and k.
48. Write user-defined BASIC functions for the formulas for h and k found in Problem 47.

In Problems 49–50, write user-defined BASIC functions for the quadratic functions.

49. $y = 2x^2 - 3x + 4$
50. $y = -4x^2 - 3x + 11$

4.4 Exponential Functions

Up to this point, we have only dealt with functions in which the base was a variable and the exponent was a constant (for example, x^2). In this section, we will consider functions in which the exponent itself is a variable.

exponential function

Definition 4.5 An **exponential function** is a function of the form

$$f(x) = a^x \qquad a > 0.$$

In Definition 4.5, a is a constant and x is a variable. For example, $f(x) = 2^x$ is the exponential function with base 2. The base can be *any* real number larger than zero.

Why can't we use $a = 0$ or $a < 0$? These questions will be posed in the problem set (Problems 19–22).

Exponential functions are used to describe such things as population growth, radioactive decay, interest rates, and logarithms (see Section 4.5).

EXAMPLE 4.12 Graph $y = 2^x$ and $y = (\frac{1}{2})^x$ on the same graph.

Solution The points are tabulated in the following tables. The graphs are shown in Figure 4.4.

x	$y = 2^x$
0	$2^0 = 1$
1	$2^1 = 2$
-1	$2^{-1} = \frac{1}{2}$
2	$2^2 = 4$
-2	$2^{-2} = \frac{1}{4}$

x	$y = (\frac{1}{2})^x$
0	$(\frac{1}{2})^0 = 1$
1	$(\frac{1}{2})^1 = \frac{1}{2}$
-1	$(\frac{1}{2})^{-1} = 2$
2	$(\frac{1}{2})^2 = \frac{1}{4}$
-2	$(\frac{1}{2})^{-2} = 4$

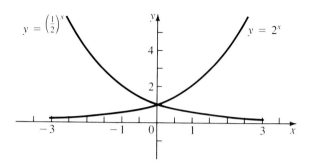

Figure 4.4

Some general observations can be made about the graphs of exponential functions.

Graphs of Exponential functions $(y = a^x)$

1. The domain is all real numbers.
2. The range is $\{y \mid y > 0\}$.
3. $f(0) = 1$.
4. If $a > 1$, the graph is always increasing from left to right.
5. If $0 < a < 1$, the graph is always decreasing from left to right.

What happens when $a = 1$? This event is explored in Problem 17.

Since $a > 0$, a^x will always be positive. The negative exponents a^{-x} are simply $1/a^x$ and will still be positive.

The rules for exponents that we learned in Chapter 1 still hold with variable exponents.

■ EXAMPLE 4.13 Simplify the expression:

   a. $2^x \cdot 2^y$ **b.** $3^x \cdot 4^x$

Solution $2^x \cdot 2^y = \underline{2^{x+y}}$ $3^x \cdot 4^x = (3 \cdot 4)^x = \underline{12^x}$

   c. $\dfrac{7^{2x}}{7^{4x}}$ **d.** $(5^{x+1})^{3x}$

Solution $\dfrac{7^{2x}}{7^{4x}} = 7^{2x-4x} = \underline{7^{-2x}}$ $(5^{x+1})^{3x} = 5^{(x+1)(3x)} = \underline{5^{3x^2 + 3x}}$

   e. $\dfrac{9^{3-y}}{3^{3-y}}$

Solution $\dfrac{9^{3-y}}{3^{3-y}} = \underline{3^{3-y}}$

4.4 Exponential Functions

Many computers and calculators have built-in functions to handle exponential functions. For example, BASIC has a built-in function EXP(x), which returns the value of e^x, where $e \approx 2.71828$. The number e is irrational and is very important in higher mathematics. In the next section, we will see how EXP(x) can be used to find values of a^x for *any* $a > 0$.

After studying exponential functions, we can consider variations of $y = a^x$ where the exponent is an expression in x ($y = 4^{2x}$) or where there is a coefficient in front of a^x ($y = 2 \cdot 3^x$).

EXAMPLE 4.14 Graph $y = 2 \cdot 3^{-x}$.

Solution The points are tabulated in the following table. The graph is shown in Figure 4.5.

x	$2 \cdot 3^{-x}$
0	$2 \cdot 3^0 = 2 \cdot 1 = 2$
1	$2 \cdot 3^{-1} = 2 \cdot \frac{1}{3} = \frac{2}{3}$
2	$2 \cdot 3^{-2} = 2 \cdot \frac{1}{9} = \frac{2}{9}$
-1	$2 \cdot 3^1 = 2 \cdot 3 = 6$
-2	$2 \cdot 3^2 = 2 \cdot 9 = 18$

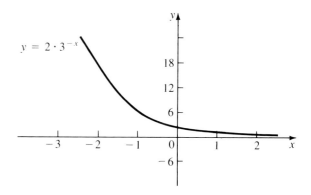

Figure 4.5

EXAMPLE 4.15 Graph $y = 4^{2x}$.

Solution The points are tabulated in the following table. The graph is shown in Figure 4.6.

x	4^{2x}
0	$4^0 = 1$
$\frac{1}{2}$	$4^1 = 4$
1	$4^2 = 16$
2	$4^4 = 256$
$-\frac{1}{2}$	$4^{-1} = \frac{1}{4}$
-1	$4^{-2} = \frac{1}{16}$
-2	$4^{-4} = \frac{1}{256}$

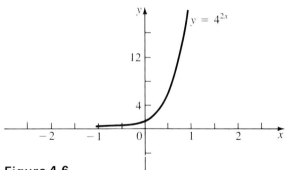

Figure 4.6

Problem Set 4.4

In Problems 1–16, plot at least five points, and graph.

1. $y = 3^x$
2. $y = 4^x$
3. $y = 5^x$
4. $y = 10^x$
5. $y = (\frac{1}{3})^x$
6. $y = (\frac{1}{4})^x$
7. $y = (0.2)^x$
8. $y = (0.1)^x$
9. $y = 5^{-x}$
10. $y = 10^{-x}$
11. $y = (0.2)^{-x}$
12. $y = (0.1)^{-x}$
13. $y = 2^{3x}$
14. $y = 2^{2x}$
15. $y = 3^{1/x}$
16. $y = 64^{1/x}$

17. What does the graph of $y = 1^x$ look like?
18. Why do exponential functions always go through the point $(0, 1)$?
19. What does the graph of $y = 0^x$ look like?
20. What happens when $x = 0$ in $y = 0^x$?
21. Consider $y = (-4)^x$. Find values for $x = -2, -1, 0, \frac{1}{2}, 1, 2$.
22. Consider $y = (-9)^x$. Find values for $x = -2, -1, 0, \frac{1}{2}, 1, 2$.
23. When does an exponential function equal zero?
24. When is an exponential function less than zero?
25. For what values of x is an exponential function undefined?
26. Can the graph of an exponential function ever "change direction" like a quadratic function does?

In Problems 27–38, simplify the expression (if possible).

27. $2^x \cdot 3^x$
28. $4^{-x} \cdot 4^{3x}$
29. $7^x + 7^x$
30. $\dfrac{4^x}{4^{2x}}$
31. $(5^x)^x$
32. $(3^{3x})^{x-3}$
33. $\dfrac{6^{1-x}}{3^{1-x}}$
34. $\dfrac{8^{2x}}{8^9}$
35. $2^{x-2} \cdot 2^{2-x}$
36. $12^{4x} \cdot 5^{4x}$
37. $\dfrac{e^{-2x}}{e^{1+x}}$
38. $e^x + e^y$

In Problems 39–42, for the EXP(x) function from BASIC, find (approximately) the value.

39. EXP(1)
40. EXP(0)
41. EXP(2)
42. EXP(−1)

Suppose the population of a city is given by the formula

$$P = 20{,}000(3^{0.05t})$$

where t represents the time (in months). In Problems 43–45, find the population of the city at the specified time.

43. $t = 0$ months
44. $t = 20$ months
45. $t = 40$ months

A radioactive substance decays at such a rate that the amount of the substance remaining (y) after t days is given by the formula $y = 100(4^{-0.02t})$. In Problems 46–48, find the amount of the substance left at the specified time.

46. $t = 0$ days
47. $t = 25$ days
48. $t = 50$ days
49. If P dollars are invested in an account that pays interest I compounded continually, then the amount of the investment after t years is $A = Pe^{tI}$. Write a user-defined BASIC function using EXP(x) to represent this formula.
50. If P dollars are invested in an account that pays interest I compounded m times a year, then the amount of the investment after t years is $A = P(1 + (I/m))^{tm}$. Find the amount in the account after two years if $10,000 is invested into the account that pays 8% (0.08) interest compounded twice a year.

4.5 Logarithmic Functions

In the previous section, we discussed exponential functions in which the variable x was the exponent ($y = a^x$). In this section, we will discuss a similar type of function, called a **logarithmic function**, in which y is the exponent. The roles of x and y are switched in going from exponential to logarithmic functions.

logarithmic function

Consider the exponential function $y = a^x$. If we interchange x and y, we get $x = a^y$. Here, y is in the exponent. Since functions are normally solved for y, we must invent a notation to represent that y is now the exponent. The notation that has been chosen is $\log_a x$.

logarithm

Definition 4.6 $y = \log_a x$ if and only if $x = a^y$. The term y is the **logarithm** base a of x, where $a > 0$.

An understanding of Definition 4.6 allows us to convert exponential to logarithmic notation, and vice versa, and to solve simple logarithmic equations.

EXAMPLE 4.16 Convert to logarithmic notation.

Solution
a. $x = 4^y$ $y = \log_4 x$
b. $7 = 3^y$ $y = \log_3 7$
c. $8 = 2^3$ $3 = \log_2 8$
d. $64 = 4^3$ $3 = \log_4(64)$
e. $8 = 32^{0.6}$ $0.6 = \log_{32} 8$

EXAMPLE 4.17 Convert to exponential notation.

Solution
a. $y = \log_3 x$ $x = 3^y$
b. $y = \log_6 5$ $5 = 6^y$
c. $4 = \log_2 16$ $16 = 2^4$
d. $-3 = \log_3(\frac{1}{27})$ $\frac{1}{27} = 3^{-3}$
e. $\frac{4}{5} = \log_{32} 16$ $16 = 32^{4/5}$

EXAMPLE 4.18

Use Definition 4.6 to solve for x.

a. $x = \log_2 8$

Solution
$$8 = 2^x$$
$$2^3 = 2^x$$
$$x = 3$$

b. $-2 = \log_x\left(\dfrac{1}{100}\right)$

Solution
$$\dfrac{1}{100} = x^{-2}$$
$$\dfrac{1}{100} = \dfrac{1}{x^2}$$
$$\dfrac{1}{10^2} = \dfrac{1}{x^2}$$
$$10^2 = x^2$$
$$x = 10$$

Actually, there are two values of x that satisfy the equation $10^2 = x^2$: 10 and -10. Since the base must be positive, x cannot be -10.

c. $4 = \log_3 x$

Solution
$$x = 3^4$$
$$x = 81$$

We can now graph logarithmic functions by using Definition 4.6 and plotting points until the shape of the graph becomes clear.

EXAMPLE 4.19

Graph $y = \log_2 x$.

Solution Since $y = \log_2 x$, then $x = 2^y$. The points are tabulated in the following table.

x	$x = 2^y$	y
1	$1 = 2^y$	0
2	$2 = 2^y$	1
4	$4 = 2^y$	2
8	$8 = 2^y$	3
16	$16 = 2^y$	4
0.5	$0.5 = 2^y$	-1
0.25	$0.25 = 2^y$	-2

Since x is a power of 2, x cannot be zero or negative. The graph of the function is shown in Figure 4.7.

4.5 Logarithmic Functions

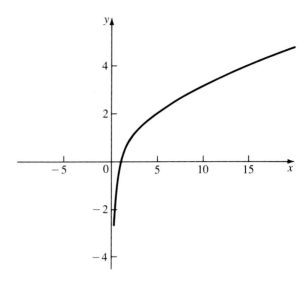

Figure 4.7

The graph in Example 4.19 is similar to the graph of $y = 2^x$ (see Figure 4.4) except that all of the points have been interchanged [(0, 1) becomes (1, 0), and so on]. This result leads to some general observations about the graphs of logarithmic functions.

Graphs of Logarithmic Functions ($y = \log_a x$)

1. The domain is $\{x | x > 0\}$.
2. The range is all real numbers.
3. $f(1) = 0$.
4. If $a > 1$, the graph is always increasing from left to right.
5. If $0 < a < 1$, the graph is always decreasing from left to right.

Compare these observations for logarithmic functions with the ones for exponential functions in the previous section. These two types of functions are closely related.

The base of a logarithmic function can be any positive real number. Logarithms base 10 are used quite often. They are called **common logarithms**, and the base is not always written:

common logarithms

$$y = \log x$$

Logarithms were used for many years to simplify complicated calculations, and every textbook had logarithmic tables in the back. With the advent of calculators and computers, logarithmic tables are no longer needed. Most computer languages have a built-in function to find common logarithms or *natural logarithms*

natural logarithms, which are logarithms based on the number e (≈ 2.71828), which is a very important number in calculus. The natural logarithm $\log_e x$ is commonly referred to as $\ln x$ in many mathematical textbooks.

In the BASIC programming language, there is a built-in function LOG(x), which finds the natural logarithm of x. If a different base is desired, it can be calculated by the following formula:

$$\log_a x = \frac{\log_e x}{\log_e a}$$

So one built-in logarithmic function is enough for a computer to calculate logarithms of any base (see Problems 61 and 62).

Logarithms are used for many scientific purposes and are actually used by computers to perform exponentiation of real numbers. Thus, calculating an integer power, like 2^4, is performed by repeated multiplication by a computer, but evaluating noninteger exponents requires logarithms. The following formula is used by many computers to find noninteger exponents:

$$a^x = e^{x \ln a}$$

where $\ln x$ is $\log_e x$ and $a > 0$. So one built-in exponential function is enough for a computer to calculate exponents of every base (see Problems 63 and 64).

Note that some computer languages, PASCAL, for example, have no exponential operation. Repeated multiplication is used for positive integer exponents. Any other exponentiation must be written by the programmer.

Problem Set 4.5

In Problems 1–10, convert to logarithmic notation.

1. $y = 3^x$
2. $y = 6^r$
3. $y = a^x$
4. $y = c^p$
5. $2 = 3^m$
6. $7 = 4^b$
7. $16 = 2^4$
8. $27 = \left(\frac{1}{3}\right)^{-3}$
9. $\frac{4}{9} = \left(\frac{8}{27}\right)^{2/3}$
10. $\frac{1}{y} = (4k)^u$

In Problems 11–20, convert to exponential notation.

11. $y = \log_4 x$
12. $y = \log_x 5$
13. $y = \log_{11} 52$
14. $y = \log_5 4$
15. $-7 = \log_4 t$
16. $5 = \log_9 x$
17. $2 = \log_7 49$
18. $-6 = \log_{0.5} 64$
19. $\log_3 81 = 4$
20. $\log_5 125 = 3$

In Problems 21–40, solve for x.

21. $x = \log_2 16$
22. $x = \log_7 343$
23. $x = \log_8 \left(\frac{1}{8}\right)$
24. $x = \log_2 \left(\frac{1}{4}\right)$
25. $x = \log_{10} (1000)$
26. $x = \log_{10} \left(\frac{1}{1000}\right)$
27. $x = \log_{16} 64$
28. $x = \log_9 27$
29. $2 = \log_3 x$
30. $4 = \log_2 x$
31. $-3 = \log_4 x$
32. $-5 = \log_3 x$
33. $3 = \log_{1/2} x$
34. $\frac{1}{4} = \log_{16} x$
35. $2 = \log_x 36$
36. $3 = \log_x 125$
37. $-3 = \log_x \left(\frac{1}{64}\right)$
38. $-2 = \log_x 9$
39. $\frac{1}{3} = \log_x 2$
40. $-\frac{1}{2} = \log_x 7$

In Problems 41–50, graph the logarithmic function.

41. $y = \log_3 x$
42. $y = \log_4 x$
43. $y = -\log_5 x$
44. $y = -\log_6 x$

4.6 Sequences

45. $y = \log_{1/2} x$
46. $y = \log_{1/3} x$
47. $y = \log_{10} 2x$
48. $y = \log_3(-x)$
49. $-y = \log_2 x$
50. $-y = \log_6 x$

51. What does the graph of $y = \log_1 x$ look like?
52. Why do logarithmic functions always go through the point $(1, 0)$?
53. What does the graph of $y = \log_a x$ look like if we are allowed to let $a = 0$.
54. What happens when $y = 0$ in $y = \log_0 x$?
55. When will a logarithmic function equal zero?
56. When will a logarithmic function be negative?
57. For what values of x is a logarithmic function undefined?
58. Can the graph of a logarithmic function ever "change direction" like a quadratic function does?
59. Graph $y = 5^x$ and $y = \log_5 x$ on the same graph. Do they intersect?
60. Graph $y = 2^x$ and $y = \log_2 x$ on the same graph. Do they intersect?
61. Define a function in BASIC to use LOG(x) to find the common logarithm of x.
62. Define a function in BASIC using LOG(x) to find the logarithm base 16 of x.
63. Define a function in BASIC to use LOG(x) and EXP(x) to find 2^x.
64. Define a function in BASIC to find 8^x.

How does the function EXP(x) find e^x? To approximate e^x, we often use a formula similar to

$$e^x \approx 1 + x + \frac{x^2}{2} + \frac{x^3}{6} + \frac{x^4}{24}$$

In Problems 65–68, use this formula to approximate the expression.

65. e^2
66. e
67. $e^{1.5}$
68. $e^{0.5}$

Animal populations can often be approximated by logarithms. Suppose that the number of penguins (y) in a certain area t months after they are introduced into the area is given by the formula $y = 100 \log_{10}(3t + 10)$. Use this information in Problems 69–71.

69. How many penguins were originally introduced into the area ($t = 0$)?
70. When will there be 200 penguins in the area?
71. Write a user-defined BASIC function to describe the penguin population.

Sound (in decibels) is given by the formula $d = \log_{10}(A/B)^{10}$, where A is the output power and B is the input power. Use this information in Problems 72–74.

72. If input and output power are equal, find the sound.
73. How many decibels will be produced if the output power is ten times the input power?
74. Write a user-defined BASIC function to calculate sound in decibels.

■ 4.6 Sequences

In this section, we will consider a special kind of function called a sequence.

infinite sequence

Definition 4.7 An **infinite sequence** is a function whose domain is N.

The range of an infinite sequence can be any real number. Since the domain of all sequences is $\{1, 2, 3, 4, \ldots\}$, the domain is usually not written. To write a sequence, we will usually list the elements of the range in order. For example, the sequence $f(n) = 2n$ is written as $2, 4, 6, \ldots$ or, in general, $f(1), f(2), f(3), \ldots$.

Sequences will usually be denoted by lowercase letters with a subscript of n

in brackets. For example, $\{a_n\}$ is the sequence $a_1, a_2, a_3, \ldots$, where $a_n = f(n)$.
Here are some examples of infinite sequences:

$$1, 3, 5, 7, \ldots \quad f(n) = 2n - 1 = a_n$$
$$1, 4, 9, 16, \ldots \quad f(n) = n^2 = a_n$$
$$3, 6, 9, 12, \ldots \quad f(n) = 3n = a_n$$
$$-1, 1, -1, 1, \ldots \quad f(n) = (-1)^n = a_n$$

The last sequence is an example of an alternating sequence.

alternating sequence

Definition 4.8 An **alternating sequence** is a sequence in which the sign alternates between $+$ and $-$.

The formula for an alternating sequence will always involve a power of -1.

explicit formula

Definition 4.9 An **explicit formula** for a sequence is a formula for the nth term of the sequence, a_n, in terms of n.

The formulas in the previous examples were explicit. An explicit formula allows us to find the nth term of a sequence directly from n without knowing any of the previous terms of the sequence. For example, the 94th term of the sequence $1, 3, 5, 7, \ldots = \{a_n\} = \{2n - 1\}$ is $2(94) - 1$, or 187.

recursive formula

Definition 4.10 A **recursive formula** for a sequence is a formula for the nth term of the sequence, a_n, in terms of previous terms.

In order to use a recursive formula to find, for example, the 94th term of a sequence, we must know one or more of the previous terms of the sequence. For the sequence $1, 3, 5, 7, \ldots = \{a_n\} = \{2n-1\}$, the recursive formula is $a_n = a_{n-1} + 2$. Recursive formulas are used often in programming and will be referred to again in Chapter 11.

Some sequences have simple explicit formulas but complicated recursive formulas. For example, the sequence $1, 4, 9, 16, \ldots$ has a simple explicit formula ($a_n = n^2$) but a complicated recursive formula [$a_n = a_{n-1} + (2n-1)$].

Fibonacci sequence

The **Fibonacci sequence** is the sequence generated by starting with $a_1 = 1$ and $a_2 = 1$ and the rule that any other term is the sum of the two previous terms. So we get $1, 1, 2, 3, 5, 8, 13, \ldots$. The recursive formula is simply $a_n = a_{n-1} + a_{n-2}$, but the explicit formula is too complicated to be practical.

■ **EXAMPLE 4.20** Find recursive and explicit formulas for the sequence.

a. $1, 2, 3, 4, \ldots$

Solution

recursive formula: $a_n = a_{n-1} + 1$
explicit formula: $a_n = n$

4.6 Sequences

b. $2, 4, 8, 16, 32, \ldots$

Solution

recursive formula: $a_n = a_{n-1} \cdot 2$
explicit formula: $a_n = 2^n$

c. $1, 2, 6, 24, 120, 720, \ldots$

Solution

recursive formula: $a_n = a_{n-1} \cdot n$
explicit formula: $a_n = n(n-1)(n-2) \cdot \ldots \cdot 3 \cdot 2 \cdot 1$

---

arithmetic sequence

common difference

Definition 4.11 An **arithmetic sequence** is a sequence in which each term differs from the previous term by the same number. This number is called the **common difference** d.

Here are some arithmetic sequences:

$$
\begin{array}{ll}
1, 2, 3, 4, \ldots & d = 1 \\
3, 6, 9, 12, \ldots & d = 3 \\
4, -1, -6, -11, \ldots & d = -5 \\
-4, 6, 16, 26, \ldots & d = 10
\end{array}
$$

The recursive formula for an arithmetic sequence is simply

$$a_n = a_{n-1} + d$$

The explicit formula for an arithmetic sequence is given by

$$a_n = a_1 + d(n-1)$$

geometric sequence

common ratio

Definition 4.12 A **geometric sequence** is a sequence in which each term differs from the previous term by the same ratio. This number is called the **common ratio** r.

Here are some geometric sequences:

$$
\begin{array}{ll}
2, 4, 8, 16, \ldots & r = 2 \\
-1, 1, -1, 1, \ldots & r = -1 \\
2, 6, 18, 54, \ldots & r = 3 \\
1, \tfrac{1}{3}, \tfrac{1}{9}, \ldots & r = \tfrac{1}{3}
\end{array}
$$

The recursive formula for a geometric sequence is simply

$$a_n = a_{n-1} \cdot r$$

NONLINEAR FUNCTIONS

The explicit formula for a geometric sequence is given by

$$a_n = a_1 \cdot r^{n-1}$$

Being able to recognize arithmetic and geometric sequences makes working with these types of sequences easy, since formulas can be used.

EXAMPLE 4.21 Find the 87th term of the sequence.
a. 3, 7, 11, 15, ...

Solution This is an arithmetic sequence with $d = 4$, so we can use the explicit formula $a_n = a_1 + d(n-1)$ with $n = 87$:

$$\begin{aligned} a_{87} &= 3 + 4(87-1) \\ &= 3 + 4(86) = 3 + 344 \\ &= \underline{347} \end{aligned}$$

b. $-2, 4, -8, 16, -32, \ldots$

Solution This is a geometric sequence with $r = -2$, so we can use the explicit formula $a_n = a_1 \cdot r^{n-1}$ with $r = -2$:

$$\begin{aligned} a_{87} &= -2 \cdot (-2)^{87-1} \\ &= -2 \cdot (-2)^{86} \\ &= \underline{(-2)^{87}} \end{aligned}$$

If we had wished to use recursive formulas in Example 4.21, we would have needed to know a_{86}.

EXAMPLE 4.22 Find x and y in each sequence.
a. $x, 2, y, 10, \ldots$ (arithmetic sequence)

Solution Since we are told that this is an arithmetic sequence, we know there is a common difference d. So $y - 2 = 10 - y$. From this equation, we can solve for y:

$$\begin{aligned} 2y &= 12 \\ y &= \underline{6} \end{aligned}$$

Now, we have $x, 2, 6, 10, \ldots$, so we know that $d = 4$.
Therefore, $2 - x = 4$, or $\underline{x = -2}$.

b. $3, x, y, 24, \ldots$ (geometric sequence)

Solution Since we are told that this is a geometric sequence, we know there is a common ratio r. So $3r = x$, $xr = y$, and $yr = 24$. By substitution, we get $3r^3 = 24$, so $r^3 = 8$, or $r = 2$. Thus,

4.6 Sequences

$$x = 3r \qquad y = \frac{24}{r}$$
$$= 3 \cdot 2 \qquad = \frac{24}{2}$$
$$= \underline{6} \qquad = \underline{12}$$

We will encounter sequences again when we discuss programming logic in Chapter 11. Computers are frequently used to repeat calculations with different values of the variables. For example, a *counter* is used in programming to count how many times a procedure is performed. A counter is simply an arithmetic sequence using the recursive formula $a_n = a_{n-1} + 1$.

Problem Set 4.6

In Problems 1–10, identify the sequence as arithmetic, geometric, or neither.

1. $1, 4, 7, 10, \ldots$
2. $2, 22, 222, 2222, \ldots$
3. $3, 9, 27, 81, \ldots$
4. $11, 22, 33, 44, \ldots$
5. $1, 2, 4, 7, 11, 16, \ldots$
6. $1, 8, 64, 512, \ldots$
7. $1, 1, 2, 3, 5, 8, 13, \ldots$
8. $1, 4, 9, 16, 25, \ldots$
9. $2, 0, -2, -4, -6, \ldots$
10. $4, 1, \frac{1}{4}, \frac{1}{16}, \ldots$

In Problems 11–24, find an explicit formula for the sequence.

11. $1, 4, 7, 10, \ldots$
12. $11, 22, 33, 44, \ldots$
13. $27, 21, 15, 9, \ldots$
14. $-3, 2, 7, 12, \ldots$
15. $1, 1.1, 1.2, 1.3, \ldots$
16. $-1, -1.3, -1.6, -1.9, \ldots$
17. $1, -1, 1, -1, \ldots$
18. $-3, 6, -12, 24, \ldots$
19. $1, 2, 4, 8, 16, \ldots$
20. $64, 16, 4, 1, \ldots$
21. $9, -3, 1, -\frac{1}{3}, \ldots$
22. $4, \frac{8}{7}, \frac{16}{49}, \frac{32}{343}, \ldots$
23. $1, 8, 27, 64, \ldots$
24. $1, \sqrt{2}, \sqrt{3}, 2, \sqrt{5}, \ldots$

In Problems 25–38, find a recursive formula for the sequence.

25. $1, 4, 7, 10, \ldots$
26. $11, 22, 33, 44, \ldots$
27. $27, 21, 15, 9, \ldots$
28. $-3, 2, 7, 12, \ldots$
29. $1, 1.1, 1.2, 1.3, \ldots$
30. $-1, -1.3, -1.6, -1.9, \ldots$
31. $1, -1, 1, -1, \ldots$
32. $-3, 6, -12, 24, \ldots$
33. $1, 2, 4, 8, 16, \ldots$
34. $64, 16, 4, 1, \ldots$
35. $9, -3, 1, -\frac{1}{3}, \ldots$
36. $4, \frac{8}{7}, \frac{16}{49}, \frac{32}{343}, \ldots$
37. $1, 8, 27, 64, \ldots$
38. $1, \sqrt{2}, \sqrt{3}, 2, \sqrt{5}, \ldots$

In Problems 39–46, find the 43rd term in the sequence.

39. $3, 5, 7, 9, \ldots$
40. $-2, 4, -8, 16, \ldots$

41. $3, \frac{1}{3}, \frac{1}{27}, \frac{1}{81}, \ldots$

42. $12, 1, -10, -21, \ldots$

43. $1, 0.1, 0.01, 0.001, \ldots$

44. $3, 3.5, 4, 4.5, 5, \ldots$

45. $2, 1.4, 0.8, 0.2, \ldots$

46. $12, 36, 108, 324, \ldots$

In Problems 47–50, find x and y in the sequence.

47. $x, y, 4, 6, \ldots$

48. $3, x, 10, y, \ldots$

49. $8, x, y, -1, \ldots$

50. $4, 1, x, y, \ldots$

51. Find an example of a sequence that is both geometric and arithmetic (if possible).

52. If $2, x, 8, y, 32$ is a geometric sequence, can you uniquely find x and y?

In Problems 53–56, find the first five terms of the sequence.

53. $a_1 = 4, a_n = a_{n-1} + 7$

54. $a_1 = -2, a_n = a_{n-1} - 3$

55. $a_1 = -3, a_n = a_{n-1} \cdot -2$

56. $a_1 = 1, a_n = a_{n-1} \cdot \frac{1}{2}$

57. A computer has been programmed to calculate and print out all terms of the Fibonacci sequence less than 100, starting with 1 and 1 and using the recursive formula $a_n = a_{n-1} + a_{n-2}$. List all of the terms that will be printed out.

58. A computer is to be programmed to calculate and print out all powers of 2, starting with 2^0 and ending at 2^{100}. Write the recursive formula that is used to accomplish this task.

CHAPTER SUMMARY

In this chapter, we have looked at several types of nonlinear functions, including the following:

Quadratic functions	$f(x) = ax^2 + bx + c$
Exponential functions	$f(x) = a^x$
Logarithmic functions	$f(x) = \log_a x$
Sequences	$y = f(x), x \in N$

Two methods were used to solve quadratic functions: completing the square and the quadratic formula:

$$x = \frac{-b \pm \sqrt{b^2 - 4ac}}{2a}$$

Rewriting quadratic functions in the form $f(x) = a(x-h)^2 + k$ is an aid in graphing since, in this form, the vertex is (h, k) and the axis of symmetry is $x = h$.

We studied two particular types of sequences:

Arithmetic sequence
- Common difference d
- Recursive formula: $a_n = a_{n-1} + d$
- Explicit formula: $a_n = a_1 + d(n-1)$

Geometric sequence
- Common ratio r
- Recursive formula: $a_n = a_{n-1} \cdot r$
- Explicit formula: $a_n = a_1 \cdot r^{n-1}$

Computers are set up to handle many different types of functions. We have looked at just a few of them in this chapter.

REVIEW PROBLEMS

Computers are set up to handle many different types of functions. We have looked at just a few of them in this chapter.

In Problems 1–10, solve the quadratic equation.

1. $x^2 + 4x - 5 = 0$
2. $x^2 - 7x - 8 = 0$
3. $x^2 + 8x - 2 = 0$
4. $x^2 - 6x + 1 = 0$
5. $2x^2 - 4x + 10 = 0$
6. $3x^2 - 15x + 18 = 0$
7. $(3x - 4)^2 = 9$
8. $(2x + 1)^2 + 1 = 5$
9. $3x^2 + 4x - 5 = x^2 + 2x + 1$
10. $x^2 - 7x = 7x^2 + 8x + 7$

In Problems 11–18, find the vertex and axis of symmetry of the equation.

11. $y = 4(x - 4)^2 + 5$
12. $y = -2(x + 3)^2 - 12$
13. $y = x^2 + 8x - 4$
14. $y = x^2 - 12x + 10$
15. $y = 3x^2 + 12x - 4$
16. $y = 5x^2 + 5x - 2$
17. $y = (x + 3)(3x - 5)$
18. $y = 2x^2 - 13x + 11$

In Problems 19–32, graph the equation.

19. $y = (x - 2)(x + 4)$
20. $y = (2x + 1)(x - 3)$
21. $y = x^2 + 5x$
22. $y = x^2 - 36$
23. $y = x^2 + 6x - 7$
24. $y = x^2 - 2x + 2$
25. $y = 7^x$
26. $y = 11^x$
27. $y = 2^{-2x}$
28. $y = (\frac{1}{3})^{-x}$
29. $y = \log_8 x$
30. $y = \log_{1/4} x$
31. $y = -\log_2 x$
32. $y = \log_3(2x)$

In Problems 33–38, solve for x.

33. $2 = \log_x 4$
34. $-3 = \log_x 64$
35. $x = \log_4 (\frac{1}{16})$
36. $x = \log_{1/2} 8$
37. $-2 = \log_5 x$
38. $\frac{1}{2} = \log_4 x$

In Problems 39–42, find explicit formulas for the sequence.

39. $2, 7, 12, 17, 22, \ldots$
40. $-3, -5, -7, -9, \ldots$
41. $4, -20, 100, -500, \ldots$
42. $72, 36, 18, 9, \ldots$

In Problems 43–46, find recursive formulas for the sequence.

43. $-3, -12, -48, -192, \ldots$
44. $4, 1, -2, -5, \ldots$
45. $13, 7, 1, -5, \ldots$
46. $7, 35, 175, 875, \ldots$

In Problems 47–48, find the 112th term of the sequence.

47. $4, 2, 1, \frac{1}{2}, \ldots$
48. $1, 18, 35, 52, \ldots$

In Problems 49–50, write user-defined BASIC functions for the function.

49. $y = 3e^{2x+1}$
50. $y = 3x^2 - 7x + 14$

NUMBER SYSTEMS

CHAPTER **5**

decimal number system

The number system that you are most familiar with is the **decimal number system**, base 10, that uses the 10 familiar symbols 0, 1, 2, 3, 4, 5, 6, 7, 8, and 9. This number system has been in use for over a thousand years, but it is not well suited for modern computers. In this chapter, we will look closer at the decimal number system and introduce three more number systems: binary, octal, and hexadecimal, which are used extensively in computer operations.

5.1 The Decimal Number System

People have found the decimal number system to be very convenient. But there is nothing special about this number system, and it is rather inconvenient for computers.

Let's examine this number system in more detail. If we want to write whole numbers larger than 9, rather than invent new symbols, we simply use the same symbols over again in appropriate places within the number. Thus, 347 represents 3 hundreds and 4 tens and 7 ones. We know this representation because of the *positions* of the symbols 3, 4, and 7. Each digit in a decimal number has a **face value** (the digit itself) and a **place value** (a power of 10). These concepts are illustrated in the following example.

face value, place value

EXAMPLE 5.1 Find the face value and place value for each digit in the number 347.

Solution

Face Value	Place Value
3	$100 = 10^2$
4	$10 = 10^1$
7	$1 = 10^0$

This idea can be extended to the right of the decimal point by using negative exponents, as shown in the next example.

EXAMPLE 5.2 Find the face value and place value for each digit in the number 1359.24.

Solution

Face Value	Place Value
1	$1000 = 10^3$
3	$100 = 10^2$
5	$10 = 10^1$
9	$1 = 10^0$
2	$.1 = 10^{-1}$
4	$.01 = 10^{-2}$

5.1 The Decimal Number System

expanded form

A decimal number can be written as a sum of the face value of each digit times its corresponding place value. This form is called the **expanded form** of the number. The expanded form is illustrated in the next example.

EXAMPLE 5.3 Write the number in expanded form.
  a. 347

Solution
$$347 = (3 \cdot 10^2) + (4 \cdot 10^1) + (7 \cdot 10^0)$$
$$= (3 \cdot 100) + (4 \cdot 10) + (7 \cdot 1)$$

  b. 1359.24

Solution
$$1359.24 = (1 \cdot 10^3) + (3 \cdot 10^2) + (5 \cdot 10^1)$$
$$+ (9 \cdot 10^0) + (2 \cdot 10^{-1}) + (4 \cdot 10^{-2})$$
$$= (1 \cdot 1000) + (3 \cdot 100) + (5 \cdot 10)$$
$$+ (9 \cdot 1) + (2 \cdot .1) + (4 \cdot .01)$$

Expanded form can be written by using the exponents (10^2, for example) or by using the number (100, for example). But in either case, note that each place value is a *power* of 10.

When mechanical counting devices were first invented, they naturally used the decimal number system (for instance, adding machines and speedometers). However, as machines got more sophisticated, this number system proved less and less desirable. Thus, another number system was needed, one more in line with modern computer equipment. This alternative number system is the subject of the next section.

Problem Set 5.1

In Problems 1–10, write the face value and place value of each digit of the decimal number.

1. 25
2. 129
3. 1789
4. 12,304
5. 13,007
6. 50,004
7. 26.71
8. 2.109
9. 453.1208
10. 123.00567

In Problems 11–20, write the number in expanded form.

11. 43
12. 598
13. 1298
14. 1249
15. 120,007
16. 30,063
17. 31.14
18. 1.9087
19. 543.6708
20. 3110.547

21. What is the largest five-digit decimal number?
22. What is the largest seven-digit decimal number?
23. What is the smallest five-digit decimal number?
24. What is the smallest seven-digit decimal number?
25. Write out all integer powers of 10 between 10^{-10} and 10^{10}.
26. What do all integer powers of 10 look like?

5.2 The Binary Number System

binary number system

The **binary number system**, base 2, is perhaps the simplest number system to use, since it consists of only two symbols. For convenience, the first two symbols of the decimal number system, namely 0 and 1, are used in this system.

Consider how appropriate this number system is for electronic computers:

- An electronic current can either be on (1) or off (0).
- A computer card can either have a hole punched in a certain spot (1) or no hole in that spot (0).
- A spot on a magnetic tape or disk can either be magnetized (1) or not magnetized (0).

All modern computers use the binary number system internally.

The binary system is an easy number system to use, especially once you understand that it follows the same rules as the decimal number system. For example, to write a number larger than one, rather than invent new symbols, we simply use the same symbols over again in appropriate positions within the number. Some important definitions for the binary system are given next.

bit
binary term, byte

Definition 5.1 A single binary digit is a **bit**. A group of eight bits is a **binary term** or a **byte**.

So a bit is simply a 0 or a 1, and a byte is an eight-digit binary number. Each digit has a face value and a place value. The only difference between decimal and binary numbers is that now the place value is a power of 2, not a power of 10.

EXAMPLE 5.4 Find the face value and place value for each digit in the binary number 1011.

Solution

Face Value	Place Value
1	$8 = 2^3$
0	$4 = 2^2$
1	$2 = 2^1$
1	$1 = 2^0$

Note: We will be referring to different number systems throughout the book; so to avoid confusion, we will use a subscript to represent the base from now on if there is any doubt. Thus, 101.1_2 is in the binary number system, and 345.8_{10} is in the decimal number system.

The idea of place value can again be extended to the right of the point (a binary point in base 2) by using negative exponents.

EXAMPLE 5.5 Find the face value and place value for each digit in the number 101.11_2.

5.2 The Binary Number System

Solution

Face Value	Place Value
1	$4 = 2^2$
0	$2 = 2^1$
1	$1 = 2^0$
1	$\frac{1}{2} = .5 = 2^{-1}$
1	$\frac{1}{4} = .25 = 2^{-2}$

Binary numbers can also be written in expanded form. We will use expanded form in the next section to convert binary numbers to decimal numbers.

EXAMPLE 5.6 Write the number in expanded form.

a. 1011_2

Solution
$$1011_2 = (1 \cdot 2^3) + (0 \cdot 2^2) + (1 \cdot 2^1) + (1 \cdot 2^0)$$
$$= (1 \cdot 8) + (0 \cdot 4) + (1 \cdot 2) + (1 \cdot 1)$$

b. 101.11_2

Solution
$$101.11_2 = (1 \cdot 2^2) + (0 \cdot 2^1) + (1 \cdot 2^0) + (1 \cdot 2^{-1}) + (1 \cdot 2^{-2})$$
$$= (1 \cdot 4) + (0 \cdot 2) + (1 \cdot 1) + (1 \cdot .5) + (1 \cdot .25)$$

With an understanding of this number system, we should be able to count in binary. The first two numbers are 0 and 1. There is no symbol for 2, so we must use two digits. Thus, 2 is 10_2: one 2 and zero 1s. Similarly, 3 is 11_2: one 2 and one 1. The following listing compares the first 12 numbers in the decimal and binary systems.

Decimal	Binary	Decimal	Binary
0_{10}	0_2	7_{10}	111_2
1_{10}	1_2	8_{10}	1000_2
2_{10}	10_2	9_{10}	1001_2
3_{10}	11_2	10_{10}	1010_2
4_{10}	100_2	11_{10}	1011_2
5_{10}	101_2	12_{10}	1100_2
6_{10}	110_2		

Notice that 1 is the highest single digit that exists in this number system, so we need more digits to write a number in binary than we needed in decimal numbers. Whenever we have a string of 1s, like 111 for example, the next highest number must include another digit, just as we add a digit when going from 999 to 1000 base 10. We will refer to the binary number system often in the next two chapters.

Problem Set 5.2

In Problems 1–10, find the face value and place value for each digit in the number.

1. 1001_2
2. 1110_2
3. 110111_2
4. 1010110_2
5. 110.1_2
6. $1.11.01_2$
7. 11011.1101_2
8. 1101.11_{10}
9. 101.11_{10}
10. 10111.11011_2

In Problems 11–20, write the number in expanded form.

11. 1110_2
12. 10110_2
13. 101110_2
14. 1101111_2
15. 11.011_2
16. 1.0111_2
17. 11.0101_{10}
18. 1101.011_{10}
19. 1010.0101_2
20. 11101.01101_2

21. Continue the comparison table for all numbers from 13_{10} to 24_{10}.
22. Write all powers of 2 from 2^{10} to 2^{-10} in decimal form.
23. Write all powers of 2 from 2^{10} to 2^{-10} in binary form.
24. What is the largest five-digit binary number?
25. What is the largest seven-digit binary number?
26. What is the smallest five-digit binary number?
27. What is the smallest seven-digit binary number?
28. How many binary numbers can be written by using no more than four digits?
29. How many binary numbers can be written by using no more than one byte?
30. How can you tell if a binary number is even or odd?
31. Which is the larger binary number: 10010_2 or 10100_2?
32. Which is the smaller binary number: 1100101_2 or 1000100_2?
33. By considering place values, determine the decimal difference between 1101_2 and 1001_2.
34. By considering place values, determine the decimal difference between 10101_2 and 11101_2.
35. Use expanded form to find the decimal equivalent of 1110_2 (Problem 11).
36. Use expanded form to find the decimal equivalent of 10110_2 (Problem 12).

5.3 The Octal Number System

The decimal number system is used by people, and the binary number system is used by computers. In this section and the next, we will introduce two more

5.3 The Octal Number System

number systems which serve as intermediate steps between decimal and binary numbers.

octal number system

The **octal number system** consists, not too surprisingly, of the eight symbols 0, 1, 2, 3, 4, 5, 6, and 7. Other than the number of symbols used, the rules for this number system are the same as for the previous systems. Again, we have face values and place values, but now the place values are powers of 8.

EXAMPLE 5.7 Find the face value and place value of each digit in the number 305.12_8.

Solution

Face Value	Place Value
3	$64 = 8^2$
0	$8 = 8^1$
5	$1 = 8^0$
1	$\frac{1}{8} = .125 = 8^{-1}$
2	$\frac{1}{64} = .015625 = 8^{-2}$

Now, consider counting in the octal number system:

Decimal	Octal	Decimal	Octal
0_{10}	0_8	7_{10}	7_8
1_{10}	1_8	8_{10}	10_8
2_{10}	2_8	9_{10}	11_8
3_{10}	3_8	10_{10}	12_8
4_{10}	4_8	11_{10}	13_8
5_{10}	5_8	12_{10}	14_8
6_{10}	6_8		

Note that when we run out of symbols, we go to two digits, as in the other systems.

Octal numbers can then easily be written in expanded form, as the next example shows.

EXAMPLE 5.8 Write 305.12_8 in expanded form.

Solution
$$305.12_8 = (3 \cdot 8^2) + (0 \cdot 8^1) + (5 \cdot 8^0) + (1 \cdot 8^{-1}) + (2 \cdot 8^{-2})$$
$$= (3 \cdot 64) + (0 \cdot 8) + (5 \cdot 1) + (1 \cdot .125) + (2 \cdot .015625)$$

Although the suitability of the octal number system for use in computers may not seem evident yet, it lies in the close relationship between the binary and octal systems. This relationship will be explored in detail later in this chapter (also see Problem 34).

Problem Set 5.3

In Problems 1–10, find the face value and place value for each digit in the number.

1. 437_8
2. 516_8
3. 1207_8
4. 3454_8
5. 35.75_8
6. 44.571_8
7. 1101.101_2
8. 1001.01101_2
9. 77.777_8
10. 766.667_8

In Problems 11–20, write the number in expanded form.

11. 237_8
12. 616_8
13. 1475_8
14. 55432_8
15. 123.4_8
16. 45.76_8
17. 101101.101_2
18. 100110.1001_2
19. 1234.567_8
20. 7.77077_8

21. Continue the comparison table for all numbers from 13_{10} to 24_{10}.
22. Write all powers of 8 from 8^5 to 8^{-5} in decimal form.
23. Write all powers of 8 from 8^5 to 8^{-5} in octal form.
24. What is the largest five-digit octal number?
25. What is the largest seven-digit octal number?
26. What is the smallest five-digit octal number?
27. What is the smallest seven-digit octal number?
28. How many octal numbers can be written by using no more than four digits?
29. How many octal numbers can be written by using no more than five digits?
30. What octal number comes after 7777_8?
31. What octal number comes after 77777_8?
32. What octal number comes just before 100_8?
33. What octal number comes just before 1000_8?
34. List the first 12 powers of 2 and the first 4 powers of 8. Can you find a relationship between the two systems? It is this relationship that makes octal numbers important for computers.
35. Use expanded form to convert 237_8 to decimal (Problem 11).
36. Use expanded form to convert 616_8 to decimal (Problem 12).
37. How can you tell if an octal number is even or odd?
38. By considering place value, determine the decimal difference between 623_8 and 633_8.
39. By considering place value, determine the decimal difference between 1247_8 and 1447_8.
40. Which number is larger: 2377_8 or 11034_8?

5.4 The Hexadecimal Number System

hexadecimal number system

The last number system that we will consider is the **hexadecimal number system**, base 16. This system consists of 16 different symbols. We are only accustomed to 10 symbols, which we can use for the first 10 symbols in the hexadecimal system. However, the numbers 10 through 15 in the decimal number system are single-digit hexadecimal numbers. By convention, the symbols A, B, C, D, E, and F are used to represent these numbers base 16.

Counting in this system is quite different from counting in the other systems that we have studied:

5.4 The Hexadecimal Number System

Decimal	Hexadecimal	Decimal	Hexadecimal
0_{10}	0_{16}	11_{10}	B_{16}
1_{10}	1_{16}	12_{10}	C_{16}
2_{10}	2_{16}	13_{10}	D_{16}
3_{10}	3_{16}	14_{10}	E_{16}
4_{10}	4_{16}	15_{10}	F_{16}
5_{10}	5_{16}	16_{10}	10_{16}
6_{10}	6_{16}	17_{10}	11_{16}
7_{10}	7_{16}	18_{10}	12_{16}
8_{10}	8_{16}	19_{10}	13_{16}
9_{10}	9_{16}	20_{10}	14_{16}
10_{10}	A_{16}		

As this listing shows, single-digit numbers are used up to the base number (in this case, 16), as in the other systems. All other numbers involve more than one digit.

The only differences between any of these four number systems are the number of symbols that are used and the power (base). The rules are the same for all number systems.

EXAMPLE 5.9 Find the face value and place value of each digit in the number $2A7.3C_{16}$.

Solution

Face Value	Place Value
2	$256 = 16^2$
$A(10_{10})$	$16 = 16^1$
7	$1 = 16^0$
3	$\frac{1}{16} = .0625 = 16^{-1}$
$C(12_{10})$	$\frac{1}{256} = .00390625 = 16^{-2}$

EXAMPLE 5.10 Write the number $2A7.3C_{16}$ in expanded form.

Solution
$$2A7.3C_{16} = (2 \cdot 16^2) + (A \cdot 16^1) + (7 \cdot 16^0) + (3 \cdot 16^{-1}) + (C \cdot 16^{-2})$$
$$= (2 \cdot 256) + (10 \cdot 16) + (7 \cdot 1) + (3 \cdot .0625) + (12 \cdot .00390625)$$

This number system is also well suited for use in computers because of its close relationship to the binary number system. We will explore this relationship later in this chapter (also see Problem 34).

Problem Set 5.4

In Problems 1–10, find the face value and place value for each digit in the number.

1. 56_{16}
2. 49_{16}
3. $14E_{16}$
4. $3CD_{16}$
5. $3A.E4_{16}$
6. $78C.B_{16}$
7. 1011.101_2
8. 543.26_8
9. $A79.CDF_{16}$
10. $E4.783A_{16}$

In Problems 11–20, write the number in expanded form.

11. 134_{16}
12. 279_{16}
13. $23ED_{16}$
14. $45E8_{16}$
15. 74.9_{16}
16. $356.EF_{16}$
17. 345.77_8
18. 11011.11011_2
19. $AD.ECB_{16}$
20. $FFE.12D_{16}$

21. Continue the comparison table for all numbers from 21_{10} to 34_{10}.
22. Write all powers of 16 from 16^3 to 16^{-3} in decimal form.
23. Write all powers of 16 from 16^3 to 16^{-3} in hexadecimal form.
24. What is the largest five-digit hexadecimal number?
25. What is the largest seven-digit hexadecimal number?
26. What is the smallest five-digit hexadecimal number?
27. What is the smallest seven-digit hexadecimal number?
28. How many hexadecimal numbers can be written by using no more than two digits?
29. How many hexadecimal numbers can be written by using no more than three digits?
30. What hexadecimal number comes after $FFFF_{16}$?
31. What hexadecimal number comes after $FFFFFF_{16}$?
32. What hexadecimal number comes just before 100_{16}?
33. What hexadecimal number comes just before 30_{16}?
34. List the first 12 powers of 2 and the first 3 powers of 16. Can you find a relationship between the two systems?
35. Use expanded form to convert 134_{16} to decimal (Problem 11).
36. Use expanded form to convert 279_{16} to decimal (Problem 12).
37. How can you tell if a hexadecimal number is even or odd?
38. Which is larger: $C798_{16}$ or $B6021_{16}$?
39. By considering place value, determine the decimal difference between $A976_{16}$ and $AA76_{16}$.
40. By considering place value, determine the decimal difference between 777_{16} and $7B7_{16}$.

5.5 Converting from Any Base to Decimal

Being able to change back and forth between the number systems not only gives a deeper understanding of these systems but also gives an insight into the operations a computer must perform to solve a simple problem. For example, all of the calculations inside the computer must be performed in the binary number system. But we usually want the results printed out in decimal form. The computer must perform this conversion. The two methods of conversion presented in this section will work for *any* base.

Method 1

This method is a relatively simple procedure. We write the number in expanded form, multiply, and add.

5.5 Converting from Any Base to Decimal

EXAMPLE 5.11 Use Method 1 to convert the number to decimal.

a. 1011.1_2

Solution
$$\begin{aligned}1011.1_2 &= (1 \cdot 2^3)+(0 \cdot 2^2)+(1 \cdot 2^1)+(1 \cdot 2^0)+(1 \cdot 2^{-1})\\ &= (1 \cdot 8)+(0 \cdot 4)+(1 \cdot 2)+(1 \cdot 1)+(1 \cdot .5)\\ &= 8+0+2+1+.5\\ &= \underline{11.5_{10}}\end{aligned}$$

b. 305.12_8

Solution
$$\begin{aligned}305.12_8 &= (3 \cdot 8^2)+(0 \cdot 8^1)+(5 \cdot 8^0)+(1 \cdot 8^{-1})+(2 \cdot 8^{-2})\\ &= (3 \cdot 64)+(0 \cdot 8)+(5 \cdot 1)+(1 \cdot .125)+(2 \cdot .015625)\\ &= 192+0+5+.125+.03125\\ &= \underline{197.15625_{10}}\end{aligned}$$

c. $2A7.C_{16}$

Solution
$$\begin{aligned}2A7.C_{16} &= (2 \cdot 16^2)+(A \cdot 16^1)+(7 \cdot 16^0)+(C \cdot 16^{-1})\\ &= (2 \cdot 256)+(10 \cdot 16)+(7 \cdot 1)+(12 \cdot .0625)\\ &= 512+160+7+.75\\ &= \underline{679.75_{10}}\end{aligned}$$

Method 2 (Integer Conversions Only)

To change an integer from any base to base 10, another method may be used. Multiply the leftmost digit by the base, add the next digit to the right to the product, and multiply the sum by the base again. Continue to add and multiply in this manner until the last digit has been added. The result is the decimal equivalent of the number.

EXAMPLE 5.12 Use Method 2 to convert the number to decimal.

a. 1101_2

Solution

Multiply the leftmost digit (1) by 2	$1 \cdot 2 = 2$
Add the next digit (1)	$2+1 = 3$
Multiply by 2	$3 \cdot 2 = 6$
Add the next digit (0)	$6+0 = 6$
Multiply by 2	$6 \cdot 2 = 12$
Add the next (last) digit (1)	$12+1 = 13$

$1101_2 = \underline{13_{10}}$

b. 345_8

Solution Multiply the leftmost digit (3) by 8 $\quad 3 \cdot 8 = 24$
Add the next digit (4) $\quad 24 + 4 = 28$
Multiply by 8 $\quad 28 \cdot 8 = 224$
Add the next (last) digit (5) $\quad 224 + 5 = 229$
$345_8 = \underline{229_{10}}$

c. $6E7_{16}$

Solution Multiply the leftmost digit (6) by 16 $\quad 6 \cdot 16 = 96$
Add the next digit (E) $\quad 96 + 14 = 110$
Multiply by 16 $\quad 110 \cdot 16 = 1760$
Add the next (last) digit (7) $\quad 1760 + 7 = 1767$
$6E7_{16} = \underline{1767_{10}}$

This method has the advantage that you never multiply by a number larger than the base. A disadvantage is that this method only works for integer conversions. In contrast, Method 1 works for fractions also.

Why does this method work? Consider Example 5.12b. The digit 3 is multiplied by 8 once to begin the problem and a second time after adding the next digit ($3 \cdot 8^2$). The digit 4 is multiplied by 8 only one time ($4 \cdot 8$). And the last digit, 5, is not multiplied by 8 at all ($5 \cdot 1$). So this method is the same as using expanded form.

Problem Set 5.5

In Problems 1–20, use Method 1 to convert the number to decimal.

1. 1110_2
2. 110111_2
3. 1.011_2
4. 110.11_2
5. 11011.0011_2
6. 100011.00101_2
7. 453_8
8. 732_8
9. 616.7_8
10. 434.23_8
11. 4112.345_8
12. 7754.63_8
13. $D39_{16}$
14. $17F_{16}$
15. 637.5_{16}
16. $143A.C_{16}$
17. $4352.E6_{16}$
18. $1ADC.B3_{16}$
19. 1232_4
20. 23.23_4

In Problems 21–40, use Method 2 to convert the number to decimal.

21. 1001_2
22. 10110_2
23. 1101101_2
24. 1100111_2
25. 11001101_2
26. 10000111_2
27. 301_8
28. 477_8
29. 2147_8
30. 3323_8
31. 2345_8
32. 6543_8
33. $7A_{16}$
34. $F9_{16}$
35. $23D_{16}$
36. $7CB_{16}$
37. $13EF_{16}$
38. $4FCA_{16}$
39. 1331_4
40. 3231_4

5.6 Converting from Decimal to Any Base

In Problems 41–46, use either method (when appropriate) to convert the number to decimal.

41. 111111_2
42. 11111111_2
43. 77_8
44. FF_{16}
45. FFF_{16}
46. 777_8

47. What is the largest number that can be written in one byte? Answer in binary form.

48. What is the largest number that can be written in one byte? Answer in decimal form.

49. Compare the answers to Problems 41 and 43. Can you reach any conclusions about a relationship between base 2 and base 8?

50. Compare the answers to Problems 42 and 44. Can you reach any conclusions about a relationship between base 2 and base 16?

5.6 Converting from Decimal to Any Base

A programmer usually enters numbers in decimal form since that is the system most familiar to people. In order for the computer to perform any calculations with these numbers, they must be changed into binary form. Changing from decimal to another base is the focus of this section.

Two different procedures must be followed: one for converting decimal integers and one for converting decimal fractions. We begin with converting integers.

Converting from a Decimal Integer to Any Base

There are several methods that may be used. We present two methods here.

Method 1 This method has the advantage that you not only can follow it but also can see *why* it works. The steps of this method follow.

Step 1 Find the largest power of the base that is not larger than the number. For example, the largest power of 2 not larger than 43 is 32 (2^5).

Step 2 By division, determine how many times this power goes into the number. This result becomes the leading digit in the converted number. Now, all that needs to be converted is the remainder from the division problem.

Step 3 Repeat steps 1 and 2 by using the remainder each time until the entire number is converted. If powers are skipped, zeros should be inserted as placeholders.

A few examples should make this process clear.

EXAMPLE 5.13 Convert the decimal integer to the given base.

a. $43_{10} = \underline{}_2$

Solution The highest power of 2 not larger than 43 is 32. And 32 divides into 43 once, with a remainder of 11. So far, we've got

32	16	8	4	2	1
1					

The highest power of 2 not larger than 11 is 8, which divides into 11 once, with a remainder of 3. So we have

32	16	8	4	2	1
1	0	1			

Since 16 was skipped, a zero was inserted.

The highest power of 2 not larger than 3 is 2, which divides into 3 once, with a remainder of 1. Thus,

32	16	8	4	2	1
1	0	1	0	1	1

Therefore, $43_{10} = \underline{101011_2}$.

When the remainder is less than the base you are converting to, put the remainder in the rightmost column, and the procedure is finished.

b. $152_{10} = \underline{}_8$

Solution The highest power of 8 not larger than 152 is 64, which divides into 152 twice, with a remainder of 24. The highest power of 8 not larger than 24 is 8, which divides into 24 *exactly* 3 times. So we get

64	8	1
2	3	0

Therefore, $152_{10} = \underline{230_8}$.

c. $202_{10} = \underline{}_{16}$

Solution The highest power of 16 not larger than 202 is 16 (the next power is $16^2 = 256$). And 16 divides into 202 twelve times (remember that 12 base 16 is the single-digit symbol C), with a remainder of 10 (A). So we get

16	1
C	A

Therefore, $202_{10} = \underline{CA_{16}}$.

Method 2 This procedure may be easier to use, but to see why it works is harder. The procedure is as follows: Repeatedly divide the number to be converted by the base to be converted to, keeping track of the remainders. When

5.6 Converting from Decimal to Any Base

you are left with zero, the converted number is simply the remainders arranged from left to right in the opposite order that they were found.

We will look at a few examples, and then we will try to understand why this procedure works.

EXAMPLE 5.14 Convert the decimal integer to the given base.

a. $57_{10} = \underline{}_2$

Solution Repeatedly divide 57 by 2, keeping track of the remainders:

Division	Remainder
2)57	1
2)28	0
2)14	0
2) 7	1
2) 3	1
2) 1	1
0	

So $57_{10} = \underline{111001}_2$.

b. $212_{10} = \underline{}_8$

Solution Repeatedly divide 212 by 8, keeping track of the remainders:

Division	Remainder
8)212	4
8)26	2
8)3	3
0	

So $212_{10} = \underline{324}_8$.

c. $333_{10} = \underline{}_{16}$

Solution Repeatedly divide 333 by 16, keeping track of the remainders:

Division	Remainder
16)333	13
16)20	4
16)1	1
0	

So $333_{10} = \underline{14D}_{16}$.

Why does this procedure work? Consider Example 5.14c. When the original number is first divided by 16, the remainder tells us how much of another 16 we are lacking. Since the remainder is less than 16, it goes into the 1s column. The 20 then represents how many groups of 16s are in the number 333. If we divide the 20 by 16, we are really dividing the original number by 16 twice, or 256. We get a 1 for this division, which tells us that 256 divides into 333 one time, so we need a 1 in the 256 column. The remainder 4 tells us how many 16s we have left to account for, and this number goes in the 16s column.

Converting from a Decimal Fraction to Any Base

The decimal number system was not designed for computers, and the procedures for conversion from decimal to the other bases we've talked about are not simple. Furthermore, a different procedure is needed for decimal fractions.

First, we will describe the procedure and then see, through an example, *why* it works. The procedure is as follows: Starting with the decimal fraction, multiply it by the base. Whatever digit is now to the left of the decimal becomes the first digit to the right of the point in the converted answer. Drop this leading digit, and multiply the result by the base. Continue the process until the converted answer stops or repeats or until you have a predetermined number of digits.

EXAMPLE 5.15 Convert the decimal fraction to the given base:

$$0.7_{10} = \underline{}_2$$

Solution

$$0.7 \cdot 2 = \cancel{1}.4$$
$$0.4 \cdot 2 = 0.8$$
$$0.8 \cdot 2 = \cancel{1}.6$$
$$0.6 \cdot 2 = \cancel{1}.2$$
$$0.2 \cdot 2 = 0.4$$
$$0.4 \cdot 2 = 0.8$$

The pattern 0110 repeats forever. To indicate that it repeats, a bar is drawn over it:

$$0.7_{10} = .1\overline{0110}_2$$

Why does this procedure work? In the example, the first digit to the right of the point measures halves. That is, 0.7 is the same as 1.4/2, or 1.4 halves. Since this number is more than one half, a one is put in the first place to the right of the point. That leaves 0.4/2 to still be accounted for, and 0.4/2 is the same as 0.8/4. So the next place to the right represents fourths. Since 0.8/4 is less than one fourth, a 0 is put in the second place to the right of the point. And so the procedure continues as far as necessary.

5.6 Converting from Decimal to Any Base

EXAMPLE 5.16 Convert the decimal fraction to the given base.

a. $0.15_{10} = \underline{}_8$

Solution
$0.15 \cdot 8 = 1.2$
$0.2 \cdot 8 = 1.6$
$0.6 \cdot 8 = 4.8$
$0.8 \cdot 8 = 6.4$
$0.4 \cdot 8 = 3.2$
$0.2 \cdot 8 = 1.6$

The pattern 1463 repeats. So

$0.15_{10} = .1\overline{1463}_8$

b. $0.75_{10} = \underline{}_{16}$

Solution
$0.75 \cdot 16 = 12.0 \quad (12_{10} = C_{16})$
$0.0 \cdot 16 = 0.0$

Since we get 0.0, the problem is finished. Thus,

$0.75_{10} = .C_{16}$

As you can see, base 10 does not convert easily to the other number systems. To convert a mixed number like 435.6_{10} to another base, you must break the problem down into two parts: one to find the integer conversion and one to find the fractional conversion.

EXAMPLE 5.17 Convert 52.8_{10} to a binary number.

Solution This problem must be done in two steps.

Step 1 We convert 52_{10} to binary. The highest power of 2 not larger than 52 is 32, which divides into 52 once, with a remainder of 20. So

32	16	8	4	2	1
1					

The highest power of 2 not larger than 20 is 16, which divides into 20 once, with a remainder of 4. Thus,

32	16	8	4	2	1
1	1				

The highest power of 2 not larger than 4 is 4, which divides into 4 once, with a remainder of 0. Thus,

32	16	8	4	2	1
1	1	0	1	0	0

So $52_{10} = 110100_2$.

Step 2 Now, convert 0.8_{10} to binary

$$0.8 \cdot 2 = \cancel{1}.6$$
$$0.6 \cdot 2 = \cancel{1}.2$$
$$0.2 \cdot 2 = 0.4$$
$$0.4 \cdot 2 = 0.8$$
$$0.8 \cdot 2 = \cancel{1}.6$$

The pattern 1100 repeats forever, so

$$0.8_{10} = .\overline{1100}_2$$

We can now conclude that $52.8_{10} = 110100.\overline{1100}_2$.

Problem Set 5.6

In Problems 1–20, perform the conversion by using Method 1.

1. $57_{10} = \underline{}_2$
2. $142_{10} = \underline{}_2$
3. $266_{10} = \underline{}_2$
4. $476_{10} = \underline{}_2$
5. $872_{10} = \underline{}_2$
6. $1259_{10} = \underline{}_2$
7. $37_{10} = \underline{}_8$
8. $85_{10} = \underline{}_8$
9. $213_{10} = \underline{}_8$
10. $452_{10} = \underline{}_8$
11. $533_{10} = \underline{}_8$
12. $1044_{10} = \underline{}_8$
13. $46_{10} = \underline{}_{16}$
14. $95_{10} = \underline{}_{16}$
15. $179_{10} = \underline{}_{16}$
16. $432_{10} = \underline{}_{16}$
17. $983_{10} = \underline{}_{16}$
18. $1453_{10} = \underline{}_{16}$
19. $87_{10} = \underline{}_4$
20. $134_{10} = \underline{}_4$

In Problems 21–40, perform the conversion by using Method 2.

21. $49_{10} = \underline{}_2$
22. $101_{10} = \underline{}_2$
23. $206_{10} = \underline{}_2$
24. $513_{10} = \underline{}_2$
25. $947_{10} = \underline{}_2$
26. $1336_{10} = \underline{}_2$
27. $55_{10} = \underline{}_8$
28. $97_{10} = \underline{}_8$
29. $148_{10} = \underline{}_8$

30. $344_{10} = \underline{}_8$
31. $747_{10} = \underline{}_8$
32. $1048_{10} = \underline{}_8$
33. $75_{10} = \underline{}_{16}$
34. $97_{10} = \underline{}_{16}$
35. $214_{10} = \underline{}_{16}$
36. $522_{10} = \underline{}_{16}$
37. $937_{10} = \underline{}_{16}$
38. $2314_{10} = \underline{}_{16}$
39. $65_{10} = \underline{}_4$
40. $215_{10} = \underline{}_4$

In Problems 41–70, perform the conversion.

41. $0.3_{10} = \underline{}_2$
42. $0.6_{10} = \underline{}_2$
43. $0.45_{10} = \underline{}_2$
44. $0.85_{10} = \underline{}_2$
45. $0.125_{10} = \underline{}_2$
46. $0.625_{10} = \underline{}_2$
47. $0.1_{10} = \underline{}_8$
48. $0.9_{10} = \underline{}_8$
49. $0.875_{10} = \underline{}_8$
50. $0.5_{10} = \underline{}_8$
51. $0.55_{10} = \underline{}_8$
52. $0.95_{10} = \underline{}_8$
53. $0.2_{10} = \underline{}_{16}$
54. $0.7_{10} = \underline{}_{16}$
55. $0.35_{10} = \underline{}_{16}$
56. $0.85_{10} = \underline{}_{16}$
57. $0.25_{10} = \underline{}_{16}$
58. $0.375_{10} = \underline{}_{16}$
59. $0.4_{10} = \underline{}_4$
60. $0.75_{10} = \underline{}_4$
61. $37.1_{10} = \underline{}_2$
62. $17.2_{10} = \underline{}_2$
63. $458.7_{10} = \underline{}_2$
64. $55.95_{10} = \underline{}_2$
65. $62.3_{10} = \underline{}_8$
66. $123.8_{10} = \underline{}_8$
67. $45.45_{10} = \underline{}_8$
68. $779.5_{10} = \underline{}_8$
69. $44.9_{10} = \underline{}_{16}$
70. $235.2_{10} = \underline{}_{16}$

71. Convert the decimal number 256 to binary, octal, and hexadecimal. Can you reach a conclusion about 256?
72. Convert the decimal number 255 to binary, octal, and hexadecimal. Can you reach a conclusion about 255?

5.7 Converting from Binary to Octal or Hexadecimal

Computers use the binary number system; people use the decimal number system. These two number systems are not very compatible (as the previous two sections illustrate), but there is a strong relationship between binary, octal, and hexadecimal numbers. Notice that 8 and 16 are both powers of 2: $2^3 = 8$ and $2^4 = 16$. So every third power of 2 is a power of 8, and every fourth power of 2 is a power of 16:

Powers of 2	1	2	4	8	16	32	64	128	256
Powers of 8	1			8			64		
Powers of 16	1				16				256

This relationship continues to the right of the point.

To convert a binary number to an octal number, simply break the number into groups of three digits each on either side of the point. Then convert each group (ranging from 000 to 111) to its corresponding octal digit (ranging from 0 to 7).

If the last group to the left of the point does not consist of exactly three digits, the missing digits can be ignored (1 is the same as 001). If the last group to the right of the point does not consist of exactly three digits, the missing zeros must be written in (110 is different from 11).

EXAMPLE 5.18 Convert the binary number to octal.

a. 1111101.10111_2

Solution Break the number into groups of three on either side of the point:

    1   111   101 . 101   110

Change each group to an octal number:

    1   7   5 . 5   6

So $1111101.10111_2 = \underline{175.56_8}$.

b. 10001101.0010111_2

Solution

Binary	10	001	101	.	001	011	100
Octal	2	1	5	.	1	3	4

So $10001101.0010111_2 = \underline{215.134_8}$.

Similarly, to convert a binary number to a hexadecimal number, break the number into groups of four digits each on either side of the point. Then convert each group (ranging from 0000 to 1111) to its corresponding hexadecimal digit (ranging from 0 to F).

EXAMPLE 5.19 Convert the binary numbers to hexadecimal.

a. 1111101.10111_2

Solution Break the number into groups of four on either side of the point:

    111   1101 . 1011   1000

Change each group to a hexadecimal number:

    7   D . B   8

So $1111101.10111_2 = \underline{7D.B8_{16}}$.

5.7 Converting from Binary to Octal or Hexadecimal

b. 100100001001.001011101_2

Solution	Binary	1001	0000	1001	.	0010	1110	1000
	Hexadecimal	9	0	9	.	2	E	8

So $100100001001.001011101_2 = \underline{909.2E8_{16}}$.

Binary integers are usually stored in a computer by using a certain number of bytes. If two bytes are available to store an integer, then 1010111101_2 is stored as 0000001010111101. The topic of storage of numbers will be covered in detail in Chapter 7.

Problem Set 5.7

In Problems 1–10, convert the binary number to an octal number.

1. 1101_2
2. 100101_2
3. 1101111_2
4. 1101101110_2
5. 110001001001_2
6. 111101100110101_2
7. 110.101101_2
8. 1001110.101_2
9. 11.1101000101_2
10. 111111.111001011_2

In Problems 11–20, convert the binary number to a hexadecimal number.

11. 1110_2
12. 11001010_2
13. 1001101101011_2
14. 1110011001110_2
15. 1001110.101_2
16. 110.101101_2
17. 110111.111011011_2
18. 11110001.011111_2
19. 110001010101.00110101_2
20. $101111000101.00011001101_2$

21. Compute powers of 2 from 2^{-1} to 2^{-6} and powers of 8 from 8^{-1} to 8^{-2}. Are there any similarities?

22. Compute powers of 2 from 2^{-1} to 2^{-8} and powers of 16 from 16^{-1} to 16^{-2}. Are there any similarities?

23. Describe a rule for converting from binary to base 4.

24. Describe a rule for converting from binary to base 32.

25. Change 110010101.101_2 to base 4.

26. Change 110010101.101_2 to base 32.

27. Convert 101101.11011_2 to base 8 and base 10. Which conversion is easier?

28. Convert 11011011.1111101_2 to base 16 and base 10. Which conversion is easier?

29. If an integer is stored by using four bytes, how many hexadecimal digits may be necessary to represent the number?

30. If an integer is stored by using four bytes, how many octal digits may be necessary to represent the number?

5.8 Converting from Octal or Hexadecimal to Binary

Converting from binary to octal or hexadecimal simply involved grouping the binary digits into groups of three or four. To convert the other way, from octal or hexadecimal to binary, take each digit and convert it to the corresponding three or four binary digits.

EXAMPLE 5.20 Convert the number to binary.

a. 720.35_8

Solution

Octal	7	2	0	.	3	5
Binary	111	010	000	.	011	101

So $720.35_8 = \underline{111010000.011101}_2$.

Note that each octal digit must be converted to three binary digits, even if there are leading zeros.

b. 1437.562_8

Solution

Octal	1	4	3	7	.	5	6	2
Binary	001	100	011	111	.	101	110	010

So $1437.562_8 = \underline{1100011111.10111001}_2$.

Note that leading zeros and trailing zeros of the converted number are not usually included in the final answer.

c. $59.D_{16}$

Solution

Hexadecimal	5	9	.	D
Binary	0101	1001	.	1101

So $59.D_{16} = \underline{1011001.1101}_2$.

Note that each hexadecimal digit must be converted to four binary digits, even if there are leading zeros.

d. $A49.EC_{16}$

Solution

Hexadecimal	A	4	9	.	E	C
Binary	1010	0100	1001	.	1110	1100

So $A49.EC_{16} = \underline{101001001001.111011}_2$.

Notice how easy conversions are between binary and these two number systems, especially compared with decimal conversions. For this reason, octal and hexadecimal numbers are sometimes used by computers. To write a 24-digit binary number in hexadecimal requires only 6 digits. This form is easier on the eyes of the programmer than a series of 0s and 1s and is quite easy for the computer to convert.

Usually, octal *or* hexadecimal numbers will be used to describe memory locations inside the computer. A programmer will not have to convert between octal and hexadecimal. However, the conversion procedure is simple and is included here for completeness.

5.8 Converting from Octal or Hexadecimal to Binary

The easiest way to convert between octal and hexadecimal is by way of binary numbers. Convert the number to binary by using the methods in this section. Then convert the binary number to the desired base by using the methods of the previous section.

EXAMPLE 5.21 Perform the conversion.

a. $35.16_8 = $ _____$_{16}$

Solution

Octal	3	5	.	1	6
Binary	011	101	.	001	110

So we have 11101.00111.

Binary	1	1101	.	0011	1000
Hexadecimal	1	D	.	3	8

So $35.16_8 = \underline{1D.38}_{16}$.

b. $457.134_8 = $ _____$_{16}$

Solution

Octal	4	5	7	.	1	3	4
Binary	100	101	111	.	001	011	100

So we have 100101111.0010111.

Binary	1	0010	1111	.	0010	1110
Hexadecimal	1	2	F	.	2	E

So $457.134_8 = \underline{12F.2E}_{16}$.

c. $4A.7_{16} = $ _____$_8$

Solution

Hexadecimal	4	A	.	7
Binary	0100	1010	.	0111

So we have 1001010.0111.

Binary	001	001	010	.	011	100
Octal	1	1	2	.	3	4

So $4A.7_{16} = \underline{112.34}_8$.

d. $A74.1F_{16} = $ _____$_8$

Solution

Hexadecimal	A	7	4	.	1	F
Binary	1010	0111	0100	.	0001	1111

So we have 101001110100.00011111.

Binary	101	001	110	100	.	000	111	110
Octal	5	1	6	4	.	0	7	6

So $A74.1F_{16} = \underline{5164.076}_8$.

As mentioned earlier, hexadecimal and octal numbers are used to describe memory locations inside the computer. All memory locations inside the computer are numbered from 0 up to the limit of the particular computer. A *memory map* is usually provided with each computer, detailing where different functions are stored and what locations are open to the programmer.

Problem Set 5.8

In Problems 1–20, convert the number to binary.

1. 634_8
2. 721_8
3. 32145_8
4. 77514_8
5. 33.75_8
6. 634.56_8
7. 44.56007_8
8. 432.4213_8
9. 14567.40213_8
10. 23567.0014_8
11. 76_{16}
12. $5A_{16}$
13. $4B7_{16}$
14. $5C9_{16}$
15. $9E.C_{16}$
16. $3A.7_{16}$
17. 163.29_{16}
18. $25E.6F_{16}$
19. $2CFE.BCD_{16}$
20. $AFD.BDE9_{16}$

In Problems 21–30, convert the number to hexadecimal.

21. 73_8
22. 45_8
23. 3452_8
24. 7732_8
25. 24.53_8
26. 37.65_8
27. 107.263_8
28. 664.731_8
29. 123.40567_8
30. 77004.0325_8

In Problems 31–40, convert the number to octal.

31. $A4_{16}$
32. $9D_{16}$
33. $176C_{16}$
34. $B4D8_{16}$
35. $8F.E_{16}$
36. $6D.B_{16}$
37. 178.34_{16}
38. $49A.5E_{16}$
39. $ED.5690E_{16}$
40. $431A.FFD_{16}$

41. Describe a rule for converting from base 4 to binary.
42. Describe a rule for converting from base 32 to binary.
43. Change 312.3_4 to binary.
44. Change 312.3_4 to octal.
45. Base 4 can be converted to base 16 by going through binary. Is there an easier way?
46. Is every power of 16 a power of 8?
47. A number is stored in memory location $4AD_{16}$. What is this location in binary? In decimal? In octal?
48. A number is stored in memory location 532_8. What is this location in binary? In decimal? In hexadecimal?
49. A computer has a total memory capacity of $FFFF_{16}$ bytes. How much is this capacity in binary? In decimal? In octal?
50. A computer has a total memory capacity of 777_8 bytes. How much is this capacity in binary? In decimal? In hexadecimal?

CHAPTER SUMMARY

In this chapter, we have discussed four number systems:

- The decimal number system (base 10), the most common number system used by people
- The binary number system (base 2), the number system used by computers
- The octal number system (base 8) and the hexadecimal number system (base 16), number systems that are related to binary and are also used in computers

The rules for all of these systems are the same. The only difference is in the number of symbols used. In discussing the conversions between these bases, we observed that octal and hexadecimal numbers are far easier to interchange with binary than decimal numbers are. However, you should be able to convert between any two of these systems without difficulty. Computers do it all the time.

REVIEW PROBLEMS

In Problems 1–40, perform the conversion.

1. $10111_2 = \underline{}_{10}$
2. $1011011_2 = \underline{}_{10}$
3. $1011.1011_2 = \underline{}_{10}$
4. $11011.1101_2 = \underline{}_{10}$
5. $435_8 = \underline{}_{10}$
6. $325.6_8 = \underline{}_{10}$
7. $43.235_8 = \underline{}_{10}$
8. $457.301_8 = \underline{}_{10}$
9. $3EC_{16} = \underline{}_{10}$
10. $ED.4_{16} = \underline{}_{10}$
11. $179.B_{16} = \underline{}_{10}$
12. $43A.F6_{16} = \underline{}_{10}$
13. $345_{10} = \underline{}_{2}$
14. $597_{10} = \underline{}_{2}$
15. $0.4_{10} = \underline{}_{2}$
16. $0.55_{10} = \underline{}_{2}$
17. $13.35_{10} = \underline{}_{2}$
18. $125.7_{10} = \underline{}_{2}$
19. $567_{10} = \underline{}_{8}$
20. $999_{10} = \underline{}_{8}$
21. $0.8_{10} = \underline{}_{8}$
22. $56.45_{10} = \underline{}_{8}$
23. $43.3_{10} = \underline{}_{8}$
24. $874_{10} = \underline{}_{16}$
25. $1346_{10} = \underline{}_{16}$
26. $0.6_{10} = \underline{}_{16}$
27. $50.65_{10} = \underline{}_{16}$
28. $77.25_{10} = \underline{}_{16}$
29. $101101101_2 = \underline{}_{8}$
30. $1111001101_2 = \underline{}_{8}$
31. $11011.101_2 = \underline{}_{8}$
32. $1101.11101_2 = \underline{}_{8}$
33. $111110111_2 = \underline{}_{16}$
34. $1000100010_2 = \underline{}_{16}$
35. $110110.11_2 = \underline{}_{16}$
36. $11.1110011_2 = \underline{}_{16}$
37. $347_8 = \underline{}_{16}$
38. $67.543_8 = \underline{}_{16}$
39. $AE.B9_{16} = \underline{}_{8}$
40. $EDC.B2_{16} = \underline{}_{8}$

41. A number is stored in location CF_{16}. What is this location in octal? In binary? In decimal?

42. A number is stored in location 437_8. What is this location in hexadecimal? In binary? In decimal?

43. Write the largest number that can be stored in four bytes in binary, octal, hexadecimal, and decimal.

44. How many different numbers can be stored by using no more than three bytes for a number? Represent this number as a power of 2, a power of 8, and a power of 16.

COMPUTER ARITHMETIC

CHAPTER **6**

By now, you should feel comfortable working in the binary, octal, and hexadecimal number systems. In this chapter, we are going to discuss the basic arithmetic operations—addition, subtraction, multiplication, and division—using these number systems. This presentation will give you not only an insight into how a computer performs arithmetic but also an appreciation of what a computer must go through to perform simple calculations.

6.1 Binary Addition

The rules for binary addition are the same as the rules for decimal addition. The only difference is that there are two symbols instead of ten.

Here is the addition table for base 2:

+	0	1
0	0	1
1	1	10

Note that $1_2 + 1_2 = 10_2$, which is equivalent to $1_{10} + 1_{10} = 2_{10}$.

The following example compares decimal and binary addition.

EXAMPLE 6.1 Perform the addition problem.

a. $324_{10} + 413_{10}$

Solution

$$\begin{array}{r} 324_{10} \\ +413_{10} \\ \hline 737_{10} \end{array}$$

b. $10100_2 + 1001_2$

Solution

$$\begin{array}{r} 10100_2 \\ +1001_2 \\ \hline 11101_2 \end{array}$$

The only "exciting" feature of decimal addition comes when the sum of any column exceeds nine. In this case, we write the rightmost digit and "carry" the other digit(s). See the next example.

6.1 Binary Addition

EXAMPLE 6.2 Add 236_{10} and 177_{10}.

Solution

$$\begin{array}{r} {}^{1\ 1} \text{(digits carried)} \\ 236_{10} \\ +177_{10} \\ \hline 413_{10} \end{array}$$

The same thing holds in binary addition, except that a carry is now necessary whenever the sum of a column exceeds 1, which (as you can imagine) happens quite often. Being able to count in binary is a valuable aid in doing these problems.

EXAMPLE 6.3 Perform the addition problem.

a. $1011_2 + 101_2$

Solution

$$\begin{array}{r} {}^{1\ 1\ 1} \\ 1011_2 \\ +\ \ 101_2 \\ \hline 10000_2 \end{array}$$

b. $101101_2 + 101011_2$

Solution

$$\begin{array}{r} {}^{1\ 1\ 1\ 1} \\ 101101_2 \\ +101011_2 \\ \hline 1011000_2 \end{array}$$

Note that in Example 6.3b, one column contained three 1s ($1_2 + 1_2 + 1_2$). The sum of the column then becomes 3_{10}, or 11_2. So a 1 is written, and a 1 is carried.

EXAMPLE 6.4 Add 1011.0_2, 110.1_2, and 1011.1_2.

Solution

$$\begin{array}{r} {}^{1\ 1\ 0\ 1\ 1} \\ 1011.0_2 \\ 110.1_2 \\ +1011.1_2 \\ \hline 11101.0_2 \ \ \text{or} \ \ 11101_2 \end{array}$$

Note that in Example 6.4, one column contained four 1s ($1_2 + 1_2 + 1_2 + 1_2$). The sum of the column then becomes 4_{10}, or 100_2. The rightmost 0 is written, and a two-digit number (10_2) is carried. Another way to do this problem is to carry the 1 over two columns and then 0 over one column.

There is another way to add a column of 1s. Every two 1s in a column add up to 10_2, so a 1 can be put in the next column for every two 1s encountered in the current column. In this way, no large sums have to be added.

The previous example included a binary point. The presence of a point will never complicate an addition problem as long as all of the points are lined up before addition takes place.

EXAMPLE 6.5 Perform the addition.

a. $101.101_2 + 110.11_2 + 1101.011_2$ **b.** $1111.11_2 + .01_2$

Solution

$$
\begin{array}{r}
\overset{1\,0\,1\,1\;\;1\,1\,1}{101.101_2} \\
110.11_2 \\
+\,1101.011_2 \\
\hline
11001.110_2
\end{array}
\qquad
\begin{array}{r}
\overset{1\,1\,1\,1\;\;1}{1111.11_2} \\
+\qquad.01_2 \\
\hline
10000.00_2 \quad \text{or} \quad 10000_2
\end{array}
$$

Within a computer, numbers are usually stored by using a certain number of bytes. If one byte is used to store integers, for example, then 101_2 is stored as 00000101. Although using eight bits when only three bits are necessary may seem like a waste of storage, this uniformity of storage is convenient for many reasons which will be explored further in the next chapter.

EXAMPLE 6.6 If one byte is used to store each integer, then find the sum.

a. $10101_2 + 110110_2$ **b.** $101_2 + 11_2$

Solution

$$
\begin{array}{r}
\overset{1\,1\;\;1}{00010101_2} \\
+\,00110110_2 \\
\hline
01001011_2
\end{array}
\qquad
\begin{array}{r}
\overset{1\,1\,1}{00000101_2} \\
+\,00000011_2 \\
\hline
00001000_2
\end{array}
$$

Problem Set 6.1

In Problems 1–20, perform the addition.

1. $\begin{array}{r} 1011_2 \\ +\ \ \ 100_2 \end{array}$
2. $\begin{array}{r} 1101_2 \\ +\ \ \ \ 10_2 \end{array}$
3. $\begin{array}{r} 11011_2 \\ +\ \ \ 101_2 \end{array}$
4. $\begin{array}{r} 10010_2 \\ +\ \ \ 110_2 \end{array}$

6.1 Binary Addition

5. $\begin{array}{r} 101011_2 \\ + 10111_2 \\ \hline \end{array}$

6. $\begin{array}{r} 110101_2 \\ + 11011_2 \\ \hline \end{array}$

7. $\begin{array}{r} 101.11_2 \\ + 11.01_2 \\ \hline \end{array}$

8. $\begin{array}{r} 1100.1011_2 \\ + 11.1001_2 \\ \hline \end{array}$

9. $\begin{array}{r} 1110.1001_2 \\ + 1.1_2 \\ \hline \end{array}$

10. $\begin{array}{r} 1111.111_2 \\ + 11.111_2 \\ \hline \end{array}$

11. $\begin{array}{r} 10101_2 \\ 1101_2 \\ + 10010_2 \\ \hline \end{array}$

12. $\begin{array}{r} 10011_2 \\ 11100_2 \\ + 1011_2 \\ \hline \end{array}$

13. $\begin{array}{r} 10111_2 \\ 1010_2 \\ 10111_2 \\ + 11101_2 \\ \hline \end{array}$

14. $\begin{array}{r} 11101_2 \\ 1111_2 \\ 10010_2 \\ + 111_2 \\ \hline \end{array}$

15. $\begin{array}{r} 101.11_2 \\ 110.1_2 \\ + 111.11_2 \\ \hline \end{array}$

16. $\begin{array}{r} 1011.101_2 \\ 101.11_2 \\ + 1100.011_2 \\ \hline \end{array}$

17. $\begin{array}{r} 1111.1101_2 \\ + .0011_2 \\ \hline \end{array}$

18. $\begin{array}{r} 1.0000001_2 \\ + .1111111_2 \\ \hline \end{array}$

19. $\begin{array}{r} 100.00111_2 \\ 11.10001_2 \\ 111.01111_2 \\ 100.00001_2 \\ + .01101_2 \\ \hline \end{array}$

20. $\begin{array}{r} 10111.011_2 \\ 1001.001_2 \\ 11011.101_2 \\ 1101.111_2 \\ + 1011.010_2 \\ \hline \end{array}$

In Problems 21–30, find the sum of the numbers.

21. 101101_2 and 11011_2
22. 11011011_2 and 1011_2
23. 101.111_2 and 1.101_2
24. 1.00101_2 and 111.101_2
25. 11.1_2 and 1.00001_2
26. 11011.011_2 and $.001_2$
27. 11011_2, 101101_2, and 10001_2
28. 10.01_2, 1.01111_2, and $.01_2$
29. 100011_2 and $.0001_2$
30. 100000.1111111_2 and 1.01_2

31. A computer can store a program beginning in memory location 10000_2 (each location represents one byte) and can use 11000_2 consecutive bytes for the program. What is the last memory location that can be used for the program?

32. A computer can store a program beginning in memory location 11111_2 and can use 10000_2 consecutive bytes for the program. What is the last memory location that can be used for the program?

33. If you add two three-digit numbers in base 10, will you always get a four-digit number?

34. If you add two three-digit numbers in base 2, will you always get a four-digit number?

35. Change 34_{10} and 25_{10} to binary, find their sum, and change the sum back to decimal. Did you get 59_{10}?

36. Change 27_{10} and 66_{10} to binary, find their sum, and change the sum back to decimal. Did you get 93_{10}?

37. Add 1 to every digit of 10110_2 and ignore all numbers carried. What happens?

38. Add 1 to every digit of 100011_2 and ignore all numbers carried. What happens?

39. Check your answer to Problem 21 by changing each binary number to decimal, performing decimal addition, and then converting your answer to binary.

40. Check your answer to Problem 22 by changing each binary number to decimal, performing decimal addition, and then converting your answer to binary.

41. Add 1101_2 and 1101_2. Can you reach a general conclusion about the sum of two identical binary numbers?

42. What binary number should be added to 11001_2 to obtain a sum of 110010_2?

Assume that one byte is used to store each number. In Problems 43–46, find the sum.

43. $101101_2 + 100001_2$
44. $10010110_2 + 110101_2$
45. $11_2 + 10000_2 + 11011_2$
46. $110111_2 + 11001000_2$

6.2 Octal and Hexadecimal Addition

The rules for octal and hexadecimal addition are the same as the rules for binary and decimal addition except that a carry must be performed whenever the sum of a column exceeds 7 or 15. For example, $6_8 + 3_8 = 11_8$, or 9_{10}. A carry is required in octal but not in decimal.

You should be able to verify the entries in the octal addition table in Figure 6.1.

+	0	1	2	3	4	5	6	7
0	0	1	2	3	4	5	6	7
1	1	2	3	4	5	6	7	10
2	2	3	4	5	6	7	10	11
3	3	4	5	6	7	10	11	12
4	4	5	6	7	10	11	12	13
5	5	6	7	10	11	12	13	14
6	6	7	10	11	12	13	14	15
7	7	10	11	12	13	14	15	16

Figure 6.1
Octal Addition Table

EXAMPLE 6.7 Perform the octal addition.
a. $53_8 + 23_8$

Solution

$$\begin{array}{r} 53_8 \\ +23_8 \\ \hline 76_8 \end{array}$$ (no carry necessary)

6.2 Octal and Hexadecimal Addition

b. $15.4_8 + 6.5_8$ **c.** $777.7_8 + .1_8$

Solution

$$\begin{array}{r} {\scriptstyle 1\ 1}\\ 15.4_8 \\ +\ \ 6.5_8 \\ \hline 24.1_8 \end{array} \qquad \begin{array}{r} {\scriptstyle 1\ 1\ 1}\\ 777.7_8 \\ +\ \ \ .1_8 \\ \hline 1000.0_8 \end{array}$$

We don't need to use Figure 6.1 to perform these additions. For instance, in Example 6.7b, in adding the first column, we can look at Figure 6.1 and see that $5_8 + 4_8 = 11_8$, or we can add 5 and 4 mentally in the decimal number system and get 9_{10} and then change 9_{10} to 11_8. There is another way of performing this addition: Add by groupings of eight. Each time we reach 8 in our sum, we carry 1. In this way, no large sums have to be calculated.

In hexadecimal addition, a carry is not required until the sum of a column exceeds 15 (F). For example, $7_{16} + 9_{16} = 10_{16}$, or 16_{10}; and $A_{16} + C_{16} = 16_{16}$, or 22_{10} (10 + 12). Although the hexadecimal addition table is not needed for these problems, it is given in Figure 6.2 for your reference.

Now, we will do a few examples.

EXAMPLE 6.8 Perform the hexadecimal addition.

a. $A19_{16} + 582_{16}$ **b.** $C86_{16} + 24B_{16}$

Solution

$$\begin{array}{r} A19_{16} \\ +582_{16} \\ \hline F9B_{16} \end{array} \text{(no carry necessary)} \qquad \begin{array}{r} {\scriptstyle 1}\\ C86_{16} \\ +24B_{16} \\ \hline ED1_{16} \end{array}$$

c. $79.FB_{16} + 8A.2D_{16}$

Solution

$$\begin{array}{r} {\scriptstyle 1\ 1\ 1}\\ 79.FB_{16} \\ +8A.2D_{16} \\ \hline 104.28_{16} \end{array}$$

d. $4.A7_{16} + C.2D_{16} + F.01_{16}$

Solution

$$\begin{array}{r} {\scriptstyle 1}\\ 4.A7_{16} \\ C.2D_{16} \\ +F.01_{16} \\ \hline 1F.D5_{16} \end{array}$$

Figure 6.2
Hexadecimal Addition Table

+	0	1	2	3	4	5	6	7	8	9	A	B	C	D	E	F
0	0	1	2	3	4	5	6	7	8	9	A	B	C	D	E	F
1	1	2	3	4	5	6	7	8	9	A	B	C	D	E	F	10
2	2	3	4	5	6	7	8	9	A	B	C	D	E	F	10	11
3	3	4	5	6	7	8	9	A	B	C	D	E	F	10	11	12
4	4	5	6	7	8	9	A	B	C	D	E	F	10	11	12	13
5	5	6	7	8	9	A	B	C	D	E	F	10	11	12	13	14
6	6	7	8	9	A	B	C	D	E	F	10	11	12	13	14	15
7	7	8	9	A	B	C	D	E	F	10	11	12	13	14	15	16
8	8	9	A	B	C	D	E	F	10	11	12	13	14	15	16	17
9	9	A	B	C	D	E	F	10	11	12	13	14	15	16	17	18
A	A	B	C	D	E	F	10	11	12	13	14	15	16	17	18	19
B	B	C	D	E	F	10	11	12	13	14	15	16	17	18	19	1A
C	C	D	E	F	10	11	12	13	14	15	16	17	18	19	1A	1B
D	D	E	F	10	11	12	13	14	15	16	17	18	19	1A	1B	1C
E	E	F	10	11	12	13	14	15	16	17	18	19	1A	1B	1C	1D
F	F	10	11	12	13	14	15	16	17	18	19	1A	1B	1C	1D	1E

You are encouraged to do these addition problems without using the tables. Mentally converting the digits to decimal, adding, and then converting back to hexadecimal is one way to do hexadecimal arithmetic. Thus, when adding 9_{16} and D_{16}, mentally add 9 and 13 base 10, getting 22_{10}. Then change 22_{10} to 16_{16}, writing the 6 and carrying the 1. These conversions should be easy for you at this point.

Another method you can use is to add by groupings of 16. Whenever a sum of 16 is reached in a column, carry a 1 to the next column. In this way, no large sums need to be calculated.

Problem Set 6.2

In Problems 1–20, perform the addition.

1. 203_8
   $+ \ 54_8$

2. 532_8
   $+ 241_8$

3. 614_8
   $+ 244_8$

4. 734_8
   $+ 106_8$

6.2 Octal and Hexadecimal Addition

5. 1407_8
   $+ 2356_8$

6. 3356_8
   $+ 2443_8$

7. 23.17_8
   $+ 4.72_8$

8. 303.127_8
   $+ 23.4_8$

9. 2347_8
   6601_8
   $+ 5435_8$

10. 7777.77_8
    234.11_8
    $+ 1456.05_8$

11. $AD7_{16}$
    $+ 214_{16}$

12. $DE4_{16}$
    $+ 209_{16}$

13. $CE2_{16}$
    $+ 849_{16}$

14. $29D_{16}$
    $+ FA4_{16}$

15. $CAD7_{16}$
    $+ 4A6E_{16}$

16. $29.7C_{16}$
    $+ 8.FA_{16}$

17. $A61.128_{16}$
    $+ 77.B_{16}$

18. $FFFF.FF_{16}$
    $+ FF.FF_{16}$

19. $ABCD_{16}$
    $EF12_{16}$
    $+ 3456_{16}$

20. 678.6_{16}
    $54.C_{16}$
    $+ EDA.D_{16}$

In Problems 21–30, find the sum of the numbers.

21. 4567_8 and 762_8
22. 423_8 and 4.52_8
23. 456.732_8 and $.0456_8$
24. 777.7_8 and 1.01_8
25. 43.21_8, 4.563_8, and 451_8
26. $AE5_{16}$ and 592_{16}
27. $34.8D_{16}$ and $57.EA_{16}$
28. $12.54A_{16}$ and $7EC.8_{16}$
29. $ED.FFC_{16}$ and 345.79_{16}
30. 67_{16}, 45.3_{16}, and $D.F_{16}$
31. A computer can store a program beginning in location 456_8 and use 777_8 bytes for the program. What is the last memory location that can be used for the program?
32. A computer can store a program beginning in location $CD4_{16}$ and use FFF_{16} bytes for the program. What is the last memory location that can be used for the program?
33. A computer has a memory of $FFFF_{16}$ bytes. Extra memory can be purchased in units of 1200_{16} bytes. If four units of extra memory are purchased, what is the total memory size of the computer?
34. A computer has a memory of 6754_8 bytes. Extra memory can be purchased in units of 540_8 bytes. If three units of extra memory are purchased, what is the total memory size of the computer?
35. Change 567_8 and 431_8 to binary, find their sum, and change the sum back to octal. Add the two numbers in octal to check your answer.
36. Change $4A_{16}$ and $56C_{16}$ to binary, find their sum, and change the sum back to hexadecimal. Add the two numbers in hexadecimal to check your answer.
37. Check your answer to Problem 21 by changing each octal number to decimal, performing decimal addition, and then converting your answer to octal.
38. Check your answer to Problem 26 by changing each hexadecimal number to decimal, performing decimal addition, and then converting your answer to hexadecimal.
39. What number must be added to 12456_8 to get 100000_8?
40. What number must be added to $8AD3_{16}$ to get 10000_{16}?
41. If two bytes are used to store an integer, how many hexadecimal digits are required to represent an integer?
42. If three bytes are used to store an integer, how many octal digits are required to represent an integer?

6.3 Decimal Subtraction by Complement Addition

complement addition

Addition can be performed relatively easily by computers. However, subtraction is more difficult. Since electronic circuitry exists to perform addition, computers use a technique known as **complement addition** to change subtraction to addition. This technique will first be illustrated by using the decimal number system.

9's complement

The **9's complement** of a three-digit decimal number is 999 minus the number. For example, the 9's complement of 537 is $999 - 537 = 462$. For a four-digit number, the 9's complement is 9999 minus the number; and so on. Another way of finding a 9's complement is to find what number must be added to each digit of 537 to change that digit to 9. Thus, $5 + 4 = 9$, $3 + 6 = 9$, and $7 + 2 = 9$. So the 9's complement is 462.

10's complement

In addition to a 9's complement, there is also a **10's complement**, which is simply the 9's complement plus 1 (or the place value of the rightmost digit). So the 10's complement of 537 is $462 + 1$, or 463. Complements are illustrated in the next example.

EXAMPLE 6.9 Find the 9's complement and 10's complement of the number.

a. 1034_{10}

Solution

9's complement $\quad 9999_{10} - 1034_{10} = \underline{8965_{10}}$

10's complement $\quad 8965_{10} + 1 = \underline{8966_{10}}$

b. 6732.7_{10}

Solution

9's complement $\quad 9999.9_{10} - 6732.7_{10} = \underline{3267.2_{10}}$

10's complement $\quad 3267.2_{10} + 0.1 = \underline{3267.3_{10}}$

The rightmost digit has a place value of 0.1, so 0.1 is added to get the 10's complement.

Subtraction can be performed by using either 9's complement or 10's complement addition. Both will be illustrated.

Subtraction using the 9's complement follows these steps:

Step 1. Replace the subtrahend with its 9's complement.
Step 2. Change the subtraction sign to an addition sign.
Step 3. Perform the addition.
Step 4. If there is an extra digit—a *leading* 1, as it is called—cross it out and add it to the sum. If there is no extra digit, then the answer is negative and is found by taking the 9's complement of the sum.

These steps are illustrated in the following decimal subtraction example.

6.3 Decimal Subtraction by Complement Addition

EXAMPLE 6.10 Subtract by using the 9's complement:

$$792_{10}$$
$$-348_{10}$$

Solution The first step of the procedure is to find the 9's complement of the subtrahend (348). The second step is to change the subtraction sign to a plus sign:

$$792_{10}$$
$$+651_{10}$$

What we have really done in this step is *add* 999 to -348.

Now, the third step is to perform the addition:

$$792_{10}$$
$$+\ 651_{10}$$
$$\overline{1443_{10}}$$

For the fourth step, we notice that there is an extra digit in the sum. We cross it out and add it to the sum:

$$\cancel{1}443_{10}$$
$$+\ \ \ 1_{10}$$
$$\overline{444_{10}}$$

What we have really done in this step is *subtract* 1000 (cross out the 1) and *add* 1. Altogether, we have added 999 and 1 and subtracted 1000, so we haven't really changed the problem. The answer is 444_{10}.

Subtraction using 10's complement follows these steps:

Step 1. Replace the subtrahend with its 10's complement.
Step 2. Change the subtraction sign to an addition sign.
Step 3. Perform the addition.
Step 4. If there is an extra digit, cross it out. If there is no extra digit, the answer is negative. In this case, take the 10's complement of the sum.

These steps are illustrated in the following decimal subtraction example.

EXAMPLE 6.11 Subtract by using 10's complement:

$$792_{10}$$
$$-348_{10}$$

Solution The 10's complement of 348 is 651 + 1, or 652. Now, change the subtraction sign to addition, and add:

$$792_{10}$$
$$+652_{10}$$
$$\overline{1444_{10}}$$

Ignoring the extra digit gives the answer $\underline{444_{10}}$. Again, we are adding 1000 and subtracting 1000, so we aren't changing the original problem.

The only difference between using 9's complement and using 10's complement occurs when the 1 is added. Computers use complement addition because it simplifies the circuitry necessary to subtract. Subtraction is still necessary to find a complement, but *borrowing* is never needed. Furthermore, many computers store negative numbers in complement form. The 10's complement is more in line with the way modern computers store numbers, so that method will be used from now on.

EXAMPLE 6.12 Perform the subtraction.

a. $3247_{10} - 1480_{10}$

Solution

$$\begin{array}{r} 3247_{10} \\ -1480_{10} \end{array} \rightarrow \begin{array}{r} 3247_{10} \\ +8520_{10} \\ \hline 11767_{10} \end{array} \quad \text{or} \quad \underline{1767_{10}}$$

b. $2796_{10} - 483_{10}$

Solution

$$\begin{array}{r} 2796_{10} \\ -483_{10} \end{array} \rightarrow \begin{array}{r} 2796_{10} \\ +9517_{10} \\ \hline 12313_{10} \end{array} \quad \text{or} \quad \underline{2313_{10}}$$

Since 483 has fewer digits than 2796, the missing digit(s) are taken as zeros. So the complements become 9.

Now, consider the case where we are subtracting a larger number from a smaller number. The answer to such a problem will be negative. There will be no leading 1 in the sum when complement addition is performed. In this case, we simply find the complement of the sum (either 9's or 10's complement, depending on how we are working the problem) and put a minus sign in front of it. Example 6.13 illustrates this procedure.

6.3 Decimal Subtraction by Complement Addition

EXAMPLE 6.13 Perform the subtraction.

a. $364_{10} - 517_{10}$

Solution We will use 10's complement for this problem. The 10's complement of 517_{10} is $482_{10} + 1_{10}$, or 483_{10}. We now perform complement addition:

$$\begin{array}{r} 364_{10} \\ +483_{10} \\ \hline 847_{10} \end{array}$$

There is no leading 1, so the number is negative. We now take the 10's complement of 847_{10} and put a minus sign in front of it:

$$-153_{10}$$

In this case, we are simply adding and subtracting the same number (1000 or 999).

b. $1726_{10} - 2485_{10}$

Solution Again, we use 10's complement.

$$\begin{array}{r} 1726_{10} \\ -2485_{10} \\ \hline \end{array} \rightarrow \begin{array}{r} 1726_{10} \\ +7515_{10} \\ \hline 9241_{10} \end{array} \quad \text{or} \quad -0759_{10}$$

c. $346.5_{10} - 749.7_{10}$

Solution Again, we use 10's complement.

$$\begin{array}{r} 346.5_{10} \\ -749.7_{10} \\ \hline \end{array} \rightarrow \begin{array}{r} 346.5_{10} \\ +250.3_{10} \\ \hline 596.8_{10} \end{array} \quad \text{or} \quad -403.2_{10}$$

Note that decimal points do not complicate the process as long as they are lined up.

Problem Set 6.3

In Problems 1–10, find the 9's complement of the number.

1. 35_{10}
2. 76_{10}
3. 123_{10}
4. 432_{10}
5. 23406_{10}
6. 43079_{10}
7. 45.87_{10}
8. 5.031_{10}
9. 123.09456_{10}
10. 4567.8904_{10}

COMPUTER ARITHMETIC

In Problems 11–20, find the 10's complement of the number.

11. 78_{10}
12. 52_{10}
13. 195_{10}
14. 378_{10}
15. 23460_{10}
16. 56703_{10}
17. 5.9054_{10}
18. 45.0097_{10}
19. 332.7896_{10}
20. 129.87004_{10}

In Problems 21–30, perform the subtraction by using 9's complement.

21. $793_{10} - 208_{10}$
22. $52243_{10} - 1760_{10}$
23. $8.843_{10} - 1.9_{10}$
24. $53.92_{10} - 53.91_{10}$
25. $678_{10} - 678_{10}$
26. $741.5_{10} - 741.5_{10}$
27. $176_{10} - 944_{10}$
28. $278.6_{10} - 333.1_{10}$
29. $57_{10} - 241_{10}$
30. $919.54_{10} - 999.99_{10}$

In Problems 31–44, perform the subtraction by using 10's complement.

31. $546_{10} - 435_{10}$
32. $712_{10} - 538_{10}$
33. $42315_{10} - 4590_{10}$
34. $56.789_{10} - 7.888_{10}$
35. $5.541_{10} - 5.541_{10}$
36. $6789_{10} - 6789_{10}$
37. $678_{10} - 945_{10}$
38. $1.203_{10} - 4.97_{10}$
39. $45.307_{10} - 67.8_{10}$
40. $55.9054_{10} - 99.9999_{10}$
41. $67.893_{10} - 0_{10}$
42. $0_{10} - 78.54_{10}$
43. $0_{10} - 345.8_{10}$
44. $567.3_{10} - 0_{10}$

45. Is it possible to get a "leading 2"?
46. Why is no borrowing necessary in complement addition?
47. A computer has a total storage capacity of 4000_{10} bytes. If 1234_{10} bytes are reserved for a program, how much memory is left (use complement addition)?
48. A computer has a total storage capacity of 2^{10} bytes. If 2^8 bytes are reserved for the operating system, how much memory is left (use complement addition)?
49. A computer can store instructions in all memory locations between 2345_{10} and 3476_{10}, inclusive. How many locations are in this capacity (use complement addition)?
50. Memory locations between 1873_{10} and 3429_{10}, inclusive, are reserved for programmer use. How many locations are available to the programmer (use complement addition)?

■ 6.4 Binary Subtraction

1's complement

2's complement

All arithmetic inside the computer must be performed in the binary number system. As mentioned in the previous section, the method of complement addition allows subtraction to take place without borrowing.

Similar to the 9's complement base 10, we have a **1's complement** base 2. The 1's complement of the binary number 10011_2 is $11111_2 - 10011_2 = 01100_2$. Since the binary number system consists of only two symbols, the 1's complement can be found by simply changing all 0s to 1s and all 1s to 0s. The **2's complement** then becomes the 1s complement plus 1. Another way to find the 2s complement of

6.4 Binary Subtraction

10011_2 is to subtract it from 100000_2. This method is not used in computers because it involves considerable borrowing.

Here are some examples of the 1's and 2's complements.

Number	1's complement	2's complement
1011_2	0100_2	0101_2
110_2	001_2	010_2
00.01_2	11.10_2	11.11_2

The procedure for binary subtraction is the same as the procedure for subtraction given in the previous section, using either the 1's complement or the 2's complement. We will use the 1's complement for the first example and the 2's complement from there on.

EXAMPLE 6.14 Perform the binary subtraction.

a. $1011_2 - 1001_2$

Solution

$$\begin{array}{r} 1011_2 \\ -1001_2 \end{array} \rightarrow \begin{array}{r} 1011_2 \\ +\ 0110_2 \\ \hline 10001_2 \end{array} \text{(1's complement)}$$

$$\begin{array}{r} \cancel{1}0001_2 \\ +\ \ \ \ \ 1_2 \\ \hline 0010_2 \end{array} \quad \text{or} \quad \underline{10_2}$$

b. $11011_2 - 101_2$

Solution

$$\begin{array}{r} 11011_2 \\ -\ \ \ 101_2 \end{array} \rightarrow \begin{array}{r} 11011_2 \\ +\ 11011_2 \\ \hline 110110_2 \end{array} \text{(2's complement)} \quad \text{or} \quad \underline{10110_2}$$

Leading zeros are changed to 1s.

c. $10.011_2 - 1.100_2$

Solution

$$\begin{array}{r} 10.011_2 \\ -\ 1.100_2 \end{array} \rightarrow \begin{array}{r} 10.011_2 \\ +\ 10.100_2 \\ \hline 100.111_2 \end{array} \text{(2's complement)} \quad \text{or} \quad \underline{.111_2}$$

d. $1001_2 - 1100_2$

Solution

$$\begin{array}{r} 1001_2 \\ -1100_2 \end{array} \rightarrow \begin{array}{r} 1001_2 \\ +0100_2 \\ \hline 1101_2 \\ -11_2 \end{array} \text{(2's complement)} \quad \text{or} \quad 0010_2 + 1_2 = \underline{11_2}$$

This number must be negative since there is no leading 1.

e. $101.11_2 - 110.01_2$

Solution

$$\begin{array}{r}101.11_2\\-110.01_2\end{array} \rightarrow \begin{array}{r}101.11_2\\+001.11_2\\\hline 111.10_2\\\hline -.1_2\end{array} \quad \text{(2's complement)}$$

or $000.01_2 + .01_2 = 000.10_2$

We mentioned earlier that inside the computer, numbers are stored by using a certain number of bytes. This storage procedure can have an effect on complement addition. For example, if one byte is used to store numbers, then 1100_2 is stored as 00001100. The 2's complement then becomes 11110100. However, if two bytes are used to store the same number 1100_2, the number is stored as 0000000000001100, and the 2's complement becomes 1111111111110100. The results of complement addition are the same in either case (see Problems 55 and 56).

EXAMPLE 6.15 Assuming one byte for storage of each integer, perform the binary subtraction.

a. $1101_2 - 110_2$

Solution

$$\begin{array}{r}00001101_2\\-00000110_2\end{array} \rightarrow \begin{array}{r}00001101_2\\+11111010_2\\\hline 100000111_2\end{array} \text{ or } \underline{00000111_2}$$

b. $1001_2 - 1100_2$

Solution

$$\begin{array}{r}00001001_2\\-00001100_2\end{array} \rightarrow \begin{array}{r}00001001_2\\+11110100_2\\\hline 11111101_2\end{array} \text{ or } \underline{-00000011_2}$$

Problem Set 6.4

In Problems 1–10, find the 1's complement of the number.

1. 1_2
2. 0_2
3. 101_2
4. 1001_2
5. 11011_2
6. 101.11_2
7. 1101.11101_2
8. 101101.001_2
9. 111.11_2
10. 0.00_2

In Problems 11–20, find the 2's complement of the number.

11. 1110_2
12. 10001_2
13. 1010101_2
14. 1.00101_2
15. 1101.110111_2
16. 11.1111_2
17. 00.000_2
18. 1.111_2
19. 11000101.00101_2
20. 100101.0011011_2

6.5 Octal and Hexadecimal Subtraction

In Problems 21–26, perform the binary subtraction by using 1's complement.

21. $110_2 - 100_2$
22. $11001_2 - 10010_2$
23. $1.0011_2 - 1.0001_2$
24. $111.1_2 - 1.1_2$
25. $101_2 - 1111_2$
26. $1010.1_2 - 1100.0_2$

In Problems 27–40, perform the binary subtraction by using 2's complement.

27. $1001_2 - 110_2$
28. $1010_2 - 1000_2$
29. $11.01_2 - 10.11_2$
30. $101.101_2 - 10.11_2$
31. $1001_2 - 1100_2$
32. $110101_2 - 111011_2$
33. $1.001_2 - 1.1_2$
34. $110.0011_2 - 111.1_2$
35. $1_2 - 10001_2$
36. $1010_2 - 0_2$
37. $101.101_2 - 101.101_2$
38. $11.101_2 - 11.101_2$
39. $1001.11_2 - 1111.11_2$
40. $10101.01010_2 - 1010.10101_2$

41. In Problem 35, there are fewer digits in the top number than in the bottom number. Why don't the leading zeros of the top number have to be considered?
42. Is zero considered positive or negative in subtraction? Consider Problem 38.
43. Add 111111_2 to 100010_2, ignoring all carries and the extra digit. What does your answer represent?
44. A computer has a total storage capacity of 1000100_2 bytes. If 11011_2 bytes are not available to the programmer, then how many bytes are available?
45. A computer can store instructions in all locations between 11010_2 and 100000_2. How many locations are there?
46. Find the 2's complement of 101_2 and 110_2. Which complement is larger?
47. Convert 14_{10} and 9_{10} to binary, perform subtraction by 2's complement, and convert the answer to decimal.
48. Convert 7_{10} and 31_{10} to binary, perform subtraction by 2's complement, and convert the answer to decimal.
49. What binary number has a 2's complement of 11011_2?
50. Find the 2's complement of 101011_2 by subtracting it from 1000000_2, and check your answer.

In Problems 51–54, assume two bytes are used to store each integer, and perform the subtraction.

51. $111000_2 - 11001_2$
52. $1100000101_2 - 100101110_2$
53. $111001000_2 - 1110011111_2$
54. $101_2 - 110111_2$

55. Assuming two bytes are used for storage of each number, perform the subtraction of Example 6.15a, verifying that the same result is obtained.
56. Assuming two bytes are used for storage of each number, perform the subtraction of Example 6.15b, verifying that the same result is obtained.

6.5 Octal and Hexadecimal Subtraction

Subtraction in the octal number system can also be performed by using complement addition, using either the 7's complement or the 8's complement.

162 COMPUTER ARITHMETIC

7's complement, The **7's complement** of 345_8 is 432_8, since $777_8 - 345_8 = 432_8$. The **8's complement**
8's complement of 345_8 then becomes $432_8 + 1_8$, or 433_8. We will use the method of 8's complement in this section.

■ **EXAMPLE 6.16** Perform the subtraction by using 8's complement.

    a. $365_8 - 241_8$

Solution

$$\begin{array}{r} 365_8 \\ -241_8 \end{array} \rightarrow \begin{array}{r} 365_8 \\ +\,537_8 \\ \hline 1124_8 \end{array} \text{ (octal addition)}$$

$$\underline{124_8}$$

    b. $260.41_8 - 4.43_8$

Solution

$$\begin{array}{r} 260.41_8 \\ -\quad 4.43_8 \end{array} \rightarrow \begin{array}{r} 260.41_8 \\ +\,773.35_8 \\ \hline 1253.76_8 \end{array} \text{ or } \underline{253.76_8}$$

    c. $30.5_8 - 477.2_8$

Solution

$$\begin{array}{r} 30.5_8 \\ -477.2_8 \end{array} \rightarrow \begin{array}{r} 30.5_8 \\ +300.6_8 \\ \hline 331.3_8 \end{array} \text{ or } \underline{-446.5_8}$$

As you have probably guessed, hexadecimal subtraction can be performed by using the 15's complement or the 16's complement. The **15's complement** of $A73_{16}$ is $58C_{16}$ since $FFF_{16} - A73_{16} = 58C_{16}$. The **16's complement** of $A73_{16}$ then becomes $58C_{16} + 1_{16}$, or $58D_{16}$. We will use the method of 16's complement in this section.

15's complement
16's complement

■ **EXAMPLE 6.17** Perform the subtraction by using 16's complement.

    a. $F86_{16} - A73_{16}$

Solution

$$\begin{array}{r} F86_{16} \\ -A73_{16} \end{array} \rightarrow \begin{array}{r} F86_{16} \\ +58D_{16} \\ \hline 1513_{16} \end{array} \text{ (hexadecimal addition)}$$

$$\underline{513_{16}}$$

    b. $79.D6_{16} - 1.7E_{16}$

6.5 Octal and Hexadecimal Subtraction

Solution
$$\begin{array}{r} 79.D6_{16} \\ -\ 1.7E_{16} \end{array} \rightarrow \begin{array}{r} 79.D6_{16} \\ +FE.82_{16} \\ \hline 178.58_{16} \end{array} \text{ or } \underline{78.58_{16}}$$

c. $29.B_{16} - A8C.7_{16}$

Solution
$$\begin{array}{r} 29.B_{16} \\ -A8C.7_{16} \end{array} \rightarrow \begin{array}{r} 29.B_{16} \\ +573.9_{16} \\ \hline 59D.4_{16} \end{array} \text{ or } \underline{-A62.C_{16}}$$

Problem Set 6.5

In Problems 1–10, find the 8's complement of the number.

1. 32_8
2. 513_8
3. 3214_8
4. 1.234_8
5. 120.056_8
6. 707.0032_8
7. 777.77_8
8. 100000_8
9. 2361.07_8
10. 4041.37_8

In Problems 11–20, find the 16's complement of the number.

11. $D4_{16}$
12. $CE3_{16}$
13. 1894_{16}
14. $75DCE_{16}$
15. 12.34_{16}
16. $69E.CB4_{16}$
17. 1000.00_{16}
18. $FF.FFF_{16}$
19. $A89.7E_{16}$
20. $C6.9BD_{16}$

In Problems 21–30, perform the subtraction by using 8's complement.

21. $356_8 - 244_8$
22. $45760_8 - 5531_8$
23. $332.45_8 - 316.30_8$
24. $56_8 - 2137_8$
25. $53.35_8 - 53.35_8$
26. $617.5_8 - 72.7_8$
27. $765.4_8 - 0_8$
28. $2.54_8 - 7777.77_8$
29. $37.4370_8 - 65.0365_8$
30. $2631.075_8 - 0.364_8$

In Problems 31–40, perform the subtraction by using 16's complement

31. $CA9_{16} - 258_{16}$
32. $12CD4_{16} - 75E8_{16}$
33. $A43.7_{16} - 850.6_{16}$
34. $A2_{16} - B71_{16}$
35. $C83.6_{16} - 9.F_{16}$
36. $AB.CD_{16} - AB.CD_{16}$
37. $9.03_{16} - DFF.FF_{16}$
38. $AB9.047_{16} - 0_{16}$
39. $8903.4B6_{16} - 22.224_{16}$
40. $CD.437B_{16} - F1.437B_{16}$

41. If 3451_8 of 5670_8 bytes are reserved for programmer use, how many bytes (in octal) are not reserved for programmer use?

42. A computer has a total memory of 4000_{16} bytes. If $2A74_{16}$ bytes are reserved for the operating system, how many bytes are left (in hexadecimal)?

43. A section of computer memory begins at location $4AD_{16}$ and ends at location 1479_{16}. How many bytes are in this section of memory?

44. A section of computer memory begins at location 704_8 and ends at location 6305_8. How many bytes are in this section of memory?

45. Convert 24_{10} and 13_{10} to octal, perform subtraction, and convert the answer to decimal.

46. Convert 65_{10} and 77_{10} to hexadecimal, perform subtraction, and convert the answer to decimal.

47. What hexadecimal number has a 16's complement of $E3.89_{16}$?

48. What octal number has an 8's complement of 34.760_8?

49. In Problem 31, convert each number to binary, subtract, and convert the difference back to hexadecimal.

50. In Problem 22, convert each number to binary, subtract, and convert the difference back to octal.

6.6 Multiplication

The rules for multiplying in the binary, octal, and hexadecimal number systems are the same as the rules for multiplying in the decimal number system. Again, the only difference is the number of symbols used.

Binary Multiplication

Multiplication is repeated addition. In the case of binary numbers, multiplication is extremely easy. Multiplication by 0_2 is 0_2, and multiplication by 1_2 leaves the number unchanged. The multiplication table is short and simple:

·	0	1
0	0	0
1	0	1

EXAMPLE 6.18 Perform the binary multiplication.

   a. $1011_2 \cdot 101_2$

Solution

$$\begin{array}{r} 1011_2 \\ \times\ 101_2 \\ \hline 1011 \\ 0000 \\ 1011 \\ \hline 110111_2 \end{array}$$ (remember that this addition is binary addition)

6.6 Multiplication

b. $1.001_2 \cdot 1.1_2$ **c.** $110.11_2 \cdot 110.1_2$

Solution

$$\begin{array}{r} 1.001_2 \\ \times\ \ 1.1_2 \\ \hline 1001 \\ 1001 \\ \hline 1.1011_2 \end{array}$$

$$\begin{array}{r} 110.11_2 \\ \times\ \ 110.1_2 \\ \hline 11011 \\ 00000 \\ 11011 \\ 11011 \\ \hline 101011.111_2 \end{array}$$

The rule for positioning the binary point in the answer is the same as it was for decimal: The sum of the places to the right of the point in the two numbers being multiplied gives the number of places to the right of the point in the product.

Octal Multiplication

There are two ways to perform octal multiplication. One method is to construct an octal multiplication table (see Problem 1) and use the table to do the problems. Filling out this table should not be too difficult. For example, to fill in 6_8 times 4_8, multiply in decimal, getting 24_{10}, and mentally convert to 30_8. If you can fill in the table, then you have enough knowledge to do these problems without using the table.

The second method is to multiply each pair of digits in decimal and then convert to octal. So 6_8 times 4_8 is 24_{10}, or 30_8. The 0_8 is written and the 3_8 is carried. The only reason this multiplication is harder than multiplying in decimal is that you have memorized the decimal multiplication table but not the octal multiplication table.

EXAMPLE 6.19   Perform the octal multiplication.

a. $42_8 \cdot 53_8$ **b.** $3.05_8 \cdot 21.4_8$

Solution

$$\begin{array}{r} 42_8 \\ \times\ \ 53_8 \\ \hline 146 \\ 252 \\ \hline 2666_8 \end{array} \text{(octal addition)}$$

$$\begin{array}{r} 3.05_8 \\ \times\ \ 21.4_8 \\ \hline 1424 \\ 305 \\ 612 \\ \hline 65.674_8 \end{array}$$

Hexadecimal Multiplication

Hexadecimal multiplication can be performed by using the hexadecimal multiplication table (see Problem 2) or by multiplying each pair of digits in decimal and then converting to hexadecimal. For example, D_{16} times 5_{16} is $13_{10} \cdot 5_{10}$, which is 65_{10}, or 41_{16}. Write the 1_{16} and carry the 4_{16}.

EXAMPLE 6.20 Perform the hexadecimal multiplication.

a. $D7_{16} \cdot 28_{16}$ **b.** $39.1B_{16} \cdot 4.0A_{16}$

Solution

$$\begin{array}{r} D7_{16} \\ \times\ 28_{16} \\ \hline 6B8 \\ 1AE \\ \hline 2198_{16} \end{array}$$

$$\begin{array}{r} 39.1B_{16} \\ \times\ \ 4.0A_{16} \\ \hline 23B0E \\ 00000 \\ E46C \\ \hline E6.A70E_{16} \end{array}$$

Problem Set 6.6

1. Construct the octal multiplication table.
2. Construct the hexadecimal multiplication table.

In Problems 3–32, perform the multiplication.

3. 10_2 × 10_2
4. 101_2 × 11_2
5. 110101_2 × 110_2
6. 100011_2 × 101_2
7. 100.01_2 × 1.11_2
8. 1.1101_2 × 11.01_2
9. 11.011_2 × 1.001_2
10. 111111_2 × 111_2
11. 1011.1001_2 × 100.011_2
12. 10.000111_2 × 1.1101_2
13. 17_8 × 34_8
14. 56_8 × 31_8
15. 145_8 × 37_8
16. 704_8 × 53_8
17. 23.32_8 × 1.04_8
18. 404.5_8 × 20.14_8
19. 4004.06_8 × 23.76_8
20. 777.7_8 × 20.3_8
21. 333.0045_8 × 23.567_8
22. 100.003_8 × 3.0456_8
23. 45_{16} × $2A_{16}$
24. $7E_{16}$ × $2D_{16}$
25. $4C7_{16}$ × $B8_{16}$
26. $DD3_{16}$ × CA_{16}
27. $C.08_{16}$ × 3.7_{16}
28. $7.9C_{16}$ × $9.F_{16}$
29. $2E.CF_{16}$ × $3.9D_{16}$
30. 456.7_{16} × 95.2_{16}
31. $FF.005_{16}$ × $2.05E_{16}$
32. $407.8AD_{16}$ × $23.7D_{16}$

33. Calculate $453_8 \cdot 23_8$ and $23_8 \cdot 453_8$ to illustrate that octal multiplication is commutative.
34. Calculate $4E_{16} \cdot 21_{16}$ and $21_{16} \cdot 4E_{16}$ to illustrate that hexadecimal multiplication is commutative.
35. Extra memory units can be purchased in groups of $4D6_{16}$ bytes. If nine units are purchased, what is the total memory purchased (in hexadecimal)?
36. If six units of memory are full and each unit consists of 2345_8 bytes, then how many bytes are used (in octal)?
37. How does a binary number change when it is multiplied by 10_2?
38. How does a binary number change when it is multiplied by 0.1_2?

6.7 Division

39. How does an octal number change when it is multiplied by 512 (8^3)?
40. How does a hexadecimal number change when it is multiplied by 256?
41. What is the inverse for multiplication of 1000_2?
42. What is the inverse for multiplication of 100_8?
43. Find the product of 26_{10} and 43_{10} by converting each number to binary, performing binary multiplication, and converting the result to decimal. (This procedure is what a computer must do.)
44. Find the product of 17_{10} and 103_{10} by converting each number to binary, performing binary multiplication, and converting the result to decimal.

6.7 Division

Division can be considered as repeated subtraction. The subtraction can be performed by complement addition within the computer. So addition again becomes the main operation involved, simplifying the circuitry needed inside the computer.

Binary Division

In the case of binary numbers, division is easy because the divisor always divides into the part of the dividend under consideration either once (if the divisor is not the larger of the two numbers) or not at all (if the divisor is the larger of the two numbers). Computer division is performed in binary.

EXAMPLE 6.21 Perform the binary division.

a. $110_2 / 101_2$

Solution
```
        1
101₂)110₂
     101
     ---
       1
```

The answer is 1 with a remainder of 1, or simply $1_2 \underline{R1}$.

b. $10110_2 / 110_2$

Solution
```
         11
110₂)10110₂
     110
     ----
     1010
      110
      ---
      100
```

$11_2 \, R100$

c. $110111_2 / 1001_2$ d. $1010001_2 / 110_2$

Solution

$$\begin{array}{r} 110 \\ 1001_2 \overline{)110111_2} \\ \underline{1001} \\ 1001 \\ \underline{1001} \\ 01 \\ \\ 110_2 \text{ R}1 \end{array}$$

$$\begin{array}{r} 1101 \\ 110_2 \overline{)1010001_2} \\ \underline{110} \\ 1000 \\ \underline{110} \\ 100 \\ \underline{0} \\ 1001 \\ \underline{110} \\ 11 \\ \\ 1101_2 \text{ R}11 \end{array}$$

Octal Division

In decimal division, sometimes you guess how many times the divisor goes into the dividend, then multiply and check to see whether you were right. The same thing is necessary in octal division. For example, how many times does 45_8 go into 237_8? We guess 5 times, multiply 5_8 times 45_8, and get 271_8, which is too big. So we can try 4_8 times 45_8 and get 224_8. We didn't have this problem in binary division, since the answer was always 0 or 1. Now, the answer can be any number from 0_8 to 7_8. Doing these problems will give you more of an understanding of octal arithmetic.

Another way of doing the octal division problem is to convert to decimal, getting 37_{10} for 45_8 and 159_{10} for 237_8. Then find how many times 37_{10} divides into 159_{10}. We can also repeatedly subtract 45_8 from 237_8 until the difference is less than 45_8. Any of these methods will work.

EXAMPLE 6.22 Perform the octal division.

a. $35_8 / 7_8$ b. $573_8 / 24_8$

Solution

$$\begin{array}{r} 4 \\ 7_8 \overline{)35_8} \\ \underline{34} \\ 1 \\ \\ 4_8 \text{ R}1 \end{array}$$

$$\begin{array}{r} 22 \\ 24_8 \overline{)573_8} \\ \underline{50} \\ 73 \\ \underline{50} \\ 23 \\ \\ 22_8 \text{ R}23 \end{array}$$

Hexadecimal Division

Hexadecimal division is similar to octal division, except that the arithmetic involved is more complicated. For example, how many times does $2F_{16}$ divide into 78_{16}? We can make an intelligent guess, change each number to decimal, repeatedly add $2F_{16}$ until we get close to 78_{16}, or repeatedly subtract $2F_{16}$ from 78_{16} until the difference is less than $2F_{16}$. All the methods work.

EXAMPLE 6.23 Perform the hexadecimal division.

a. $78_{16} / 2F_{16}$ b. $142A_{16} / 59_{16}$

Solution

$$2F_{16} \overline{)78_{16}} \quad \begin{array}{r} 2 \\ \underline{5E} \\ 1A \end{array}$$
$$2_{16} \text{ R } 1A$$

$$59_{16} \overline{)142A_{16}} \quad \begin{array}{r} 3A \\ \underline{10B} \\ 37A \\ \underline{37A} \\ 0 \end{array}$$
$$3A_{16} \text{ R } 0$$

Problem Set 6.7

In Problems 1–32, perform the division.

1. $101_2 \overline{)1010_2}$
2. $110_2 \overline{)10110_2}$
3. $1101_2 \overline{)1001101_2}$
4. $11_2 \overline{)10110_2}$
5. $1010_2 \overline{)111011_2}$
6. $11011_2 \overline{)111011001_2}$
7. $1011_2 \overline{)110101_2}$
8. $1000_2 \overline{)1010101_2}$
9. $10001_2 \overline{)101100101_2}$
10. $11010_2 \overline{)1101001_2}$
11. $10111_2 \overline{)1101101_2}$
12. $1101001_2 \overline{)10001000_2}$
13. $110001_2 \overline{)10110111_2}$
14. $1010011_2 \overline{)111111111_2}$
15. $4_8 \overline{)25_8}$
16. $7_8 \overline{)36_8}$
17. $12_8 \overline{)453_8}$
18. $37_8 \overline{)175_8}$
19. $35_8 \overline{)546_8}$
20. $44_8 \overline{)5650_8}$
21. $457_8 \overline{)34506_8}$
22. $4043_8 \overline{)556703_8}$
23. $332_8 \overline{)56702_8}$
24. $12034_8 \overline{)110544_8}$
25. $D_{16} \overline{)57_{16}}$
26. $E_{16} \overline{)9A_{16}}$
27. $C4_{16} \overline{)E67_{16}}$
28. $16_{16} \overline{)44C9_{16}}$
29. $E4_{16} \overline{)56B_{16}}$
30. $9A8_{16} \overline{)406E_{16}}$
31. $67C_{16} \overline{)5547DC_{16}}$
32. $667_{16} \overline{)3320F_{16}}$

33. Convert the two numbers in Problem 7 to decimal, perform the division, and convert the answer to binary.
34. Convert the two numbers in Problem 16 to decimal, perform the division, and convert the answer to octal.

35. A computer has 11010_2 bytes of memory per unit of storage, and the total storage capacity is 10110110_2 bytes. How many units of storage are there in this computer?
36. What is the result of dividing a binary number by 10_2?
37. What is the result of dividing an octal number by 0.1_8?
38. What is the reciprocal of 100_{16}?
39. What is the reciprocal of 1000_2?
40. A computer has a total storage capacity of 770_8 bytes. If a total of eight programs completely uses all memory, what is the average number of bytes used per program?
41. Convert the two numbers in Problem 25 to binary, perform the division, and convert the result to hexadecimal.
42. Convert the two numbers in Problem 26 to binary, perform the division, and convert the result to hexadecimal.
43. Perform the division in Problem 3 by using complement addition to do the necessary subtraction.
44. Perform the division in Problem 2 by using complement addition to do the necessary subtraction.
45. Divide 763_{10} by 27_{10} by first converting the numbers to binary, performing binary division, and converting your answer to decimal.
46. Divide 982_{10} by 33_{10} by first converting the numbers to binary, dividing, and converting your answer to decimal.

CHAPTER SUMMARY

In this chapter, we have studied the four basic operations of arithmetic—addition, subtraction, multiplication, and division—and how they are performed in the binary, octal, and hexadecimal number systems by computers.

Addition is the main operation performed by computers, and the other operations are all performed by using variations of addition. Subtraction, for example, is performed by using complement addition, since negative numbers are stored by their complements inside computers. Multiplication is repeated addition, and division can be considered repeated subtraction (complement addition). All arithmetic performed by the computer is done in the binary number system.

REVIEW PROBLEMS

In Problems 1–36, perform the indicated arithmetic operation.

1. 10110_2
   $+\ 1110_2$

2. 101.101_2
   $+\ 110.11_2$

3. 11.01011_2
   $+\ 10.111_2$

4. 1011.011_2
   $+\ 1101.111_2$

5. 1011.0111_2
   100.0101_2
   1110.111_2
   $+\ 100.011_2$

6. 101101.1_2
   110011.0_2
   1010_2
   $+\ 111000.1_2$

7. 34.75_8
   $+\ 2.06_8$

8. 507.4_8
   $+\ 304.6_8$

Review Problems

9. $B4.9E_{16}$
   $+C2.07_{16}$

10. $709.4EA_{16}$
    $+ 8E.632_{16}$

11. 12306_{10}
    $- 9871_{10}$

12. 34.789_{10}
    $- 45.021_{10}$

13. 1011.011_2
    $- 111.001_2$

14. 101101_2
    $- 11111_2$

15. 101.0001_2
    $- 111.0111_2$

16. 1011.101_2
    $- 100.001_2$

17. 405.03_8
    $- 377.76_8$

18. 44.503_8
    $- 56.703_8$

19. $E47.9C_{16}$
    $-FF3.27_{16}$

20. $109.7D_{16}$
    $- 99.AB_{16}$

21. 101.11_2
    $\times\ 10.1_2$

22. 1001.1_2
    $\times\ 101_2$

23. 1101101_2
    $\times\ 1.0111_2$

24. 101.011_2
    $\times\ 11.11_2$

25. 404.56_8
    $\times\ 35.7_8$

26. 5506.7_8
    $\times\ 45.36_8$

27. $5E.C3_{16}$
    $\times\ 34.F_{16}$

28. $457.0E_{16}$
    $\times\ 7E.96_{16}$

29. $1001_2 \overline{)11101101_2}$

30. $1011_2 \overline{)1011100_2}$

31. $1111_2 \overline{)10011001_2}$

32. $11_2 \overline{)110011_2}$

33. $23_8 \overline{)4507_8}$

34. $406_8 \overline{)30054_8}$

35. $A8_{16} \overline{)3D7_{16}}$

36. $4E_{16} \overline{)109C_{16}}$

In Problems 37–42, perform the operations that a computer would go through.

37. Convert 5.75_{10} and 16.5_{10} to binary, find their sum, and convert the sum to decimal.

38. Convert 98_{10} and 102_{10} to binary, find their difference ($98_{10} - 102_{10}$), and convert the difference to decimal.

39. Convert 101_{10} and 22_{10} to binary, find their product, and convert the product to decimal.

40. Convert 55_{10} and 17_{10} to binary, find their quotient ($55_{10} / 17_{10}$), and convert the quotient to decimal.

41. Convert the storage locations 405_8 and 563_8 to binary, find their difference, and convert the difference to octal.

42. Convert the storage locations $A37_{16}$ and 45_{16} to binary, find their difference, and convert the difference to hexadecimal.

COMPUTER CONSIDERATIONS

CHAPTER **7**

In the previous chapters, we have examined number systems used by computers and arithmetic using these number systems. Now, it is time to examine these topics from a computer standpoint to see what goes on inside the computer and what special problems might be encountered.

We must remember that a computer is a device of limited ability which must follow very precise rules. The concepts of this chapter will help you understand what a computer is capable of doing and what its limitations are.

7.1 Significant Digits, Accuracy, and Precision

Numbers cannot be stored in the computer with commas. Thus, 7,000 must be stored as 7000. For this reason, commas will not be used in any numbers throughout this chapter.

Consider the following sentence: "A crowd of 7000 people had gathered to watch the parade." Does this statement mean that *exactly* 7000 people saw the parade? Probably not. The figure 7000 may be an approximation to the nearest thousand.

Now, consider the statement "There were 6243 applications for marriage licenses received in this office last year." In this case, the figure 6243 means that exactly this many applications were received.

The two numbers just mentioned both contain four digits, but the number of important, or "significant," digits in these numbers is different. That is, 7000 has only one significant digit (7), but 6243 has four significant digits. The following rules will help you to determine the significant digits in a number.

Rules for Significant Digits

1. All nonzero digits are significant.
2. A zero is a significant digit if it is between two significant digits.
3. Leading 0s are not significant.
4. Trailing 0s are significant only if the number contains a decimal point.

EXAMPLE 7.1 Find the number of significant digits in the number.

Solution
a. 527 — 3
b. −5032 — 4 (0 is between 5 and 3 and is significant)
c. 3.68 — 3

174

7.1 Significant Digits, Accuracy, and Precision

d.	−.0019	2 (leading 0s are not significant)
e.	15009	5
f.	23.004	5
g.	−27500	3 (since there is no decimal point in the number, the trailing 0s are not significant)
h.	3000.00	6 (the trailing 0s are significant since the number includes a decimal point)

Significant digits leads to the concept of accuracy.

accuracy

Definition 7.1 The **accuracy** of a number is the number of significant digits in the number.

The number $\pi = 3.14$ is accurate to 3 significant digits; $\sqrt{2} = 1.14142$ is accurate to 6 significant digits. In Example 7.1, we were really finding the accuracy of those numbers.

Now, consider the following sentence: "Joe's daughter, Debbie, made $4.17 selling lemonade last year, but Joe earned $125000 last year." Both figures have the same accuracy. However, we know Debbie's figure to the nearest penny, but we only know her father's figure to the nearest $1000. This example leads to the idea of precision.

precision

Definition 7.2 The **precision** of a number is the place value of the rightmost significant digit.

By definition, precision will always be a power of 10.

EXAMPLE 7.2 Find the accuracy and precision of the number.

Solution

Number	Accuracy	Precision
3.02	3	$.01 = 10^{-2}$
−1470	3	$10 = 10^1$
−.006043	4	$.000001 = 10^{-6}$
23	2	$1 = 10^0$
1000000	1	$1000000 = 10^6$

If a distance can be measured to the nearest inch, then it has a precision of 1 inch, regardless of how many significant digits are in the number. The numbers 3.02, −21.74, and 33021.79 all have the same precision (.01); but 331, −.00204, and 356000 all have the same accuracy (3).

Understanding accuracy and precision will become important later in this chapter when we discuss how numbers are represented in computers.

Problem Set 7.1

In Problems 1–20, find the accuracy and precision of the number.

1. 3
2. −7
3. −14
4. 270
5. 2.3
6. 31.9
7. 0.2
8. −4.9
9. −216.4
10. 0.8
11. .0003
12. .00219
13. 10.0
14. 1470.00
15. −36.0040
16. 2000.001
17. 33267.174
18. 999.999
19. 55000
20. −6300000

21. The elevation of Mt. LeConte in the Great Smoky Mountains has been measured as 6593 feet. What is the accuracy and precision of this elevation?

22. The distance between Big Rapids, Michigan, and Okemos, Michigan, has been measured as 115.9 miles. What is the accuracy and precision of this distance?

23. When is 0 a significant digit?
24. When is 0 not a significant digit?
25. When is 7 not a significant digit?
26. When is 7 a significant digit?
27. Which number has a better precision: 23.617 or 131349.24?
28. Which number has a better precision: 3271.9 or −.0002?
29. Which number is more accurate: 23.617 or 131349.24?
30. Which number is more accurate: 3271.9 or −.0002?
31. Which is more precise: $17236.19 or $1.04?
32. Which is more accurate: $17236.19 or $1.04?
33. Convert 134_{10} to binary. How many significant binary digits are there?
34. Convert 255_{10} to binary. How many significant binary digits are there?

7.2 Scientific Notation

Before we can examine how computers store real numbers, we must first discuss scientific notation. Consider the distance from the earth to the moon in centimeters: approximately 46000000000. This measurement is important for scientists when they study the earth's gravitational pull.

Now, consider the time it takes light to travel 1 inch: approximately .00000000008 seconds. Measurements like this one are important to scientists studying physics. However, these measurements require too many digits to be stored in a computer the way they are written here. So another way of writing numbers is needed—a form called scientific notation.

The first step in writing a number in scientific notation is to move the decimal point so that the number is 1 or larger but less than 10. In other words, there is exactly one nonzero digit to the left of the decimal point.

Moving the decimal point is equivalent to multiplying the number by a power of 10. If n is a positive integer, then multiplying a number by 10^n moves the decimal point n places to the right. For example, $3.45 \times 10^3 = 3450$ and $1.763 \times 10^2 = 176.3$. Similarly, multiplying a number by 10^{-n} moves the decimal

7.2 Scientific Notation

point n places to the left. So $2.76 \times 10^{-2} = .0276$ and $4.53 \times 10^{-3} = .00453$.

The second step in writing a number in scientific notation is to multiply the number from the first step by the appropriate power of 10 so that the product equals the original number.

scientific notation

Thus, we have the following definition: **Scientific notation** results in a number written in the form

(a number between 1 and 10) $\times 10^n$, $\quad n \in I$

EXAMPLE 7.3 Write the number in scientific notation.

a. 3476

Solution $\underline{3.476 \times 10^3}$

The decimal point has been moved 3 places to the left to get a number between 1 and 10. So we must multiply by 10^3 to move the point back 3 places to the right so that our result is equal to the given number. Although 3476 also equals 34.76×10^2, this form is not considered scientific notation.

b. .00023

Solution $\underline{2.3 \times 10^{-4}}$

Since .00023 is less than 1, the exponent is negative.

c. 3.71

Solution $3.71 \times 10^0 = 3.71 \times 1 = \underline{3.71}$

If the original number is between 1 and 10, then it is already in scientific notation, and 10^0 is usually not written.

d. .0000001437

Solution $\underline{1.437 \times 10^{-7}}$

In this case, the decimal point was moved 7 places to the right to get a number between 1 and 10. So we must multiply by 10^{-7} to move the point back 7 places to the left.

e. 56421000

Solution $\underline{5.6421 \times 10^7}$

Only the significant digits need be written in scientific notation. So the last three zeros are not written.

f. -42.3407

Solution $\underline{-4.23407 \times 10^1}$

The minus sign in the original number does not complicate the problem. It simply makes the number in scientific notation negative.

g. 10000

Solution $1 \times 10^4 = \underline{10^4}$

If a number is a power of 10, then there is no need to write the 1. Simply write 10^n.

Another way of writing a number in scientific notation is to count the number of places that the decimal point has been moved to achieve a number between 1 and 10 and multiply by 10 to that power. If the original number is 10 or larger, then the exponent is positive. If the original number is less than 1, then the exponent is negative.

Taking a number out of scientific notation is accomplished by moving the decimal point the specified number of places (the power of 10) to the right if the exponent is positive and to the left if the exponent is negative. Remember that multiplying a number by a power of 10 merely moves the decimal point and doesn't change the digits.

EXAMPLE 7.4 Write the number in decimal notation.

a. 4.04×10^0

Solution $4.04 \times 10^0 = 4.04 \times 1 = \underline{4.04}$

b. 1×10^{-5}

Solution $1 \times 10^{-5} = \underline{.00001}$

c. 3.1426×10^3

Solution $3.1426 \times 10^3 = \underline{3142.6}$

d. 1.174×10^{-4}

Solution $1.174 \times 10^{-4} = \underline{.0001174}$

e. 2.35×10^{11}

Solution $2.35 \times 10^{11} = \underline{235000000000}$

Even though the trailing zeros are not significant digits, they, of course, must be written down.

f. 5.4127×10^{-6}

Solution $5.4127 \times 10^{-6} = \underline{.0000054127}$

g. 345.009×10^2

Solution $345.009 \times 10^2 = \underline{34500.9}$

Even though this number was not given in scientific notation, the rules for moving the decimal point still apply.

Problem Set 7.2

In Problems 1–20, write the number in scientific notation.

1. 100
2. 100000
3. 31.2
4. 55.3
5. 457
6. 296
7. .012
8. .0023
9. 3.098
10. 9.0054
11. -35
12. -93
13. 32617
14. 45104
15. .000004
16. .000027
17. -3141.59
18. -2631.794
19. 0
20. 4

21. The president of a large corporation has an annual salary of $140000. Write this salary in scientific notation.
22. The elevation of Mt. LeConte is 6593 feet above sea level. Write this elevation in scientific notation.

In Problems 23–42, write the number in decimal notation.

23. 1×10^9
24. 1×10^{-4}
25. 2.1×10^0
26. 3.9×10^1
27. 3.12×10^3
28. 4.09×10^5
29. 3.8×10^{-1}
30. 4×10^{-4}
31. 1.002×10^{-7}
32. 3.334×10^{-8}
33. 1.87×10^{12}
34. 3.08×10^{-12}
35. -6.54×10^{-3}
36. -5.04×10^6
37. 34.56×10^5
38. 53.09×10^{-6}
39. 1234×10^{-3}
40. 2005.6×10^2
41. $.00324 \times 10^6$
42. $-.34005 \times 10^1$

43. A *googol* is defined to be 1 followed by 100 zeros. Write this number in scientific notation.
44. A *googolplex* is defined to be 1 followed by a googol of zeros (see Problem 43). Write this number in scientific notation.
45. Write the number $(10^{10})^{10}$ in scientific notation.
46. Write the number $(10^{-10})^{10}$ in scientific notation.
47. Change 111001.11_2 to decimal, and write the answer in scientific notation.
48. Change 307.4_8 to decimal, and write the answer in scientific notation.

7.3 Integers and Reals

In Chapter 1, we studied the different sets of numbers: N, I, Q, and R. Most computers don't distinguish between all four of these sets, but they do make a distinction between numbers that include a decimal point (3.14) and numbers that don't include a decimal point (35). In this chapter, we will call numbers that include the decimal point *real numbers* and numbers that don't include the decimal point *integers*.

These two types of numbers are stored in different ways inside the computer.

COMPUTER CONSIDERATIONS

The computer integer 35 is different from the computer real number 35., even though they both represent the same number.

Every integer is written without a decimal point. Integers are stored inside the computer in binary form. If one byte (eight bits) is available to store an integer, then 29_{10} is stored as 00011101, the binary equivalent of 29_{10}.

How should negative integers be stored? They *could* be stored by using the first bit to indicate the sign (0 for +, 1 for −). The next example illustrates.

EXAMPLE 7.5 If one byte is available to store an integer, with the first bit used for the sign, then determine how the integer should be stored.

a. 58_{10}

Solution 00111010

The first bit (0) indicates that the number is positive.

b. 114_{10}

Solution 01110010

c. -37_{10}

Solution 10100101

The first bit (1) indicates that this is a negative number, and the other bits represent the number 37_{10}.

d. -127_{10}

Solution 11111111

e. 127_{10}

01111111

A problem with this storage scheme is that special circuitry must be used to interpret the sign. Another method, and the one used by many computers, is to store the negative numbers in 2's complement form. If one byte is available for an integer, then 77_{10} is stored as 01001101, and -77_{10} is stored as 11111111 $-01001101 + 1$, or 10110011, the 2's complement of 77_{10}. In this method, addition is straightforward.

EXAMPLE 7.6 Determine how the number is stored when using one byte and 2's complement form for negatives.

Solution   a. 104_{10} 01101000

b. -41_{10} $11111111 - 00101001 + 1 = \underline{11010111}$

7.3 Integers and Reals

    c. -1_{10} $11111111 - 00000001 + 1 = \underline{11111111}$

    d. 0_{10} $\underline{00000000}$ (0 is considered nonnegative).

    e. -42_{10} $11111111 - 00101010 + 1 = \underline{11010110}$

EXAMPLE 7.7   Perform these additions by using 2's complements.

    a. $104_{10} + (-41_{10})$

Solution

$$\begin{array}{r} 01101000 \\ +\ 11010111 \\ \hline \mathit{1}00111111 \\ \underline{00111111} \end{array} \quad \text{or} \quad \underline{63_{10}}$$

    b. $-41_{10} + (-1_{10})$

Solution

$$\begin{array}{r} 11010111 \\ +\ 11111111 \\ \hline \mathit{1}\,11010110 \\ \underline{11010110} \end{array} \quad \text{or} \quad \underline{-42_{10}}$$

In part b, since the first digit of the answer is 1, this number is negative. Taking the 2's complement of the remaining bits yields 42_{10}, so the answer is -42_{10}.

Negative integers are stored as "higher" numbers than positive integers, with -1 (11111111) at the upper end. This process may seem to be a strange way to store integers, but it makes arithmetic far simpler for the computer.

Real numbers are stored by a totally different method. Many computers store real numbers in a form of scientific notation. For instance, the number 186000 is written as 1.86×10^5 in scientific notation. Some computers use the notation 1.86E + 05 for this number, where the E stands for exponent and the next number gives the power of 10. Other computers use a **normalized notation** where the decimal point is moved one place to the immediate left of the leftmost significant digit. In normalized notation, 186000 is written as .186E + 06. Thus, in normalized notation, the number is always written as a number between 0 and 1 followed by E and the power of 10 necessary to give the original number.

normalized notation

EXAMPLE 7.8   Write the number as a computer would in scientific notation.

Solution

    a.   3500 $\underline{3.5\text{E}+03}$

    b. -2217.4 $\underline{-2.2174\text{E}+03}$

    c.   .0043 $\underline{4.3\text{E}-03}$

    d.   2.97 $\underline{2.97\text{E}+00}$

EXAMPLE 7.9

Write the number as a computer would in normalized notation.

Solution

a.	3500	.35E+04
b.	−2217.4	−.22174E+04
c.	.0043	.43E−02
d.	2.97	.297E+01

Although many storage schemes are possible, the following scheme is representative of the way computers store real numbers. Suppose that four bytes (32 bits) are used to store a real number. The first bit represents the sign of the number (0 for +, 1 for −). The next 7 bits represent the exponent, and the remaining 24 bits are used to store the number.

The number is stored in normalized notation in binary form. The exponent is also stored in binary form and represents a power of 2, but some modification is necessary. We would like to include negative exponents. So we need either a bit to hold the sign of the exponent or a different storage scheme where the sign of the exponent is not necessary. Many computers use the latter method, as explained next.

Using 7 bits to hold an exponent allows exponents from 0 to 127 (1111111_2). We would like half of these exponents to be negative. To achieve this goal, we add 64 to each exponent before storing it in the computer. For example, if the exponent is 14, then we store $14+64$, or 78, as the exponent. If -35 is the exponent, then we store $-35+64$, or 29, as the exponent. In this way, we can have negative exponents without having to worry about the sign of the exponent, which will always be positive. We can now store exponents ranging from -64 to $+63$.

One of the advantages of using normalized notation is that the point is always to the left of the number. And since this format is known, the point itself does not have to be stored.

EXAMPLE 7.10

Represent the real number by using the storage scheme just described.

a. $25._{10}$

Solution

In binary form	$25_{10} = 11001_2$
In normalized binary form	$.11001 \times 2^5$
The sign of the number is +	0
The stored exponent	$5+64 = 69_{10}$
The exponent in binary form	1000101_2

The final storage is

0 1000101 110010000000000000000000

7.3 Integers and Reals

b. -34.2_{10}

Solution

In binary form	$34.2_{10} = 100010.\overline{0011}_2$
In normalized binary form	$.100010\overline{0011}_2 \times 2^6$
The sign of the number is $-$	1
The stored exponent	$6 + 64 = 70_{10}$
The exponent in binary form	1000110_2

The final storage is

<u>1 1000110 100010001100110011001100</u>

Since the number repeats forever, it must be cut off, or *truncated*, after 24 bits. This topic will be covered in detail in the next section.

c. $.025_{10}$

Solution

In binary form	$.025_{10} = .0000\overline{0011}_2$
In normalized binary form	$.\overline{1100}_2 \times 2^{-5}$
The sign of the number is $+$	0
The stored exponent	$-5 + 64 = 59_{10}$
The exponent in binary form	0111011_2

The final storage is

<u>0 0111011 110011001100110011001100</u>

Now, consider 25 (integer) and 25. (real). The number 25 can be stored by using four bytes as 00000000 00000000 00000000 00011001, but 25. may be stored as 01000101 11001000 00000000 00000000. Thus, we see that the same number can be stored two completely different ways, depending on whether or not the point is written. We will continue this discussion in Section 7.5, where we will examine arithmetic operations for integers and reals.

Problem Set 7.3

In Problems 1–10, represent the integer by using one byte with the first bit for the sign.

1. 23
2. 17
3. -31
4. -23
5. 1
6. 0
7. 120
8. -86
9. -45
10. 118

In Problems 11–20, represent the number by using one byte and 2's complement for negatives.

11. 27
12. 14
13. 0
14. 1
15. -50
16. -72
17. 116
18. -93
19. -58
20. 126

In Problems 21–26, perform the addition by using the results of Problems 11–20. Give your answer in binary.

21. $27_{10} + (-50_{10})$
22. $14_{10} + (-72_{10})$
23. $-50_{10} + 116_{10}$
24. $(-72_{10}) + (-93_{10})$
25. $(-58_{10}) + (-50_{10})$
26. $126_{10} + (-93_{10})$

In Problems 27–36, write the number as a computer would in scientific notation. Give your answer in decimal.

27. 3.17
28. -2.903
29. $-14.$
30. 106.
31. $-.0125$
32. .1307
33. 156.123
34. -14.9
35. 1000000.1
36. 22000.004

In Problems 37–46, write the number as a computer would in normalized notation. Give your answer in decimal.

37. -1.351
38. 2.04
39. 51.0
40. $-47.$
41. .023
42. .01491
43. -153.6
44. 26.232
45. 33004.97
46. 2200000.03

In Problems 47–56, represent the storage of the real number by using four bytes and the storage scheme described in this section.

47. $12._{10}$
48. $31._{10}$
49. -7.0_{10}
50. -154.0_{10}
51. 35.8_{10}
52. 17.45_{10}
53. $.02_{10}$
54. $.004_{10}$
55. $-.125_{10}$
56. $-.0625_{10}$

57. What is the largest positive integer that can be represented by using one byte with the first bit for the sign?

58. What is the smallest negative integer that can be represented by using one byte with the first bit for the sign?

▰ 7.4 Truncation, Rounding, and Conversion Error

No matter how large a computer is, it still has a *limited* amount of storage. Consider the fraction $\frac{2}{3} = .6666\overline{6}$. If a computer can store six decimal digits for a number, for example, $\frac{2}{3}$ will be stored as .666666 or .666667, depending on how the computer cuts off the number. No computer can accurately store $\frac{2}{3}$ because it repeats indefinitely, and the computer must cut the number off at some point. Neither .666666 nor .666667 exactly equals $\frac{2}{3}$.

truncate

To **truncate** a number means to simply ignore the extra digits that the computer can't store and use as many of the actual digits for the number as possible. For example, .2349 truncated to three decimal digits is .234. The 9 is ignored. And $\frac{2}{3}$ truncated to six decimal digits is .666666.

Many computers truncate digits from the right (least significant digits) if the numbers include too many digits to store in memory. Real numbers are written in scientific or normalized notation and truncated on the right.

What about integers? Suppose a computer allows six decimal digits to be stored for an integer. How will the computer handle a number like 1234567, for example? This number is too large to be stored as 1234567. Truncation on the left

7.4 Truncation, Rounding, and Conversion Error

leaves 234567, which eliminates the most significant digit. Truncation on the right leaves 123456, but the decimal point is one place beyond the 6. In such cases, the computer normally gives an **overflow error**, indicating that the number is simply too large to be stored in the computer.

overflow error

EXAMPLE 7.11 Truncate the decimal number to the given number of significant digits (which is specified in parentheses).

Solution

a.	2.53 (2)	2.5
b.	.1874 (3)	.187
c.	2.15 (2)	2.1
d.	−35.98 (3)	−35.9
e.	.00012935 (4)	.0001293

rounding

An alternative to truncation is **rounding**, where the last digit is "adjusted" to give a more accurate representation of the number.

Rule for Rounding When a number is to be rounded to n digits, if the $(n+1)$st digit is greater than or equal to 5, then the nth digit is increased by 1. Otherwise, it isn't changed.

EXAMPLE 7.12 Round the number to the given number of significant digits.

Solution

a.	2.53 (2)	2.5
b.	17.948 (3)	17.9
c.	−.002463 (2)	−.0025
d.	$.66\overline{6}$ (6)	.666667
e.	$.1\overline{73}$ (5)	.17374

No matter whether truncation or rounding is used to store numbers, errors are inevitable. Consider, for example, storing $\frac{1}{9}$ by using five decimal digits. Here, $\frac{1}{9}$ is stored as 0.11111, which does not exactly equal $0.1\overline{1}$.

Since computers use binary numbers, we must also consider binary examples. For instance, 0.1_{10} is $.000\overline{1100}_2$. Repeating fractions are very common in decimal-to-binary conversions, as we discovered in Chapter 5. Assume that seven bits are available to store this number. Then $.0001100_2$ is stored for $.1_{10}$. Changing this number back to decimal gives $.09375_{10}$, which does not equal $.1_{10}$.

Note that no arithmetic was performed in this example. We simply changed $.1_{10}$ to seven binary digits and changed it back to decimal. The resulting error is called **conversion error** and is quite common in decimal-to-binary conversions in computers.

conversion error

COMPUTER CONSIDERATIONS

If more bits had been used for storage, the result would have been closer to $.1_{10}$, but it would never be exactly equal to 0.1_{10}. You should be aware that conversion error exists and understand why you don't always get the number you should get.

EXAMPLE 7.13 A student needs a grade point average (GPA) of 3.6 or higher to make the honor roll. Joe has a GPA of exactly 3.6. If a computer can use seven bits for a number, what will the computer get for Joe's GPA?

Solution

$3.6_{10} = 11.1\overline{0011}_2$
11.10011 (7 bits)
$11.10011_2 = \underline{3.59375}$

According to the computer, Joe's average is *not* 3.6 or higher. So a student who should be on the honor roll isn't (if a computer is in charge).

One way to cut down on conversion error is for a computer to store numbers in a decimal format. Numbers must be represented in binary form inside the computer, but they do not have to be represented in the true binary number system. There are many methods of storing decimal numbers in binary form. We will consider one example.

binary-coded decimal

Computers sometimes use a four-bit, **binary-coded decimal** (BCD) format, where each decimal digit is changed to four bits. The decimal number is not converted to a binary number, but each digit is converted to its binary equivalent using exactly four bits. The next example illustrates the technique.

EXAMPLE 7.14 Write the number in BCD format.

a. 327

Solution

3   2   7
0011   0010   0111
$\underline{001100100111}$

b. 39845

Solution

3   9   8   4   5
0011   1001   1000   0100   0101
$\underline{0011100110000100 0101}$

In the BCD format, the numbers are written in binary form, but not as true binary numbers. The leading and trailing zeros must be written in this code. While this conversion is similar to the conversion from hexadecimal to binary, we

7.4 Truncation, Rounding, and Conversion Error

are only using the decimal digits 0 through 9. In BCD format, there is a byte preceding the number containing such information as the sign of the number and the placement of the decimal point.

Conversion error will still occur with repeating *decimals* ($7.\bar{4}_{10}$, for example) but not with numbers like $.1_{10}$. Some disadvantages of this system are that it may use more storage than true binary numbers, and arithmetic must be performed differently.

In scientific calculations, small conversion errors can be tolerated; but in business applications, figures must be exact. An error of even one penny is not acceptable. Consider Example 7.13 again, but now suppose that the 3.6 represents Joe's hourly salary. If the computer rounds or truncates to the nearest penny, it will obtain the result 3.59. Joe will be "robbed" of one penny for every hour he works. Multiply this one penny by 10,000 employees working for a year and we are talking about a substantial amount of money. So binary-coded decimal formats similar to the BCD code are used frequently in business applications.

■ Problem Set 7.4

In Problems 1–18, truncate the number to the given number (in parentheses) of significant digits.

1. 14.38 (4)
2. −3.109 (4)
3. −.257 (2)
4. .826 (2)
5. 2614.35 (5)
6. −1000.000 (6)
7. 3.12 (2)
8. 7.03 (2)
9. −26.09 (3)
10. 73.86 (3)
11. .00023 (2)
12. −.0000147 (3)
13. −35.009 (4)
14. 176.123 (5)
15. 2768.435 (5)
16. $.\bar{1}$ (9)
17. $.\bar{9}$ (7)
18. $-.\overline{29}$ (7)

In Problems 19–36, round the number to the given number (in parentheses) of significant digits.

19. 2.79 (3)
20. 14.06 (4)
21. −33.59 (3)
22. −226.77 (4)
23. 47.265 (4)
24. 95.45 (3)
25. 279.329 (4)
26. 88.848 (4)
27. $-.\bar{1}$ (9)
28. $.\bar{9}$ (5)
29. $3.\overline{25}$ (6)
30. $-17.\overline{16}$ (9)
31. $.\overline{2793}$ (7)
32. $.\overline{9748}$ (8)
33. $.4\bar{9}$ (7)
34. $-.\overline{4276}$ (7)
35. $.2003 \cdot 10^{-4}$ (3)
36. $78.9 \cdot 10^{-4}$ (2)

In Problems 37–48, change the number to binary, truncate to seven bits, and convert back to decimal. Note any conversion errors.

37. 35_{10}
38. 47_{10}
39. 73_{10}
40. 104_{10}
41. 127_{10}
42. 126_{10}
43. 0.25_{10}
44. 0.9_{10}
45. 0.3_{10}
46. 0.875_{10}
47. 26.1_{10}
48. 7.4_{10}

49. The decimal $0.1_{10} = .000\overline{1100}_2$, which is $.09375_{10}$ if seven bits are used to store the number. What is it if eight bits are used?
50. What is the decimal equivalent of 0.1_{10} if nine bits are used to store the number?

In Problems 51–56, write the number in BCD format.

51. 379_{10}
52. 2246_{10}
53. 180_{10}
54. 479_{10}
55. 35380_{10}
56. 2017_{10}

57. Write 535_{10} in BCD format and as a true binary number. How many bits does each representation take?
58. Write 1525_{10} in BCD format and as a true binary number. How many bits does each representation take?

In Problems 59–60, use the following information: A large corporation pays its employees $3.70 an hour. There are 2000 employees, who each work 4000 hours a year for the corporation.

59. What is the total amount of money that the corporation should pay its employees?
60. If the computer stored the hourly salary as $3.69 because of conversion error, then find the amount of money that the corporation does not pay the employees because of this error.

7.5 Order of Operations

Now that we have examined numeric representation inside computers, it is time again to discuss computer arithmetic. As you may have noticed, we have been using the / to represent division throughout this book, because computers use this symbol. Computers often use the asterisk (*) to represent multiplication since × or · can be mistaken for other things, so we will use it in this section and the next.

The normal notation for exponentiation also won't work with computers. All expressions must be written on one line without subscripts or superscripts. We will use the notation ˆ to indicate exponentiation. So 2^3 is written as 2ˆ3.

Consider the expression 3 + 4*2. If the addition is performed first, the result is 7*2, or 14. If the multiplication is performed first, the result is 3 + 8, or 11. Thus, we must understand the order of operations within the computer. Of course, it is the computer that performs the arithmetic, but the programmer must write the expression. So the programmer must know what the computer will do with the expressions that are written.

Since computers were designed by mathematicians, the order of operations used by many computer languages follows the same rules as the order of operations in algebra. Note, however, that some languages and software packages don't follow any precedence rules for operations but use a left-to-right evaluation scheme. Others let you choose whether you want addition or multiplication to have precedence. If addition were given precedence over multiplication or if operations were performed from left to right, 3 + 4*2, for example, would equal 14.

In this book, we will use the order of operations used in algebra. This order is listed next.

Order of Operations

1. All expressions inside parentheses are performed from the inside out.
2. All exponents are evaluated from right to left.
3. All multiplications and divisions are performed from left to right.
4. All additions and subtractions are performed from left to right.

7.5 Order of Operations

In some languages, exponents are evaluated from left to right rather than from right to left (see Problem 41).

The use of parentheses allows the programmer to alter the normal order of operations. Given 3+4*2, the computer will multiply first, getting 11. Writing (3+4)*2 forces the computer to add first, getting 14. Similarly, given 4*3^2, the computer will evaluate the exponent first, getting 4*9, or 36. The programmer can force the multiplication to be performed first by writing (4*3)^2, which is 144. The next example gives further illustrations.

EXAMPLE 7.15 Evaluate the expression.

a. 7*3+2^4−9/(3+6)

Solution

7*3+2^4−9/9	(parentheses)
7*3+16 −9/9	(exponentiation)
21 +16 −1	(multiplication and division)
36	(addition and subtraction)

b. 3+(5−(2+1))/2*5

Solution

3+(5−3)/2*5
3+2/2*5
3+1*5
3+5
8

c. 3−2*2^3^2+11

Solution

3−2*2^9+11	(exponents are evaluated from right to left)
3−2*512+11	
3−1024 +11	
−1010	

Depending on the order of operations used, there are many different possible answers to each of these expressions, but only *one* that the computer will come up with. Understanding what the computer will do with an expression is vital to good programming.

As noted earlier, some computers distinguish between integers and reals when performing arithmetic. If integer arithmetic is performed, the result will always be an integer. This procedure leads to interesting results when two integers are divided. For instance, 7/4, which is 1.75, is simply 1 in integer division. The fractional part is truncated. If arithmetic is performed involving a real number and an integer, the integer is converted to a real number and the result is real.

EXAMPLE 7.16 Evaluate the expression by using integer arithmetic.

    a. $16/5$ **b.** $4/10$

Solution $16/5 = 3.2 = \underline{3}$ $4/10 = 0.4 = \underline{0}$

    c. $-35/4$ **d.** $199/10$

Solution $-35/4 = -8.75 = \underline{-8}$ $199/10 = 19.9 = \underline{19}$

Unless specifically mentioned, we will not be doing integer arithmetic in this book.

Problem Set 7.5

In Problems 1–20, evaluate the expression.

1. $1+3*5$
2. $4*5+3$
3. $3*2-2$
4. $5-2*3$
5. $2+4/2$
6. $6/3+3$
7. $14/2-9$
8. $15-10/5$
9. $3-(4-7)$
10. $(3-4)-7$
11. $2\verb|^|3\verb|^|2$
12. $2\verb|^|2\verb|^|3$
13. $1+4*7\verb|^|2$
14. $6\verb|^|(-2)+2-4/2$
15. $3*(4-(8+3))-4$
16. $14-((8+3)*-4)/(5-3)$
17. $1-(7+2\verb|^|3*4)/30/2$
18. $5\verb|^|2-4*2*7$
19. $3*2*4/2*4*3$
20. $60/2/5/3/2$

In Problems 21–30, evaluate the expression by using integer arithmetic.

21. $12/4$
22. $20/5$
23. $17/4$
24. $24/10$
25. $33/2$
26. $21/2$
27. $-34/7$
28. $-93/11$
29. $-23/14$
30. $-79/88$

In Problems 31–40, write computer expressions to perform the arithmetic.

31. Add 3, 9, and 12
32. Find the average of 3, 9, and 12
33. Evaluate the quadratic formula with $a=1$, $b=4$, $c=2$ (two expressions)
34. Evaluate the quadratic formula with $a=3$, $b=9$, $c=3$ (two expressions)
35. 3^2+5^2
36. $4(3+7^2)$
37. $2(3+5(2-7))$
38. $(3^2)^{-5}+9\sqrt{2}$
39. Find the monthly interest (balance times yearly interest rate divided by twelve) given a balance of \$10,000 and a yearly interest rate of 0.06
40. Find the weekly salary of an employee who is paid \$4 an hour for the first 40 hours and \$6 an hour for all hours beyond 40, if the employee works 52 hours this week
41. Evaluate $(2\verb|^|2)\verb|^|3$ and $2\verb|^|(2\verb|^|3)$. What can you conclude about evaluating exponents?
42. Does it matter if $+$ and $-$ are performed from right to left? Give an example to support your answer.

43. Does it matter if * and / are performed from right to left? Give an example to support your answer.
44. Which has higher priority: * or /?
45. Which has higher priority: + or −?
46. Write an expression with no parentheses in which * is performed before /.
47. Write an expression with no parentheses in which / is performed before *.
48. Write an expression with no parentheses in which − is performed before +.
49. Write an expression with no parentheses in which + is performed before −.
50. Is it possible to write an expression without parentheses in which * is performed before ^? If so, give an example.

7.6 Assignment Statements

In the previous section, we discussed arithmetic operations and the order in which the computer performs these operations. Computers can't simply perform arithmetic. They must "do something" with the result. Usually, the computer assigns the result to a variable in an **assignment statement**.

assignment statement

In algebra, x and y are used for variables. Although the rules for variable names may differ from one language to another, many languages accept any single letter for a numeric variable. Sometimes, more than a single letter can be used, but in this book, we will usually use single letters to represent variables.

The general form of an assignment statement is a single variable on the left, an equal sign in the middle, and the expression on the right. For example, X = 5 + 2 and Y = 9 are assignment statements.

Note that assignment statements are *not* equations, even though an equal sign is used. In the assignment statement Y = 9, we are not saying that Y and 9 are equal. We are *assigning* the value of 9 to the variable Y, regardless of the previous value of Y. Some computer languages use ← or := for assignment statements instead of = to make this distinction clear.

EXAMPLE 7.17 Determine whether the statement is a proper assignment statement.

Solution
a. X = 2 — Yes
b. A = 4 + 3*2 — Yes
c. 9 = Y — No (the left side must be a variable)
d. D = X − 2 — Yes (any expressions are allowed on the right, including variables)
e. 4*A = 7*B — No (the left side must be a single variable)
f. C = C + 1 — Yes

Example 7.17f is certainly different from any equation you have ever seen. That is, in algebra, there is *no* number which equals itself plus 1. In this assignment statement, the computer takes the value of C, adds 1 to it, and assigns the result to C. So C is increased by 1. In a similar way, for instance, every year your age is increased by 1, but it is still called age (A = A + 1).

EXAMPLE 7.18 Write a computer assignment statement for the expression.

a. $b^2 - 4ac$

Solution D = B^2 − (4*A*C)

Although the parentheses are not necessary, they make the expression easier to read.

b. $ax^2 + bx + c$

Solution P = (A*X^2) + (B*X) + C

c. $\sqrt{d(d-a)(d-b)(d-c)}$

Solution A = (D*(D − A)*(D − B)*(D − C))^.5

When a computer first encounters an assignment statement, it performs the calculation on the right (if necessary). It then assigns the result of the calculation to the variable on the left.

EXAMPLE 7.19 Find the value of A after the calculation of the assignment statement has been done. Assume A has a value of 7, B has a value of 10, and C has a value of 4 when the computer gets to each statement.

a. A = 9   **b.** A = 4 + 3*5 − 1

Solution Value: 9   4 + 15 − 1
     Value: 18

c. A = B*C   **d.** A = ((A + B + C) − 5)^.5

Solution 10*4   ((7 + 10 + 4) − 5)^.5
Value: 40   (21 − 5)^.5
     16^.5
     Value: 4

e. A = A − 1

Solution 7 − 1
Value: 6

7.6 Assignment Statements

Problem Set 7.6

In Problems 1–20, determine whether the statement is a proper assignment statement. If it is not, state why not.

1. $A = 0$
2. $B = -2$
3. $C = 5 + 7$
4. $D = 7*4$
5. $E + 1 = 4$
6. $3 = F$
7. $G = G + 2$
8. $H = 2*H$
9. $I + I = I$
10. $J = J + J$
11. $17 = K$
12. $1 = -2 + 3*4\char`^2$
13. $M\char`^2 = M$
14. $N + 1 = 1 + N$
15. $P*Q = Q*P$
16. $-R = 4$
17. $-S = 2 + (6*T + 2/U)$
18. $2 = 2$
19. $1*V = V$
20. $W = -A - (2 + (3\char`^2))/18$

In Problems 21–36, write a computer assignment statement for the expression or situation.

21. $A + B + C + D$
22. $xy^3 - 2yx^4$
23. $3x^2 + 4x - 5$
24. $\frac{9}{5}C + 32$
25. $\frac{5}{9}(F - 32)$
26. $(x - 2)(x + 4)$
27. $(x + 1)(x - 3)^2$
28. $(y - 4)^2(x + 3)^3$
29. 2^x
30. 3^{x-7}
31. $\sqrt{x^3 y z^5}$
32. $\sqrt{1 + x^3 - x}$

33. The cost for attending a movie is $6 for an adult and $2 for a child. Find an expression for the total cost for a group of x adults and y children to attend the movie.

34. An employee's gross weekly pay is hours worked times hourly rate plus time and a half for any hours over 40. Find an expression for the employee's weekly pay.

35. Find an expression for monthly interest (balance times yearly interest rate divided by 12) in a savings account.

36. A salesman earns a commission of 6.5% (.065) on all of his sales. Find an expression for his earnings in terms of his total sales.

In Problems 37–48, find the value of A after the calculation of the assignment statement has been done. Assume the value of A is 5, the value of B is 7, and the value of C is -4 when the computer gets to each statement.

37. $A = -3$
38. $A = 2$
39. $A = 2 + (3 - (2*5))$
40. $A = (2*6)/(4 - 1)$
41. $A = 2*B + C$
42. $A = B\char`^2 - C$
43. $A = 3*C\char`^2 - 17 + B$
44. $A = B - A$
45. $A = A + 5$
46. $A = 7 - A$
47. $A = (A + B - C)\char`^.5$
48. $A = (A\char`^2 - B + C/2)\char`^.5$

49. Is $C = C$ a proper assignment statement?
50. What will the computer do with the statement $C = C$?

7.7 Coding Nonnumeric Data

Not only numbers but everything else must be stored in the computer in binary form. The instructions, mathematical symbols (+, −, =, and so on), or any words that are part of the program must *all* be reduced to 0s and 1s.

The computer distinguishes between numeric items, which we have been discussing for the past three chapters, and the nonnumeric items that were mentioned in the previous paragraph. Several codes have been developed to store these nonnumeric data items.

EBCDIC
ASCII–8

Two of the more widely used codes are **EBCDIC** (extended binary-coded decimal interchange code) and **ASCII–8** (American standard code for information interchange). Both of these codes use one byte (eight bits) to store each character, so 2^8, or 256, different characters can be stored in each code.

Figure 7.1 shows how the EBCDIC code represents selected characters. Figure 7.2 shows how the ASCII–8 code represents letters and numbers.

alphanumeric expressions

Nonnumeric data items are called **alphanumeric expressions** and are usually written within quotation marks, or quotes (for instance, "Salary" or "1959"). There is a big difference between the number 9 and the alphanumeric expression "9". Arithmetic cannot be performed on alphanumeric expressions.

Notice that both codes represent the letters in increasing order. In this way, expressions can be compared, with the lower representation signifying that it occurs earlier in the alphabet. A computer simply compares the codes of the two expressions to see which one comes first alphabetically.

Character	EBCDIC	Character	EBCDIC
0	11110000	F	11000110
1	11110001	G	11000111
2	11110010	H	11001000
3	11110011	I	11001001
4	11110100	J	11010001
5	11110101	K	11010010
6	11110110	L	11010011
7	11110111	M	11010100
8	11111000	N	11010101
9	11111001	O	11010110
Blank	01000000	P	11010111
.	01001011	Q	11011000
<	01001100	R	11011001
*	01011100	S	11100010
)	01011101	T	11100011
;	01011110	U	11100100
A	11000001	V	11100101
B	11000010	W	11100110
C	11000011	X	11100111
D	11000100	Y	11101000
E	11000101	Z	11101001

Figure 7.1 EBCDIC Code

Note that the blank character must be given an eight-bit representation. We will write ƀ for a blank when the symbol is necessary.

7.7 Coding Nonnumeric Data

For a computer, you can't use the *letters* l and O, as you can on a typewriter, to represent the *numbers* 0 and 1. They are stored in totally different ways by the computer. We will represent zero by Ø when dealing with alphanumeric expressions.

Character	ASCII–8	Character	ASCII–8
Ø	01010000	I	10101001
1	01010001	J	10101010
2	01010010	K	10101011
3	01010011	L	10101100
4	01010100	M	10101101
5	01010101	N	10101110
6	01010110	O	10101111
7	01010111	P	10110000
8	01011000	Q	10110001
9	01011001	R	10110010
A	10100001	S	10110011
B	10100010	T	10110100
C	10100011	U	10110101
D	10100100	V	10110110
E	10100101	W	10110111
F	10100110	X	10111000
G	10100111	Y	10111001
H	10101000	Z	10111010

Figure 7.2 ASCII–8 Code

EXAMPLE 7.20 Represent "HELP7" by using the EBCDIC code.

Solution

H	E	L	P	7
11001000	11000101	11010011	11010111	11110111

11001000110001011101001110101111111110111

EXAMPLE 7.21 Represent "ADDØ1" by using the ASCII–8 code.

Solution

A	D	D	Ø	1
10100001	10100100	10100100	01010000	01010001

1010000110100100101001000101000001010001

There are many numeric items which can be stored in alphanumeric form. For example, social security numbers, zip codes, telephone numbers, and identification numbers are numeric items which can be stored in alphanumeric form since no arithmetic is usually performed on these items.

Problem Set 7.7

In Problems 1–10, represent the expression by using the EBCDIC code.

1. "HI"
2. "BYE"
3. "3*2"
4. "3<7"
5. "5"
6. "1"
7. "GObTEAM"
8. "HIbTHERE"
9. "R2D2"
10. "MADEbINbUSA"

In Problems 11–20, represent the expression by using the ASCII–8 code.

11. "GOOD"
12. "BAD"
13. "C3PØ"
14. "1957"
15. "OCTOBER"
16. "ØØ21"
17. "2ØØ1"
18. "COMPUTER"
19. "ASCII"
20. "L1ØØ"

21. Which expression has the smaller EBCDIC representation: "123" or "ABC"?
22. Which expression has the smaller ASCII–8 representation: "123" or "ABC"?
23. Find the hexadecimal representation for each EBCDIC representation in Figure 7.1 (for example, 11110000 is FØ).
24. Find the hexadecimal representation for each ASCII–8 representation in Figure 7.2.
25. Some computers use seven-bit codes to represent characters. How many different characters can be represented by such a code?
26. Represent 2.75 as a real number in binary (normalized) notation and as an EBCDIC expression.
27. Represent 5.125 as a real number in binary (normalized) notation and as an EBCDIC expression.
28. What alphanumeric expression does the ASCII–8 code

    10100010101101011010 1101

    represent?
29. What alphanumeric expression does the ASCII–8 code

    10100100101010011010 1101

    represent?
30. Which alphanumeric expression has the smaller EBCDIC representation: "JAMESbBAY" or "JAMESAN"?

CHAPTER SUMMARY

In this chapter, we studied the special topics that must be learned to understand how the computer performs arithmetic. The computer distinguishes between integers and reals, storing real numbers in a form of scientific notation.

Because of limited storage capability, numbers may have to be truncated or rounded to a specified number of digits, which can often lead to conversion error.

The computer follows very precise rules for the order in which operations are performed: parentheses, exponentiation, multiplication and division, and finally, addition and subtraction.

Assignment statements are used to hold the results of arithmetic calculations. Even nonnumeric information must be stored in binary form. Several codes, including EBCDIC and ASCII, were studied in this chapter.

REVIEW PROBLEMS

In Problems 1–4, find the accuracy and precision of the number.

1. 5.87
2. 0.785
3. −44.009
4. −2000.0

5. The *Viking* spacecraft on Mars measured a wind velocity of 155 miles an hour. What is the accuracy and precision of this measurement?

6. The balance in Jan's checking account is $1432.06. What is the accuracy and precision of this amount?

In Problems 7–10, write the number in scientific notation.

7. 678.9
8. 1414
9. 0.000403
10. −0.000000051

11. Write the wind velocity measurement of Problem 5 in scientific notation.

12. Write the amount of money in Problem 6 in scientific notation.

In Problems 13–16, write the number in decimal notation.

13. 5.7×10^7
14. -3.087×10^6
15. 1.023×10^{-5}
16. 4.4×10^{-13}

In Problems 17–20, represent the storage of the integer by using one byte and 2's complement for negatives.

17. 49
18. 79
19. −55
20. −111

In Problems 21–24, write the number as a computer would in scientific notation. Give your answer in decimal.

21. .254
22. 33.56
23. −23.456
24. −.0006703

In Problems 25–28, write the number as a computer would in normalized notation. Give your answer in decimal.

25. 12.34
26. −.0097
27. 2.034
28. 2305.67

In Problems 29–32, truncate the number to the given number (in parentheses) of significant digits.

29. 45.087 (4)
30. 777.00 (4)
31. .234569 (3)
32. .000302 (2)

In Problems 33–36, round the number to the given number (in parentheses) of significant digits.

33. 4.96 (2)
34. 3321.092 (5)
35. −.021567 (3)
36. .00090003 (4)

In Problems 37–38, convert to binary, truncate to seven bits, and convert back to decimal. Note any conversion errors.

37. 0.4_{10}
38. 14.15_{10}

In Problems 39–40, write the number in BCD format.

39. 36708_{10}
40. 404792_{10}

41. Write an assignment statement to find the total cost of purchasing X items which cost Y dollars apiece.

42. Write an assignment statement to calculate the new salary for an employee if her present salary is S and she receives a $1000 raise.

In Problems 43–44, find the value of A after the assignment statement has been executed. Assume that A is 3, B is −2, and C is 5 when the computer gets to each statement.

43. A = B + 1*(C^(3−1))
44. A = 4*A + (B/(C−3))

45. Evaluate 23/5 by using integer arithmetic.

46. Evaluate $-16/7$ by using integer arithmetic.
47. Do assignment statements represent functions?
48. What is the difference between an assignment statement and an equation?
49. Represent "CH 1" by using the EBCDIC code.
50. Represent "GO FOR IT" by using the EBCDIC code.
51. Represent "PENGUIN" by using the ASCII–8 code.
52. Represent "R2D2" by using the ASCII–8 code.

SETS

CHAPTER **8**

We have studied how computers perform arithmetic. It is time now to see how computers can make decisions.

We mentioned the concept of sets earlier in this book. In this chapter, we will study them in detail. Sets are fundamental to many branches of mathematics and are directly related to the area of computer logic, which is our main reason for studying them.

As you proceed through the next four chapters, you should see the concepts of this chapter as the foundation for the later chapters.

8.1 Introduction

We should first review what was said about sets in Chapter 1. A set is undefined in mathematics because it is one of the fundamental building blocks of the science of mathematics, and fundamental concepts are left undefined. Informally, a **set** is merely a collection of things. In this book, sets will usually consist of numbers. There are two ways of writing sets. One method is listing the members, or **elements**, of the set:

set

elements

$$\{1, 3, 5, 7, \ldots\}$$

set-builder notation

The other method is **set-builder notation**:

$$\{x \mid x \text{ is an odd natural number}\}$$

which is read "the set of all elements x such that x is an odd natural number."

Notation: Rather than write "x is a natural number," we will use the symbol $\in$ for "is an element of" and write $x \in N$. To indicate that an element does not belong to N, we write $x \notin N$, which is read "x is not an element of set N."

Capital letters are used for sets, and lower case letters are used for elements of a set. We illustrate these basic ideas of sets in the following examples.

EXAMPLE 8.1 List the elements in the set.

    a. $A = \{x \mid x \in N \text{ and } x < 4\}$

Solution   elements: <u>1, 2, and 3</u>

    b. $B = \{x \mid x \text{ is a day of the week}\}$

Solution   elements: <u>Monday, Tuesday, Wednesday, Thursday, Friday, Saturday, Sunday</u>

8.1 Introduction

    c. $C = \{x \mid x \text{ is a binary number between } 10_2 \text{ and } 110_2\}$

Solution   elements: 11_2, 100_2, and 101_2

EXAMPLE 8.2   Write the set in set-builder notation.

    a. $A = \{a, e, i, o, u\}$

Solution $A = \{x \mid x \text{ is a vowel}\}$

    b. $B = \{\ldots, -4, -3, -2\}$

Solution $B = \{x \mid x \in I \text{ and } x < -1\}$

    c. $C = \{1_8, 10_8, 100_8, 1000_8, \ldots\}$

Solution $C = \{x \mid x \text{ is a nonnegative integer power of 8 written in octal}\}$

We now turn to the idea of comparing sets.

equal sets

Definition 8.1   Two sets are **equal sets** if and only if they contain the same elements.

EXAMPLE 8.3   Determine whether the sets are equal.

    a. $A = \{1, 2, 3\}$, $B = \{3, 2, 1\}$, and $C = \{1, 2, 1, 3\}$

Solution   The sets are all equal since they contain the elements 1, 2, and 3. Remember that order doesn't matter within a set (although we will write all sets in increasing order), and repeating an element isn't necessary.

    b. $A = \{x \mid x \in I \text{ and } x \notin N\}$ and $B = \{\ldots, -3, -2, -1, 0\}$

Solution   The sets are equal because they describe the same set and, therefore, contain the same elements.

The two sets $A = \{1, 3, 5\}$ and $B = \{2, 4, 6\}$ are obviously not equal, but they do contain the same *number* of distinct elements (3). Such sets are said to be equivalent.

Notation: The phrase "if and only if," one of the strongest phrases in mathematics, will be abbreviated as **iff**.

iff

equivalent sets

Definition 8.2   Two sets are **equivalent sets** iff they contain the same number of elements.

EXAMPLE 8.4   Determine whether the sets are equivalent.

a. $A = \{1, 3, 4, 7\}$ and $B = \{2, 4, 9, 11\}$

Solution The sets are equivalent since they both contain four elements.

b. $A = \{1, 2, 3\}$ and $B = \{2, 4, 6, \ldots\}$

Solution The sets are not equivalent since A contains three elements and B contains an infinite number of elements.

Problem Set 8.1

In Problems 1–16, list the elements in the set.

1. $\{x \mid x < 10 \text{ and } x \in N\}$
2. $\{x \mid 1 \leq x < 7 \text{ and } x \in N\}$
3. $\{x \mid x \in I \text{ and } -2 < x \leq 3\}$
4. $\{x \mid x \in I \text{ and } x > -3\}$
5. $\{x \mid x \in Q \text{ and } x = \frac{1}{n}, n \in N\}$
6. $\{x \mid x \in R \text{ and } x = \sqrt{n}, n \in N\}$
7. $\{x \mid x \text{ is a binary number less than three digits}\}$
8. $\{x \mid x \text{ is a single-digit hexadecimal number}\}$
9. The set of all integers between -2 and 3
10. The set of all natural numbers larger than 10
11. The set of all planets in the solar system
12. The set of all months that contain less than 31 days
13. The set of letters in the word *hexadecimal*
14. The set of colors in the rainbow
15. The set of all octal numbers between 12_8 and 23_8
16. The set of all three-digit binary numbers

In Problems 17–30, write the set in set-builder notation.

17. $\{1, 2, 3, 4\}$
18. $\{1, 3, 5, 7\}$
19. $\{\text{binary, octal, decimal, hexadecimal}\}$
20. $\{\text{May, June, July, August}\}$
21. $\{2, 4, 6, 8, \ldots\}$
22. $\{10_2, 11_2\}$
23. $\{1, 2, 4, 8, 16, \ldots\}$
24. $\{0.1, 0.01, 0.001, 0.0001, \ldots\}$
25. The set of all positive rational numbers
26. The set of all negative integers
27. The set of all customers overdrawn at Michigan State Bank
28. The set of all employees whose salaries are over $50,000
29. The set of all binary numbers less than 1000_2
30. The set of all negative octal numbers

In Problems 31–44, state whether the pairs of sets are equal, equivalent, both, or neither.

31. $A = \{\text{money, wealth, power}\}$
    $B = \{\text{wealth, power, prestige}\}$
32. $A = \{a, c, f, j, t\}$
    $B = \{f, t, a, j, c\}$
33. $A = \{3, 2, 3\}$
    $B = \{2, 3\}$
34. $A = \{2, 1, 3, 5\}$
    $B = \{4, 2, 6, 10\}$
35. $A = \{3, 6, 9\}$
    $B = \{9, 3, 6\}$
36. $A = \{2, 7, 10\}$
    $B = \{2, 4, 2\}$
37. $A = \{1, 3, 1, 1\}$
    $B = \{3, 1, 1, 2\}$
38. $A = \{4, 7, 9, 10\}$
    $B = \{4, 9, 10, 9, 4, 7\}$
39. $A = \{1, 3, 5, 7, \ldots\}$
    $B = \{2, 4, 6, 8, \ldots\}$
40. $A = \{\frac{1}{2}, \frac{1}{3}, \frac{1}{4}, \ldots\}$
    $B = \{\sqrt{2}, \sqrt{3}, \sqrt{4}, \ldots\}$
41. $A = \{0_2, 1_2, 10_2, 11_2\}$
    $B = \{0_8, 1_8, 2_8, 3_8\}$
42. $A = \{10_2, 100_2, 1000_2\}$
    $B = \{10_{16}, 100_{16}, 1000_{16}\}$
43. $A = \{a, b, c, d\}$
    $B = \{A, B, C, D, E\}$

44. $A = \{1, 2, 3, 4, \ldots\}$
    $B = \{1, 2, 3, 4\}$
45. Are all equal sets equivalent?
46. Are all equivalent sets equal?
47. List the elements of the set $\{x | x \in N \text{ and } x < 0\}$.
48. List the elements of the set of all rational numbers that aren't real.

8.2 Subsets

Consider the sets $A = \{1, 2, 4\}$ and $B = \{1, 2, 3, 4, 5\}$. These sets are neither equal nor equivalent. However, there is a relationship between these sets. The three elements of A are all found in B. This observation leads to the concept of a subset.

subset

Definition 8.3 The set A is a **subset** of the set B, denoted $A \subseteq B$, iff every element in A is also in B.

EXAMPLE 8.5 Determine the relationship between the sets.

a. $A = \{a, f\}, B = \{f\}$

Solution $B \subseteq A$

b. $A = \{1, 2, 3\}, B = \{1, 2, 4\}$

Solution $A \nsubseteq B$ and $B \nsubseteq A$

c. $A = \{1, 2, 3, 4\}, B = \{1, 2, 3, 4\}$

Solution We know that $A = B$, but by Definition 8.3,

$A \subseteq B$ and $B \subseteq A$

This is another way of defining equality of sets.

d. $A = \{1, 3, 5, 7, \ldots\}, B = \{1, 2, 3, 4, 5, \ldots\}$

Solution $A \subseteq B$

e. N and I, I and Q, Q and R

Solution $N \subseteq I$ $I \subseteq Q$ $Q \subseteq R$

Definition 8.4 Two sets A and B are equal iff $A \subseteq B$ and $B \subseteq A$.

Consider the following sets: $\{x | x \in N \text{ and } 2.5 < x < 2.6\}$. There are *no* elements in this set.

empty, null set

Definition 8.5 The set consisting of no elements is called the **empty**, or **null**, **set**, denoted by the symbol $\emptyset$ or the empty braces $\{\ \}$.

The empty set is just as important in set theory as the number zero is in arithmetic. There are many ways to describe the empty set. For example, $\{x|x>0$ and $x<0\}$ or $\{x|x$ is a person over 10 feet tall$\}$. If these sets have nothing in them, then they are the empty set.

Is $\emptyset$ a subset of $A = \{1, 2, 3\}$? If it isn't, then there must be an element in $\emptyset$ that isn't in A. However, $\emptyset$ has no elements. Therefore, $\emptyset \subseteq A$. A similar argument should convince you that $\emptyset$ is a subset of *every* set.

What about $\{\emptyset\}$? Many students write $\{\emptyset\}$ for the empty set, which is *not* correct. The empty set is written as $\{\ \}$ or $\emptyset$. Consider an empty bowl and an empty box. If the empty bowl is put into the empty box, then the box is no longer empty. Similarly, putting $\emptyset$ in braces results in a nonempty set.

EXAMPLE 8.6 List all of the subsets of A.

a. $A = \{1\}$ (1 element)

Solution subsets: $\underline{\emptyset \text{ and } \{1\}}$ (2 subsets)

b. $A = \{\$, \#\}$ (2 elements)

Solution subsets: $\underline{\emptyset, \{\$\}, \{\#\}, \{\$, \#\}}$ (4 subsets)

c. $A = \{1, 2, 3\}$ (3 elements)

Solution subsets: $\underline{\emptyset, \{1\}, \{2\}, \{3\}, \{1, 2\}, \{1, 3\}, \{2, 3\}, \{1, 2, 3\}}$ (8 subsets)

As you may have guessed, there is a relationship between the number of elements in a set and the number of subsets. If a set has n elements, then the set will have 2^n subsets. So $A = \{1, 2, 3, 4\}$ has $2^4 = 16$ subsets, including $\emptyset$ and A. This result is the first connection between set theory and binary numbers. There will be more.

proper subset

Definition 8.6 The set A is a **proper subset** of the set B, denoted $A \subset B$, iff $A \subseteq B$ and $B \neq A$.

The difference between a subset ($\subseteq$) and a proper subset ($\subset$) is the same as the difference between less than or equal ($\leq$) and less than ($<$). We are simply excluding the possibility of equality.

EXAMPLE 8.7 List all of the proper subsets of $A = \{1, 2, 3\}$.

Solution $\underline{\emptyset, \{1\}, \{2\}, \{3\}, \{1, 2\}, \{1, 3\}, \{2, 3\}}$

If a set has n elements, then it will have $2n - 1$ proper subsets (all subsets except the entire set).

We conclude this section with the definition of one more set.

8.2 Subsets

universal set

Definition 8.7 The **universal set** U is the set of all items under consideration.

The set U may change from problem to problem. The universal set in algebra is usually R. We must be told what the universal set is when we are performing operations on sets (see the next section). Once the universal set is chosen, every other set is a subset of U.

Problem Set 8.2

In Problems 1–18, determine whether $A \subseteq B$, $B \subseteq A$, both, or neither.

1. $A = \{a, b, c\}$
   $B = \{b, c\}$
2. $A = \{/, +, \hat{\,}, *\}$
   $B = \{\hat{\,}, +, *, /\}$
3. $A = \{\text{PRINT, INPUT, READ}\}$
   $B = \{\text{INPUT, READ, PRINT}\}$
4. $A = \{\text{BASIC, COBOL, Pascal}\}$
   $B = \{\text{Pascal, PL1, FORTRAN}\}$
5. $A = \{2, 6, 8, 14\}$
   $B = \{2, 6, 8, 14, 19\}$
6. $A = \{1, 3, 5, 7\}$
   $B = \{1, 2, 4, 7\}$
7. $A = \{1, 2, 3, \ldots\}$
   $B = N$
8. $A = \{2, 4, 6, 8, \ldots\}$
   $B = I$
9. $A = \{\ldots, -4, -2, 0, 2, 4, \ldots\}$
   $B = Q$
10. $A = R$
    $B = N$
11. $A = \{x \mid x \text{ is a power of 2}\}$
    $B = \{256, 512, 1024\}$
12. $A = \{34_8, 45_8, 107_8\}$
    $B = \{x \mid x \text{ is a power of 8}\}$
13. $A = \{1, 2, 3, 4, 5\}$
    $B = \{x \mid x \in N \text{ and } x < 6\}$
14. $A = \{-3, 4\}$
    $B = \{x \mid x \in I \text{ and } -6 < x\}$
15. $A = \{x \mid x \text{ is a power of 2}\}$
    $B = \{x \mid x \text{ is a power of 16}\}$
16. $A = \{x \mid x \text{ is a power of 8}\}$
    $B = \{x \mid x \text{ is a power of 2}\}$
17. $A = \{x \mid x \text{ is a color in the United States flag}\}$
    $B = \{\text{blue, white, green, red}\}$
18. $A = \{x \mid x \text{ is a state in the United States}\}$
    $B = \{\text{West Virginia, Michigan, Puerto Rico}\}$

In Problems 19–24, list all of the subsets of the set.

19. $A = \{4\}$
20. $B = \emptyset$
21. $C = \{\text{savings, checking}\}$
22. $D = \{\text{red, orange, yellow}\}$
23. $E = \{1, 2, 3, 4\}$
24. $F = \{4, 5, 6, 7\}$

In Problems 25–26, determine the number of subsets in the set.

25. $G = \{a, b, c, d, e, f\}$
26. $H = \{1, 2, 3, 4, 5, 6, 7\}$

In Problems 27–34, list all of the proper subsets of the set.

27. $A = \{7\}$
28. $B = \{3, 9\}$
29. $C = \{\ \}$
30. $D = \{\text{Mercury, Venus, Mars}\}$
31. $E = \{\text{east, west, north}\}$
32. $F = \{a, b, c, d, e\}$
33. $G = \{x \mid x \in N \text{ and } x < 3\}$
34. $H = \{y \mid y \in I \text{ and } \sqrt{y} - 2\}$

In Problems 35–36, determine the number of proper subsets in the set.

35. $J = \{a, b, c, d, e, f\}$
36. $K = \{-5, -4, -3, -2, -1, 0, 1, 2\}$

In Problems 37–46, use the sets $U = \{1, 2, 3, 4, 5\}$, $A = \{1, 2, 3\}$, and $B = \{2, 4\}$. **Describe the relationship between the two items by using one of** $\in$, $\notin$, $\subseteq$, $\nsubseteq$, $\subset$, **or** $\not\subset$. **For example, the relationship between 5 and** A **is** $5 \notin A$.

37. 1, A
38. A, U

39. B, A
40. A, B
41. $4, A$
42. $5, U$
43. B, U
44. U, A
45. U, B
46. $3, B$
47. In general, if $A \subseteq B$ and $B \subseteq C$, does it follow that $A \subseteq C$?
48. In general, if $A \subset B$ and $B \subset C$, does it follow that $A \subset C$?
49. If $A \subseteq B$, then can $A \subset B$?
50. If $A \subset B$, then can $A \subseteq B$?

8.3 Operations on Sets

We have four basic operations—addition, subtraction, multiplication, and division—in arithmetic. In set theory, we have three basic operations: union, intersection, and complement. Each operation is discussed in the following subsections.

Union

union

The **union** of two sets A and B, denoted $A \cup B$, is the set consisting of all elements either in A or in B or in both A and B. In set-builder notation, $A \cup B = \{x | x \in A \text{ or } x \in B\}$. To be in $A \cup B$, an element must be in *at least one* of the sets A and B.

EXAMPLE 8.8 Find the union of the sets.

a. $A = \{a, b, c\}, B = \{b, c, d\}$

Solution $A \cup B = \{a, b, c, d\}$

Here, $a \in A$ and $a \notin B$. Also, $b, c \in A$ and $b, c \in B$. Finally, $d \notin A$ and $d \in B$.

b. $C = \{1, 2, 7, 9\}, D = \{3, 5, 10\}$

Solution $C \cup D = \{1, 2, 3, 5, 7, 9, 10\}$

There are *no* elements that are in both sets, but common elements are not necessary to have a union.

c. $E = \{x | x \text{ is a state beginning with the letter V}\}$
$F = \{x | x \text{ is a state ending in the letter t}\}$

Solution $E = \{\text{Vermont, Virginia}\}$ and $F = \{\text{Connecticut, Vermont}\}$
$E \cup F = \{\text{Connecticut, Vermont, Virginia}\}$

d. $G = \{1, 3, 5, 7, \ldots\}, H = \{2, 4, 6, 8, \ldots\}$

Solution $G \cup H = N$

Intersection

intersection

The **intersection** of two sets A and B, denoted $A \cap B$, is the set consisting of all elements that A and B have in common. In set-builder notation, $A \cap B = \{x \mid x \in A \text{ and } x \in B\}$. An element must be in *both* sets to be in the intersection.

EXAMPLE 8.9 Find the intersection of the sets.

a. $A = \{+, *, /\}, B = \{*, /, \hat{\ }\}$

Solution $A \cap B = \underline{\{*, /\}}$

Only * and / appear in both sets.

b. $C = \{1, 2, 7, 9\}, D = \{3, 5, 10\}$

Solution $C \cap D = \underline{\varnothing}$

Sets C and D have *no* elements in common.

disjoint sets

Definition 8.8 Two sets A and B are said to be **disjoint sets** iff $A \cap B = \varnothing$.

Notice that sets C and D in Example 8.9b are disjoint.

EXAMPLE 8.10 Find the intersection of the sets.

a. $A = \{1, 2, 3, 4, 5\}, B = \{-1, 0, 1\}$

Solution $A \cap B = \underline{\{1\}}$

b. $C = \{4, 8, 12, 16, \ldots\}, D = \{6, 12, 18, \ldots\}$

Solution $C \cap D = \underline{\{12, 24, 36, \ldots\}}$

Complement

complement

If U is the universal set, then any set A is a subset of U. The **complement** of A, denoted $\bar{A}$, is the set consisting of all element in U that are *not* in A.

EXAMPLE 8.11 For $U = \{1, 2, 3, 4, 5, 6\}$, find the complement of the given set.

a. $A = \{1, 4\}$ b. $B = \{2, 3, 5\}$

Solution $\bar{A} = \underline{\{2, 3, 5, 6\}}$ $\bar{B} = \underline{\{1, 4, 6\}}$

c. $C = \{1, 6\}$

Solution $\bar{C} = \underline{\{2, 3, 4, 5\}}$

These three operations will be important for the rest of this chapter and will be repeated in different forms in the following two chapters.

Once the basic operations of arithmetic were studied, more complicated expressions were introduced, like $(2 \cdot 3) + (3 + 4)$, involving several operations. We will do the same thing with these set operations.

EXAMPLE 8.12 Perform the given operations for $U = \{1, 2, 3, 4, 5, 6\}$, $A = \{1, 2, 3, 6\}$, $B = \{2, 4, 6\}$, and $C = \{4, 5\}$.

a. $(A \cap B) \cup C$

Solution We first find $A \cap B$; then we take the union of this set with C.

$$A \cap B = \{2, 6\}$$
$$(A \cap B) \cup C = \{2, 6\} \cup \{4, 5\}$$
$$(A \cap B) \cup C = \underline{\{2, 4, 5, 6\}}$$

b. $(A \cup B) \cap (A \cup C)$

Solution
$$A \cup B = \{1, 2, 3, 4, 6\}$$
$$A \cup C = \{1, 2, 3, 4, 5, 6\}$$
$$(A \cup B) \cap (A \cup C) = \{1, 2, 3, 4, 6\} \cap \{1, 2, 3, 4, 5, 6\}$$
$$= \underline{\{1, 2, 3, 4, 6\}}$$

c. $\overline{A} \cap (B \cup C)$

Solution
$$\overline{A} = \{4, 5\} \quad \text{and} \quad B \cup C = \{2, 4, 5, 6\}$$
$$\overline{A} \cap (B \cup C) = \underline{\{4, 5\}}$$

d. $B \cup (\overline{A \cap C})$

Solution We must first find $A \cap C$ before we can find $\overline{A \cap C}$.

$$A \cap C = \{\ \}$$
$$\overline{A \cap C} = \{1, 2, 3, 4, 5, 6\}$$
$$B \cup (\overline{A \cap C}) = \{2, 4, 6\} \cup \{1, 2, 3, 4, 5, 6\}$$
$$= \underline{\{1, 2, 3, 4, 5, 6\}}$$

Problem Set 8.3

For Problems 1–26, $U = \{1, 2, 3, 4, 5, 6, 7, 8, 9, 10\}$, $A = \{1, 3, 5, 7, 9\}$, $B = \{2, 4, 5, 8\}$, $C = \{1, 10\}$, and $D = \{1, 3, 6, 7, 8\}$. Perform the given operations.

1. $A \cup B$
2. $B \cap C$
3. $B \cap D$
4. $C \cup D$
5. $A \cap D$
6. $D \cap B$
7. $B \cup C$
8. $A \cup D$
9. $A \cup (B \cap C)$
10. $B \cup (D \cap C)$
11. $A \cup (B \cup D)$
12. $B \cap (C \cup A)$
13. $(A \cap B) \cup (B \cap C)$

14. $(C \cap D) \cup (D \cup C)$
15. $(A \cap C) \cap (B \cup A)$
16. $(A \cup B) \cap (A \cup C)$
17. $\bar{A}$
18. $\bar{B}$
19. $\overline{A \cup B}$
20. $\overline{B \cap C}$
21. $A \cup \overline{(B \cap C)}$
22. $\overline{(B \cup C)} \cap D$
23. $\bar{C} \cup \bar{D}$
24. $\bar{B} \cap \bar{C}$
25. $\overline{A \cap (B \cup D)}$
26. $\overline{B \cap (C \cup D)}$

For Problems 27–34, $U = \{x \mid x \text{ is a letter in the word } octal\}$, $A = \{x \mid x \text{ is a vowel in the word } octal\}$, and $B = \{x \mid x \text{ is one of the last two letters in the word } octal\}$.

27. Find $\bar{A}$.
28. Find $A \cap B$.
29. Find $A \cup B$.
30. Find $\bar{B}$.
31. If $A \subset B$, then what can you say about $A \cap B$?
32. If $A \subset B$, then what can you say about $A \cup B$?
33. What is $\bar{\varnothing}$?
34. What is $\bar{U}$?

For Problems 35–44, consider *any* sets A and B. Decide when (if ever) the statement is possible.

35. $A \cap B = A \cup B$
36. $\bar{A} = A$
37. $A \cup B = B \cup A$
38. $A \cap B = B \cap A$
39. $A \cap U = \varnothing$
40. $A \cup \varnothing = U$
41. $A \cap A = A$
42. $A \cup A = A$
43. $A \cap B = \varnothing$
44. $A \cap B = U$

For Problems 45–46, $U = \{0_2, 1_2\}$, $A = \{0_2\}$, and $B = \{1_2\}$.

45. What is $\bar{B}$?
46. What is $\bar{A}$?

For Problems 47–48, we define the *difference* of two sets A and B to be $A - B = A \cap \bar{B}$.

47. Does $A - B = B - A$ in general? Give an example.
48. What is $A - A$?

8.4 Venn Diagrams

We were able to "picture" equations in x and y by use of the Cartesian coordinate system. To "picture" relationships between sets, the English mathematician John Venn (1834–1923) invented the Venn diagram.

Venn diagram

A **Venn diagram** is always drawn inside a rectangle, the rectangle representing the universal set. Individual sets are represented by circles, as illustrated in Figure 8.1.

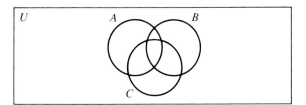

Figure 8.1

We can use Venn diagrams to illustrate the concepts of the previous two sections. To illustrate $A \subset B$, we draw set A completely inside set B, as in figure 8.2.

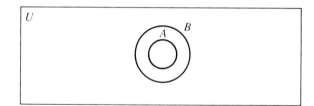

Figure 8.2
$A \subset B$

Figure 8.3 illustrates the intersection of sets A and B, the part of U inside *both* sets. We shade or hatch the part of the diagram that represents the answer.

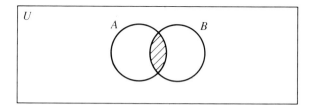

Figure 8.3
$A \cap B$

Figure 8.4 illustrates the union of sets A and B, the part of U inside one set *or* the other set or both.

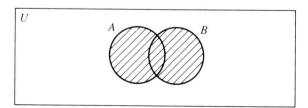

Figure 8.4
$A \cup B$

To illustrate the complement of A, we shade in everything in U that is outside A, as shown in Figure 8.5.

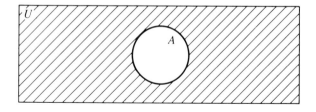

Figure 8.5
$\bar{A}$

8.4 Venn Diagrams

Disjoint sets are drawn as nonintersecting circles, as shown in Figure 8.6.

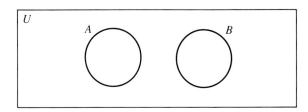

Figure 8.6
Disjoint Sets

We can also use Venn diagrams to illustrate more complicated relationships, as illustrated in the following example.

EXAMPLE 8.13 Use a Venn diagram to illustrate the relationship.

a. $(A \cup B) \cap (A \cup C)$

Solution If we were trying to find this relationship for given sets A, B, and C, we would first find $A \cup B$ and $A \cup C$. Similarly, we will hatch $A \cup B$ like this: \\\\; and $A \cup C$ like this: ////. The intersection then is that part of the diagram where the hatching lines cross. The solution is illustrated in Figure 8.7.

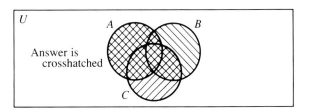

Figure 8.7
$(A \cup B) \cap (A \cup C)$

b. $\bar{A} \cup (B \cap C)$

Solution Hatch $\bar{A}$ by using ////, and hatch $B \cap C$ by using \\\\. The union is every part that is hatched in any way, as illustrated in Figure 8.8.

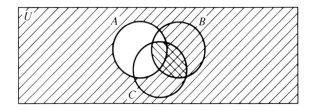

Figure 8.8
$\bar{A} \cup (B \cap C)$

When we draw sets A, B, and C in general, we draw them overlapping as in Figure 8.1. In this way, Venn diagrams can be used to illustrate relationships between *specific* sets by writing the elements inside the circles. The next example illustrates the process.

EXAMPLE 8.14 Use Figure 8.9 to find the relationship.

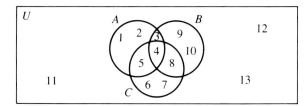

Figure 8.9

a. $A \cap B$

Solution $A \cap B = \{3, 4\}$

Only 3 and 4 can be found in both sets.

b. $\bar{A}$

Solution $\bar{A} = \{6, 7, 8, 9, 10, 11, 12, 13\}$

c. $B \cup C$

Solution $B \cup C = \{3, 4, 5, 6, 7, 8, 9, 10\}$

d. $(A \cap B) \cap C$

Solution $(A \cap B) \cap C = \{3, 4\} \cap \{4, 5, 6, 7, 8\} = \{4\}$

e. $\bar{A} \cap C$

Solution $\bar{A} \cap C = \{6, 7, 8, 9, 10, 11, 12, 13\} \cap \{4, 5, 6, 7, 8\} = \{6, 7, 8\}$

Problem Set 8.4

In Problems 1–20, for sets A, B, and C in general, draw a Venn diagram and hatch the relationship.

1. A
2. B
3. $B \cup C$
4. $A \cap C$
5. $A \cap B$
6. $A \cup B$
7. $A \cap (B \cup C)$
8. $B \cup (A \cap C)$
9. $\bar{A}$
10. $\bar{B}$
11. $A \cap (B \cap C)$
12. $A \cup (B \cup C)$
13. $\overline{A \cup C}$
14. $\overline{B \cap C}$
15. $\bar{A} \cap \bar{B}$
16. $\bar{B} \cup \bar{C}$
17. $(A \cap B) \cup (B \cap C)$
18. $(A \cup C) \cap (B \cap C)$
19. $\bar{A} \cap (\overline{B \cap C})$
20. $(\overline{A \cap B}) \cup \bar{C}$

In Problems 21–32, draw a Venn diagram to represent the relationship between A, B, and C.

21. $A \subset B, B \subset C$
22. $A \subset B, B = C$

8.5 Basic Properties of Sets

23. $A \subset B, C \subset B, A \cap C = \emptyset$
24. $A \subset B, C \subset B, A \cap C \neq \emptyset$
25. $A \subset B, B \cap C = \emptyset$
26. $(A \cap B) \subset C, A \nsubseteq C, B \nsubseteq C$
27. $A \cap B = \emptyset, B \cap C = \emptyset$
28. $A \cup B \cup C = U$
29. $\bar{A} = B$
30. $A \cup B = C$
31. $A \cap B = C$
32. $(A \cup B) \subset C$

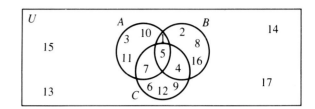

Figure 8.10

In Problems 33–44, use the Venn diagram in Figure 8.10 to find the relationship.

33. $A \cap B$
34. $B \cup C$
35. $A \cap (B \cap C)$
36. $\bar{A}$
37. $\bar{B}$
38. $A \cap C$
39. $B \cup A$
40. $A \cup (B \cup C)$
41. $\overline{A \cup (B \cup C)}$
42. $\overline{(A \cap B) \cup C}$
43. $\bar{A} \cap \bar{B}$
44. $\bar{B} \cup \bar{C}$

45. Draw a Venn diagram for $(A \cap B) \cap C$, and compare the answer with the answer for Problem 35. What can you conclude?
46. Draw a Venn diagram for $(A \cup B) \cup C$, and compare the answer with the answer for Problem 40. What can you conclude?
47. Draw Venn diagrams for $A \cup (B \cap C)$ and $(A \cup B) \cap (A \cup C)$. What can you conclude?
48. Draw Venn diagrams for $A \cap (B \cup C)$ and $(A \cap B) \cup (A \cap C)$. What can you conclude?

8.5 Basic Properties of Sets

In Chapter 1, we studied the basic properties of real numbers. Now, we will consider the basic properties of sets, and you should notice a direct relationship between the properties of sets and R.

Commutative Property For any two sets A and B:

$$A \cup B = B \cup A$$
$$A \cap B = B \cap A$$

This property follows from the definitions of union and intersection.

Associative Property For any sets A, B, and C:

$$A \cup (B \cup C) = (A \cup B) \cup C$$
$$A \cap (B \cap C) = (A \cap B) \cap C$$

The best way to illustrate this property is by using Venn diagrams. See Figure 8.11, which illustrates $A \cap (B \cap C) = (A \cap B) \cap C$.

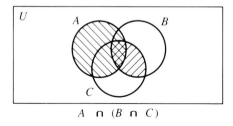
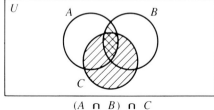

A ∩ (B ∩ C)   (A ∩ B) ∩ C

Figure 8.11

Closure Property   For any two sets $A, B \subseteq U$:

$A \cup B \subseteq U$
$A \cap B \subseteq U$

This property follows from the fact that A and B are both subsets of U, and $A \cup B$ and $A \cap B$ contain, by definition, nothing that isn't in one set or the other.

Identity Property   For every set A:

$A \cup \emptyset = A$
$A \cap U = A$

The null set $\emptyset$ is the identity for union, and U is the identity for intersection. Since $\emptyset$ has nothing to contribute to the union of $A \cup \emptyset$, the union is A. Since $A \subseteq U$, $A \cap U$ must be A.

So far, the properties have exactly paralleled those of real numbers, where $\cup$ and $\cap$ correspond to $+$ and $\cdot$, and $\emptyset$ and U correspond to 0 and 1. However, the similarity ends here. The next property of R that we studied was the property of inverses. There is no inverse property for sets:

$A \cap ? = \emptyset$
$A \cap ? = U$

If $A \neq \emptyset$, then A contains at least one element, so $A \cup ?$ contains at least one element and, therefore, cannot be empty. Similarly, if $A \neq U$, then there is at least one element in U that is not in A, so there is at least one element in U that cannot be in $A \cap ?$ Therefore, $A \cap ?$ cannot be U. So no inverses exist for $\cup$ and $\cap$.

What about the distributive property for sets? Does $A \cap (B \cup C) = (A \cap B) \cup (A \cap C)$? Figure 8.12 illustrates that the answer is yes.

8.5 Basic Properties of Sets

Figure 8.12

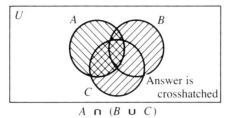
Answer is crosshatched
$A \cap (B \cup C)$

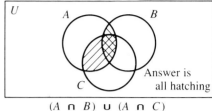
Answer is all hatching
$(A \cap B) \cup (A \cap C)$

Remember that in algebra, there was only *one* distributive property: multiplication over addition. In set theory, the "other" distributive property also works: $A \cup (B \cap C) = (A \cup B) \cap (A \cup C)$. This concept is so important, and so different from what you learned in algebra, that it is worthwhile for you to illustrate it by using Venn diagrams (see Problem 36).

Distributive Property For any sets A, B, and C:

$$A \cap (B \cup C) = (A \cap B) \cup (A \cap C)$$
$$A \cup (B \cap C) = (A \cup B) \cap (A \cup C)$$

Now, consider the sets $U = \{1, 2, 3, 4, 5, 6, 7, 8, 9, 10\}$, $A = \{1, 2, 4, 7, 10\}$, and $B = \{2, 3, 6\}$. Let's find $\overline{A} \cup \overline{B}$ and $\overline{A \cup B}$ and compare the results.

$$\overline{A} \cup \overline{B} = \{3, 5, 6, 8, 9\} \cup \{1, 4, 5, 7, 8, 9, 10\}$$
$$= \underline{\{1, 3, 4, 5, 6, 7, 8, 9, 10\}}$$
$$\overline{A \cup B} = \overline{\{1, 2, 3, 4, 6, 7, 10\}} = \underline{\{5, 8, 9\}}$$

In general, $\overline{A} \cup \overline{B} \neq \overline{A \cup B}$.
Similarly,

$$\overline{A} \cap \overline{B} = \{3, 5, 6, 8, 9\} \cap \{1, 4, 5, 7, 8, 9, 10\} = \underline{\{5, 8, 9\}}$$
$$\overline{A \cap B} = \overline{\{2\}} = \underline{\{1, 3, 4, 5, 6, 7, 8, 9, 10\}}$$

In general, $\overline{A} \cap \overline{B} \neq \overline{A \cap B}$.

However, it does appear that $\overline{A} \cup \overline{B}$ and $\overline{A \cap B}$ are the same and that $\overline{A} \cap \overline{B}$ and $\overline{A \cup B}$ are the same. These results are called DeMorgan properties.

DeMorgan Properties For any sets A and B:

$$\overline{A} \cup \overline{B} = \overline{A \cap B}$$
$$\overline{A} \cap \overline{B} = \overline{A \cup B}$$

These properties can be illustrated by using Venn diagrams (see Problems 37 and 38).

Idempotent Properties For any set A:

$$A \cup A = A$$
$$A \cap A = A$$

idempotent

Idempotent means that an operation is performed by using the same set twice and the result is the original set. This property is obviously not true in R: $1 + 1 \neq 1$ and $2 \cdot 2 \neq 2$.

Complement Properties For any set A:

$$A \cup \bar{A} = U$$
$$A \cap \bar{A} = \emptyset$$

These properties follow from the definition of complement. By definition, A and $\bar{A}$ can have nothing in common ($A \cap \bar{A} = \emptyset$), and every element in U must be either in A or in $\bar{A}$ ($A \cup \bar{A} = U$).

You should fully understand these properties. We will be referring to them in the next two chapters.

Note that some computer languages, such as Pascal, have sets and set operations built into the language. But many languages do not have built-in set operations.

Problem Set 8.5

In Problems 1–16, name the property.

1. $A \cup B = B \cup A$
2. $A \cap (B \cap C) = (A \cap B) \cap C$
3. $A \cap B \subseteq U$
4. $A \cup B \subseteq U$
5. $\overline{A \cup B} = \overline{A} \cap \overline{B}$
6. $\overline{A \cap B} = \overline{A} \cup \overline{B}$
7. $A \cap U = A$
8. $A \cup \emptyset = A$
9. $A \cup (B \cup C) = (A \cup B) \cup C$
10. $A \cap B = B \cap A$
11. $A \cap A = A$
12. $A \cup A = A$
13. $A \cup \bar{A} = U$
14. $A \cap \bar{A} = \emptyset$
15. $A \cup (B \cap C) = (A \cup B) \cap (A \cup C)$
16. $A \cap (B \cup C) = (A \cap B) \cup (A \cap C)$

In Problems 17–24, $U = \{1, 2, 3, 4, 5, 6, 7\}$, $A = \{1, 2, 3, 4, 5\}$, $B = \{1, 2, 4, 6\}$, and $C = \{3, 4, 5, 7\}$. **Show that the relationship holds.**

17. $A \cap (B \cap C) = (A \cap B) \cap C$
18. $A \cup \emptyset = A$
19. $A \cup A = A$
20. $A \cup (B \cup C) = (A \cup B) \cup C$
21. $\overline{A \cup B} = \overline{A} \cap \overline{B}$
22. $\overline{A \cap B} = \overline{A} \cup \overline{B}$
23. $A \cup (B \cap C) = (A \cup B) \cap (A \cup C)$
24. $A \cap (B \cup C) = (A \cap B) \cup (A \cap C)$
25. Describe in words what the closure properties mean.

26. Describe in words what the DeMorgan properties mean.
27. Are any numbers in R idempotent for addition?
28. Are any numbers in R idempotent for multiplication?

In Problems 29–34, consider the special case where $U = A = B = \{p, q\}$ and $C = \emptyset$. Show that the relationship holds.

29. $A \cup (B \cup C) = (A \cup B) \cup C$
30. $A \cap (B \cap C) = (A \cap B) \cap C$
31. $\overline{B} \cup \overline{C} = \overline{B \cap C}$
32. $\overline{A} \cap \overline{C} = \overline{A \cup C}$
33. $A \cap U = A$
34. $B \cup \emptyset = B$

In Problems 35–40, draw two Venn diagrams to show that the property is true.

35. The associative property for union
36. The second distributive property (union over intersection)
37. The first DeMorgan property
38. The second DeMorgan property
39. The complement property for union
40. The complement property for intersection

CHAPTER SUMMARY

In this chapter, we have studied sets, their operations and properties. There are three main operations on sets:

Union $A \cup B$ is the set of all elements either in A or in B or in both A and B

Intersection $A \cap B$ is the set of all elements that are common to both A and B

Complement $\overline{A}$ is the set of all elements in U not in A

There are eight basic properties of sets:

Commutative property	$A \cup B = B \cup A$
	$A \cap B = B \cap A$
Associative property	$A \cup (B \cup C) = (A \cup B) \cup C$
	$A \cap (B \cap C) = (A \cap B) \cap C$
Closure property	$A \cup B \subseteq U$
	$A \cap B \subseteq U$
Identity property	$A \cup \emptyset = A$
	$A \cap U = A$
Distributive property	$A \cap (B \cup C) = (A \cap B) \cup (A \cap C)$
	$A \cup (B \cap C) = (A \cup B) \cap (A \cup C)$
DeMorgan property	$\overline{A} \cup \overline{B} = \overline{A \cap B}$
	$\overline{A} \cap \overline{B} = \overline{A \cup B}$
Idempotent property	$A \cup A = A$
	$A \cap A = A$

Complement property $A \cup \bar{A} = U$
$A \cap \bar{A} = \emptyset$

We also discussed the concept of a subset, a set completely contained inside another set. And we used Venn diagrams to picture the operations and relationships on sets.

REVIEW PROBLEMS

In Problems 1–4, list the elements in the set.

1. $\{x \mid x \in I \text{ and } -7 < x < 3\}$
2. $\{x \mid x \in N \text{ and } x < -6\}$
3. The set of all single-digit octal numbers
4. The set of all binary numbers between 100_2 and 1000_2

In Problems 5–8, write the set in set-builder notation.

5. $\{1, 8, 64, 512, \ldots\}$
6. $\{a, b, c, d, e, f, g, h, i, j, k, l, m\}$
7. The set of all banks that offer IRAs
8. The set of all irrational numbers between 1 and 2

In Problems 9–12, determine whether the sets are equal, equivalent, both, or neither.

9. $A = \{2, 5, 8\}$
   $B = \{8, 5.0, \sqrt{4}\}$
10. $C = \{-3, -2, -1\}$
    $D = N$
11. $E = \{a, b, c, d, e\}$
    $F = \{x, y, z, x, a\}$
12. $G = \{10_{10}, 100_{10}\}$
    $H = \{10_8, 100_8\}$

In Problems 13–14, list all of the subsets of the set.

13. $A = \{-4, 1, 9\}$
14. $B = \{a, e, i, o, u\}$

In Problems 15–16, list all of the proper subsets of the set.

15. $C = \{x \mid x \in N \text{ and } x < 5\}$
16. $D = \{\$, \#, *\}$

For Problems 17–32, $U = \{1, 2, 3, 4, 5, 6, 7, 8\}$, $A = \{4, 5, 6, 7\}$, $B = \{1, 3, 5\}$, and $C = \{1, 8\}$. Find the relationship.

17. $A \cap C$
18. $B \cup C$
19. $A \cap (B \cup C)$
20. $B \cup (A \cap C)$
21. $\bar{B}$
22. $\bar{C}$
23. $\overline{B \cup C}$
24. $\overline{A \cap B}$
25. $(A \cap B) \cup (B \cap C)$
26. $(A \cup C) \cap (B \cup C)$
27. $C \cap \bar{B}$
28. $\bar{A} \cup \bar{C}$
29. $A \cap \overline{(B \cup C)}$
30. $\bar{B} \cup (A \cap \bar{C})$
31. $\overline{B \cup (A \cap C)}$
32. $(B \cap \bar{C}) \cup \overline{(A \cup C)}$

For Problems 33–40, draw Venn diagrams to illustrate the relationship.

33. $A \cup (B \cap C)$
34. $(B \cup A) \cap (B \cap C)$
35. $\overline{A \cap C}$
36. $B \cup \overline{(A \cap C)}$
37. $A \subseteq B, C = \emptyset$
38. $A \cap B \cap C = \emptyset$
39. $\overline{A \cup B} = C$
40. $A \cap C = \emptyset, B = U$

In Problems 41–46, name the property.

41. $A \cap B \nsubseteq U$
42. $A \cup (B \cup C) = (A \cup B) \cup C$
43. $A \cup (B \cap C) = (A \cup B) \cap (A \cup C)$
44. $A \cap B = B \cap A$
45. $A \cup \bar{A} = U$
46. $\bar{A} \cap \bar{B} = \overline{A \cup B}$

In Problems 47–48, draw a Venn diagram to illustrate the set.

47. The set of all elements in either A or B but *not* both
48. The set of all elements in both A and B or in neither A nor B
49. Find an expression to represent the set described in Problem 47.
50. Find an expression to represent the set described in Problem 48.

LOGIC

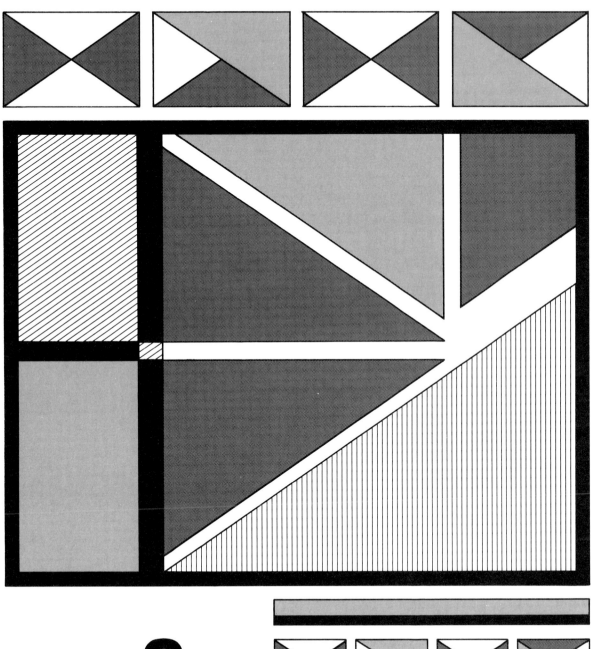

CHAPTER 9

Computers can do more than perform mathematical calculations. They can also make logical decisions. In this chapter, we will discuss the fundamentals of mathematical logic, which will lay the foundation for computer logic. You should see a great similarity between operations in logic and the set operations of the previous chapter.

Logic is the science of reasoning, where certain statements are either *true* or *false*. These two possible outcomes are easily adaptable to computers and the binary number system, using 1 for true and 0 for false.

Can computers think? This question is a difficult one, and many people do not understand it. In order to comprehend how a computer makes decisions, we must first investigate the principles behind computer decisions. This investigation is the purpose of this chapter.

9.1 Simple and Compound Statements

We start this section with a discussion of what a statement represents in logic.

statement

Definition 9.1 A **statement** is a declarative sentence that is either true or false but not both.

EXAMPLE 9.1 Determine whether the sentence is a valid statement.

    a.   The bank is open.

Solution   Yes; the statement is a declarative sentence.

    b.   When is the payment due?

Solution   No; interrogative sentences (questions) are not statements.

    c.   It snowed every day in July.

Solution   Yes; a statement need not be true.

    d.   Invest now!

Solution   No; imperative sentences (commands) are not statements.

    e.   How sweet it is!

Solution   No; exclamatory sentences are not statements.

    f.   A binary digit is either a 0 or a 1.

Solution   Yes.

9.1 Simple and Compound Statements

g. The selling price.

Solution No; a phrase is not a sentence.

In symbolic logic, letters are used to represent simple statements, usually starting with p, q, and r. We can then talk about **compound statements**, statements that are combinations of simple statements by use of **connectives**. We will consider the three basic connectives of logic in this section: conjunction, disjunction, and negation.

compound statements
connectives

conjunction

Definition 9.2 The **conjunction** of the two simple statements p and q is the compound statement p and q. The connective *and* is written by using the symbol $\wedge$, so the conjunction is written as $p \wedge q$.

disjunction

Definition 9.3 The **disjunction** of the two simple statements p and q is the compound statement p or q. The connective *or* is written by using the symbol $\vee$, so the disjunction is written as $p \vee q$.

negation

Definition 9.4 The **negation** of the simple statement p is the compound statement *not p*. The connective *not* is written by using the symbol $\sim$, so the negation is written $\sim p$.

The conjunction of p and q is true only when both p and q are true. The disjunction of p and q is true whenever at least one of p and q is true. These concepts will be explored in more detail in the next section (truth tables).

Many computer languages (BASIC, COBOL, and Pascal) use the words AND, OR, and NOT to represent the logic connectives. Others (APL) use the symbols $\wedge$, $\vee$ and $\sim$. Still other symbols are used by some languages.

The connective *not* is not considered to be a "compound" statement in English. But any statement that involves a connective is a compound statement in logic.

Now, let's explore these ideas through some examples.

EXAMPLE 9.2 Determine whether the following statements are simple or compound.

Solution

a. The moon is made of green cheese.

Simple

b. The inflation rate goes up and the savings accounts go down.

Compound (AND)

c. Dinosaurs are not smart.

Compound (NOT)

d. The days are getting shorter or the nights are getting darker.

Compound (OR)

EXAMPLE 9.3 Write the compound statement in symbolic logic.

a. The space shuttle flies and the moon sets.

Solution
p: The space shuttle flies.
q: The moon sets.
$p \wedge q$

b. The stock market is not open.

Solution
p: The stock market is open.
$\sim p$

c. The economics course is hard or the mathematics class is not hard.

Solution
p: The economics course is hard.
q: The mathematics class is hard.
$p \vee \sim q$

EXAMPLE 9.4 Write the negation of these statements.

a. The customer's payment is due now.

Solution The customer's payment is not due now.

b. Family history is not an interesting hobby.

Solution Family history is an interesting hobby.

EXAMPLE 9.5 Let p be the statement "The client has retired" and q be the statement "The client is in poor health." Write the symbolic statement in English.

a. $p \wedge q$

Solution The client has retired and the client is in poor health.

b. $\sim p \vee q$

Solution The client has not retired or the client is in poor health.

c. $\sim q \wedge \sim p$

Solution The client is not in poor health and the client has not retired.

Problem Set 9.1

In Problems 1–14, determine whether the sentence is a valid statement. If it is not, state why not.

1. The sky is blue.
2. All penguins live in Florida.
3. Close the door!
4. Joe gets paid time and a half for overtime.

9.2 Truth Tables: AND, OR, and NOT

5. What a beautiful view!
6. Is it raining?
7. Adult tickets cost $7 apiece.
8. Brush your teeth!
9. Are bulldogs fat?
10. The quick brown fox.
11. The computer disk is full.
12. What a great day!
13. The refund check.
14. This book is very interesting.

In Problems 15–20, determine whether the statement is simple or compound.

15. The sun is up or the moon is down.
16. The insurance company is losing money.
17. The X–15 was a rocket plane that went into space before the space shuttle.
18. The politicians are not very smart.
19. I missed the bus and I was late for work.
20. The client is healthy, but he is not eligible for insurance.

In Problems 21–28, write the statement in symbolic logic.

21. I invested my money in stocks, and I went bankrupt.
22. The grass is dying, or the weeds are growing.
23. The baseball team is not having a good season.
24. The soccer team is always late, or the football team loses.
25. The algebra class is boring, and the computer class is not boring.
26. Midnight-shift workers get a 10% bonus and afternoon-shift workers get a 5% bonus.
27. Either the tire is flat or the road is rough.
28. The wind doesn't blow, and the river doesn't flow.

In Problems 29–34, write the negation of the statement.

29. The moon is blue.
30. There are eight symbols in the octal number system.
31. There isn't enough time to finish the job.
32. The oak tree is not doing very well.
33. At least one person believes me.
34. No one can do the things he claims to do.

In Problems 35–42, p is the statement "Landing on the moon was dramatic," and q is the statement "Flying the space shuttle is a challenge." Write the symbolic statement in English.

35. $p \vee q$
36. $p \wedge q$
37. $\sim p$
38. $\sim q$
39. $\sim q \wedge p$
40. $\sim p \vee q$
41. $\sim (p \vee q)$
42. $\sim (p \wedge q)$

9.2 Truth Tables: AND, OR, and NOT

Consider the simple statement p. It has exactly *two* possible outcomes: true or false. When listing these possibilities, we will use T for true and F for false, and T will always precede F:

p
T
F

Now, consider two statements: *p* and *q*. There are *four* possible outcomes: both true, both false, only *p* true, or only *q* true. For convenience, we will always list these four possibilities in the following order:

p	*q*
T	T
T	F
F	T
F	F

Similarly, with the three statements *p*, *q*, and *r*, there are *eight* possible outcomes, which will always be listed in the following order:

p	*q*	*r*
T	T	T
T	T	F
T	F	T
T	F	F
F	T	T
F	T	F
F	F	T
F	F	F

truth table

We will now investigate the truth value of the compound statements defined in the previous section for *each* possible outcome. The result is a **truth table**. We begin with conjunctions.

$p \wedge q$: Conjunction

As the definition states, the conjunction is only true when both *p* and *q* are true. Suppose that you were told that school would be canceled if the following statement were true: "It snows today, and the temperature reaches 120° today." In order for school to be canceled, both simple statements must be true. It must snow today, and the temperature must reach 120° today also. This is not a very likely combination of events. If it snows and the temperature doesn't reach 120°, then the statement is not true. Similarly, if the temperature happens to reach 120° but it doesn't snow, then the statement is still false.

The truth table for a conjunction follows:

p	*q*	$p \wedge q$
T	T	T
T	F	F
F	T	F
F	F	F

9.2 Truth Tables: AND, OR, and NOT

$p \vee q$: Disjunction

The disjunction is true whenever *at least* one of the two statements is true. A disjunction is also referred to as an *Inclusive OR*. Another OR form, the *Exclusive OR*, will be considered in the next section.

Now, consider the statement "It snows today or the temperature reaches 120° today." This statement will be true if either simple statement is true. The only case in which the disjunction is false is when both statements are false. So this statement will be false only if it doesn't snow *and* the temperature doesn't reach 120°.

The truth table for disjunction follows:

p	q	$p \vee q$
T	T	T
T	F	T
F	T	T
F	F	F

$\sim p$: Negation

The negation changes the truth value of the original statement. This statement is directly related to complementing binary digits. The truth table for the negation follows:

p	$\sim p$
T	F
F	T

To find truth tables for more complicated statements, we construct a table listing all possible outcomes and including one column for each operation in the statement. Then we evaluate each operation. In order to find truth values for $\sim p \wedge \sim q$, for example, we must determine the values for $\sim p$ and $\sim q$ before the conjunction can be considered.

EXAMPLE 9.6 Find a truth table for $\sim p \wedge \sim q$.

Solution

p	q	$\sim p$	$\sim q$	$\sim p \wedge \sim q$
T	T	F	F	F
T	F	F	T	F
F	T	T	F	F
F	F	T	T	T

Note that we list the values of the conjunction in the last column, and they are derived from the values in columns 3 and 4 ($\sim p$ and $\sim q$), not columns 1 and 2. A few examples should make the process clear.

EXAMPLE 9.7 Find the truth table.

a. $\sim(p \vee \sim q)$

Solution

p	q	$\sim q$	$p \vee \sim q$	$\sim(p \vee \sim q)$
T	T	F	T	F
T	F	T	T	F
F	T	F	F	T
F	F	T	T	F

b. $(p \wedge \sim q) \vee (\sim p \wedge q)$

Solution

p	q	$\sim q$	$p \wedge \sim q$	$\sim p$	$\sim p \wedge q$	$(p \wedge \sim q) \vee (\sim p \wedge q)$
T	T	F	F	F	F	F
T	F	T	T	F	F	T
F	T	F	F	T	T	T
F	F	T	F	T	F	F

c. $p \vee (q \wedge \sim r)$

Solution

p	q	r	$\sim r$	$q \wedge \sim r$	$p \vee (q \wedge \sim r)$
T	T	T	F	F	T
T	T	F	T	T	T
T	F	T	F	F	T
T	F	F	T	F	T
F	T	T	F	F	F
F	T	F	T	T	T
F	F	T	F	F	F
F	F	F	T	F	F

Problem Set 9.2

In Problems 1–32, find the truth table for the statement.

1. $\sim(\sim p)$
2. $p \wedge p$
3. $p \wedge \sim q$
4. $q \vee \sim p$
5. $p \vee p$
6. $\sim p \vee p$
7. $p \wedge \sim p$
8. $\sim p \wedge q$
9. $p \wedge (q \vee p)$
10. $q \vee (p \wedge q)$
11. $\sim(p \wedge q)$
12. $\sim(p \vee q)$
13. $\sim(p \vee q)$
14. $\sim(p \wedge q)$
15. $p \wedge (q \wedge r)$
16. $p \vee (q \vee r)$
17. $(p \wedge q) \wedge r$
18. $(p \vee q) \vee r$
19. $p \wedge [q \wedge (p \vee \sim q)]$
20. $(p \vee \sim q) \wedge (\sim p \vee q)$
21. $(p \wedge \sim q) \wedge (\sim p \vee q)$

22. $(p \wedge q) \vee \sim(p \vee \sim q)$
23. $\sim(\sim p \wedge \sim q)$
24. $\sim p \vee \sim q$
25. $p \vee (q \wedge r)$
26. $p \wedge (q \vee r)$
27. $\sim p \wedge (q \vee \sim r)$
28. $p \wedge [q \vee (r \wedge \sim q)]$
29. $q \wedge \sim(p \vee r)$
30. $\sim(p \wedge q) \vee \sim r$
31. $(p \vee q) \wedge (p \vee r)$
32. $(p \wedge q) \vee (p \wedge r)$

33. If four statements (p, q, r, and s) are involved, how many possible outcomes are there?
34. In general, if n statements are involved, how many possible outcomes are there?
35. Compare the truth tables for Problems 15 and 17. What can you conclude?
36. Compare the truth tables for Problems 16 and 18. What can you conclude?
37. Compare the truth tables for Problems 25 and 31. What can you conclude?
38. Compare the truth tables for Problems 26 and 32. What can you conclude?
39. Compare the truth tables for Problem 13 and Example 9.6. What can you conclude?
40. Compare the truth tables for Problems 14 and 24. What can you conclude?
41. Is the conjunction commutative?
42. Is the disjunction commutative?
43. When will $p \vee q \vee r \vee s \vee t$ be true?
44. When will $p \wedge q \wedge r \wedge s \wedge t$ be true?

In Problems 45–48, describe in English when the sentence will be false.

45. The bank is closed, and today is a holiday.
46. The client is wealthy, or the client is ill.
47. Overtime workers don't get a bonus at our company.
48. The interest rate is not over 7%, and the minimum balance is $500.

9.3 Other Truth Tables: NAND, NOR, and EOR

In this section, we will consider three more compound statements which are very important in the design of computer circuits: NAND, NOR, and EOR.

NAND: Not AND ($\bar{\wedge}$)

NAND statement

The **NAND statement** is simply the negation of the conjunction: $\sim(p \wedge q)$. We will use the symbol $\bar{\wedge}$ to indicate the NAND operation. This statement is true whenever the conjunction is false. You should be able to compute the NAND truth table directly without having to first find $p \wedge q$. The truth table follows.

p	q	$p \bar{\wedge} q$
T	T	F
T	F	T
F	T	T
F	F	T

NOR: Not OR ($\bar{\vee}$)

NOR statement

The **NOR statement** is the negation of the disjunction: $\sim(p \vee q)$. We will use the symbol $\bar{\vee}$ to indicate the NOR operation. This statement is true whenever the disjunction is false, that is, when both p and q are false. The truth table follows.

p	q	$p\bar{\vee}q$
T	T	F
T	F	F
F	T	F
F	F	T

EOR: Exclusive OR ($\underline{\vee}$)

EOR statement

The **EOR statement** is true whenever *exactly one* of p and q is true: $(p \wedge \sim q) \vee (\sim p \wedge q)$. We are excluding from the disjunction the case where both p and q are true. This statement is important in binary arithmetic. That is, if T and F are replaced by 1 and 0, the Exclusive OR operation gives the same result as adding single binary digits and ignoring the carry (see Problem 38). The truth table for EOR follows.

p	q	$p\underline{\vee}q$
T	T	F
T	F	T
F	T	T
F	F	F

Although all three of these operations can be written as combinations of $\wedge$, $\vee$, and $\sim$, they are important enough to warrant their own symbols, which we have shown.

The next example illustrates statements involving the six operations covered so far.

EXAMPLE 9.8 Find the truth table for the statement.

a. $(p \bar{\wedge} q) \vee (q \bar{\vee} p)$

Solution

p	q	$p\bar{\wedge}q$	$q\bar{\vee}p$	$(p\bar{\wedge}q) \vee (q\bar{\vee}p)$
T	T	F	F	F
T	F	T	F	T
F	T	T	F	T
F	F	T	T	T

b. $(p \bar{\vee} q) \wedge (p \vee q)$

9.3 Other Truth Tables: NAND, NOR, and EOR

Solution

p	q	$p \bar{\vee} q$	$p \vee q$	$(p \bar{\vee} q) \wedge (p \vee q)$
T	T	F	T	F
T	F	F	T	F
F	T	F	T	F
F	F	T	F	F

contradiction   A statement which is false in *every* case is called a **contradiction**.

c. $\sim(p \underline{\vee} q) \vee (p \bar{\wedge} q)$

Solution

p	q	$p \underline{\vee} q$	$\sim(p \underline{\vee} q)$	$p \bar{\wedge} q$	$\sim(p \underline{\vee} q) \vee (p \bar{\wedge} q)$
T	T	F	T	F	T
T	F	T	F	T	T
F	T	T	F	T	T
F	F	F	T	T	T

tautology   A statement which is true in every case is called a **tautology**.

Contradictions and tautologies play a role in the design of computer circuits, as discussed in the next chapter.

Problem Set 9.3

In Problems 1–20, find the truth table for the statement.

1. $\sim(p \wedge q)$
2. $\sim(p \vee q)$
3. $(p \wedge \sim q) \vee (\sim p \wedge q)$
4. $p \underline{\vee} \sim q$
5. $\sim p \bar{\wedge} q$
6. $\sim(p \bar{\vee} q)$
7. $(p \vee q) \vee (p \underline{\vee} q)$
8. $(p \wedge q) \bar{\wedge} (p \vee q)$
9. $(\sim p \bar{\wedge} \sim q) \vee p$
10. $p \underline{\vee} (\sim p \wedge q)$
11. $p \bar{\wedge} p$
12. $p \underline{\vee} p$
13. $(p \bar{\vee} q) \wedge \sim(p \bar{\wedge} q)$
14. $\sim(p \underline{\vee} q) \wedge q$
15. $p \bar{\wedge} (q \underline{\vee} r)$
16. $(p \bar{\vee} q) \wedge \sim r$
17. $(p \bar{\vee} q) \wedge (p \bar{\wedge} r)$
18. $\sim(\sim p \vee \sim q)$
19. $\sim(p \bar{\wedge} \sim q) \bar{\wedge} q$
20. $(\sim p \wedge q) \bar{\vee} (q \wedge \sim p)$
21. Write a statement in p and q which is a tautology.
22. Write a statement in p and q which is a contradiction.

For Problems 23–34, suppose that t is a tautology, c is a contradiction, and p is any statement. Find the truth table for the statement.

23. $c \wedge t$
24. $c \vee t$
25. $\sim c$
26. $\sim t$
27. $c \underline{\vee} t$
28. $c \bar{\wedge} t$
29. $c \bar{\vee} t$
30. $c \vee c$
31. $p \vee t$
32. $p \vee c$
33. $p \wedge c$
34. $p \wedge t$
35. Describe in your own words what the NAND operation does.
36. Describe in your own words what the NOR operation does.
37. What is the difference between OR and Exclusive OR?
38. Change T and F to 1 and 0, and show that Exclusive OR gives the sum in binary (ignoring the carry).
39. Is the NAND operation associative?
40. Is the NOR operation associative?

9.4 Conditional and Biconditional Statements

The last two statements we will consider in this chapter are the conditional and biconditional statements.

conditional statement

The **conditional statement** is a statement of the form "If p, then q" and will be designated $p \rightarrow q$. In order for $p \rightarrow q$ to be true, q must be true whenever p is true. If p is false, there is no requirement on q. So the only time $p \rightarrow q$ is false is when p is true and q is false. The truth table follows.

p	q	$p \rightarrow q$
T	T	T
T	F	F
F	T	T
F	F	T

Some students have difficulty understanding how $p \rightarrow q$ can be true when p is false. Consider the statement "If it snows, then school is canceled," where p represents the statement "It snows" and q represents the statement "School is canceled." If it doesn't snow (p is false), the statement is not in error whether or not school is canceled. However, if it does snow (p is true), school must be canceled (q true) or else the statement is in error. Statement p being true implies that q must also be true. So $p \rightarrow q$ is also called the **implication statement**.

implication statement

The implication $p \rightarrow q$ is the only statement we will consider in which *order* makes a difference.

EXAMPLE 9.9 Find a truth table for $q \rightarrow p$.

Solution

p	q	$q \rightarrow p$
T	T	T
T	F	T
F	T	F
F	F	T

Notice that $q \rightarrow p$ is only false when q is true and p is false. Statement $q \rightarrow p$ is *not* the same as $p \rightarrow q$.

converse

Statement $q \rightarrow p$ is the **converse** of $p \rightarrow q$, and as Example 9.9 shows, the converse is not the same as the original statement. The statement $\sim p \rightarrow \sim q$ is the **inverse** of $p \rightarrow q$, and $\sim q \rightarrow \sim p$ is the **contrapositive** of $p \rightarrow q$. These statements are examined in the next example.

inverse, contrapositive

9.4 Conditional and Biconditional Statements

EXAMPLE 9.10 Find the converse, inverse, and contrapositive of the following statement:

If you work more than 40 hours, then you get extra pay.

Solution

Converse — If you get extra pay, then you work more than 40 hours. (The converse simply reverses the order of the two statements.)

Inverse — If you don't work more than 40 hours, then you don't get extra pay. (The inverse simply negates the two statements without changing their order.)

Contrapositive — If you don't get extra pay, then you don't work more than 40 hours. (The contrapositive reverses the order *and* negates the two statements.)

equivalent statements

Definition 9.5 Two statements are **equivalent statements** iff they have the same truth values for *every* possible outcome.

As you will show in the problem set (Problems 21 and 22), the converse and inverse are equivalent, and the original statement and the contrapositive are equivalent.

biconditional statement

The idea of equivalence leads to the last type of statement that we will consider in this chapter: the biconditional. The **biconditional statement** is a statement of the form "p iff q" and will be designated $p \leftrightarrow q$. Statement $p \leftrightarrow q$ is true whenever p and q have the same truth value (both true or both false). So if two statements are equivalent, then combining them with $\leftrightarrow$ will always lead to a tautology. An example of a biconditional English statement is the following: "An investment of \$1000 yields \$100 interest in a year if and only if the interest rate is 10% a year." The biconditional truth table follows.

p	q	$p \leftrightarrow q$
T	T	T
T	F	F
F	T	F
F	F	T

EXAMPLE 9.11 Show that $p \leftrightarrow q$ and $\sim(p \veebar q)$ are equivalent.

Solution

p	q	$p \leftrightarrow q$	$p \veebar q$	$\sim(p \veebar q)$	$(p \leftrightarrow q) \leftrightarrow \sim(p \veebar q)$
T	T	T	F	T	T
T	F	F	T	F	T
F	T	F	T	F	T
F	F	T	F	T	T

In Example 9.11, the Exclusive OR operation and the biconditional are shown to be negations of each other.

Another way of writing the biconditional is in terms of the conditional: $(p \rightarrow q) \land (q \rightarrow p)$, as shown in the next example.

EXAMPLE 9.12 Find the truth table for $(p \rightarrow q) \land (q \rightarrow p)$.

Solution

p	q	$p \rightarrow q$	$q \rightarrow p$	$(p \rightarrow q) \land (q \rightarrow p)$
T	T	T	T	T
T	F	F	T	F
F	T	T	F	F
F	F	T	T	T

Notice that the truth values in the last column are identical to the truth values of the biconditional. Thus, $(p \rightarrow q) \land (q \rightarrow p)$ is equivalent to the biconditional $p \leftrightarrow q$.

EXAMPLE 9.13 Find the truth table.

a. $(p \leftrightarrow q) \lor \sim(p \land \sim q)$

Solution

p	q	$p \leftrightarrow q$	$\sim q$	$p \land \sim q$	$\sim(p \land \sim q)$	$(p \leftrightarrow q) \lor \sim(p \land \sim q)$
T	T	T	F	F	T	T
T	F	F	T	T	F	F
F	T	F	F	F	T	T
F	F	T	T	F	T	T

Notice that this truth table is identical to the truth table for $p \rightarrow q$.

b. $(p \rightarrow r) \land [(p \leftrightarrow q) \veebar r]$

Solution

p	q	r	$p \rightarrow r$	$p \leftrightarrow q$	$(p \leftrightarrow q) \veebar r$	$(p \rightarrow r) \land [(p \leftrightarrow q) \veebar r]$
T	T	T	T	T	F	F
T	T	F	F	T	T	F
T	F	T	T	F	T	T
T	F	F	F	F	F	F
F	T	T	T	F	T	T
F	T	F	T	F	F	F
F	F	T	T	T	F	F
F	F	F	T	T	T	T

Problem Set 9.4

In Problems 1–20, find the truth table.

1. $p \rightarrow {\sim}q$
2. ${\sim}p \rightarrow q$
3. ${\sim}(p \rightarrow q)$
4. ${\sim}(q \rightarrow p)$
5. $(p \wedge q) \rightarrow q$
6. $q \rightarrow (p \vee q)$
7. $(p \veebar q) \rightarrow {\sim}(p \vee q)$
8. $(p \rightarrow q) \wedge (p \rightarrow {\sim}q)$
9. $p \rightarrow [{\sim}q \barwedge (r \veebar p)]$
10. $[{\sim}(q \wedge r) \veebar p] \rightarrow (p \wedge q)$
11. ${\sim}p \leftrightarrow q$
12. $p \leftrightarrow {\sim}q$
13. ${\sim}p \leftrightarrow {\sim}q$
14. ${\sim}q \leftrightarrow {\sim}p$
15. $(p \wedge q) \leftrightarrow (p \rightarrow q)$
16. $(q \rightarrow p) \leftrightarrow (p \vee q)$
17. $(p \leftrightarrow q) \wedge (p \veebar q)$
18. $(p \rightarrow q) \vee (q \leftrightarrow p)$
19. $(p \wedge {\sim}r) \rightarrow (r \leftrightarrow {\sim}q)$
20. $({\sim}q \rightarrow p) \leftrightarrow [r \vee ({\sim}q \wedge p)]$

21. Show that the converse $(q \rightarrow p)$ and the inverse $({\sim}p \rightarrow {\sim}q)$ are equivalent.
22. Show that the original statement $(p \rightarrow q)$ and the contrapositive $({\sim}q \rightarrow {\sim}p)$ are equivalent.

In Problems 23–28, find the inverse, converse, and contrapositive of the statement.

23. If the interest rates are low, then we can buy a house.
24. If my raise is not high enough, then I will quit.
25. If computers don't use binary numbers, then computers are dumb.
26. If the octal number system has a zero, then the biconditional statement is not true.
27. ${\sim}p \rightarrow q$
28. $q \rightarrow {\sim}p$

29. Find a statement in p and q that uses only $\wedge$, $\vee$, and ${\sim}$ and that is equivalent to $p \rightarrow q$.
30. Find a statement in p and q that uses only $\wedge$, $\vee$, and ${\sim}$ and that is equivalent to $p \leftrightarrow q$.
31. Does order matter in the biconditional? In other words, is $p \leftrightarrow q$ equivalent to $q \leftrightarrow p$?
32. Find the disjunction of the conditional and the biconditional. What is this disjunction equivalent to?
33. Find the conjunction of the conditional and the biconditional. What is this conjunction equivalent to?
34. How many different truth tables are there for statements p and q?

In Problems 35–40, determine whether or not the pair of statements are equivalent.

35. $p \rightarrow q$, ${\sim}(p \wedge q)$
36. $p \vee q$, ${\sim}(p \rightarrow q)$
37. $p \wedge (q \vee r)$, $p \vee (q \wedge r)$
38. ${\sim}p \wedge {\sim}q$, ${\sim}(p \wedge q)$
39. ${\sim}p \vee {\sim}q$, ${\sim}(p \wedge q)$
40. $p \vee (q \vee r)$, $(p \vee q) \vee r$

9.5 Properties of Logic

We have studied eight operations in logic in this chapter. All eight of these operations can be written by using $\wedge$, $\vee$, and ${\sim}$:

Conjunction	$p \wedge q$
Disjunction	$p \vee q$
Negation	${\sim}p$
NAND ($p \barwedge q$)	${\sim}(p \wedge q)$
NOR ($p \barvee q$)	${\sim}(p \vee q)$

Exclusive OR $(p \underline{\vee} q)$ $(\sim p \wedge q) \vee (p \wedge \sim q)$
Conditional $(p \rightarrow q)$ $\sim (p \wedge \sim q)$
Biconditional $(p \leftrightarrow q)$ $(p \wedge q) \vee (\sim p \wedge \sim q)$

For this reason, we will restrict our discussion of the properties of logic to equivalences involving these three basic operations. Remember that equivalence means that both statements yield the same truth values under every condition. It is as strong a statement as can be made in mathematical logic.

Now, we will investigate the properties of logic.

Commutative Properties

$(p \wedge q) \leftrightarrow (q \wedge p)$
$(p \vee q) \leftrightarrow (q \vee p)$

These properties follow directly from the definitions of $\wedge$ and $\vee$.

Associative Properties

$[p \wedge (q \wedge r)] \leftrightarrow [(p \wedge q) \wedge r]$
$[p \vee (q \vee r)] \leftrightarrow [(p \vee q) \vee r]$

These properties may not seem obvious, but they can be verified by using truth tables to show that the statement is a tautology.

EXAMPLE 9.14   Show that $[p \wedge (q \wedge r)] \leftrightarrow [(p \wedge q) \wedge r]$ is a tautology.

Solution

p	q	r	$q \wedge r$	$p \wedge q$	$p \wedge (q \wedge r)$	$(p \wedge q) \wedge r$	$[(p \wedge q) \wedge r] \leftrightarrow [p \wedge (q \wedge r)]$
T	T	T	T	T	T	T	T
T	T	F	F	T	F	F	T
T	F	T	F	F	F	F	T
T	F	F	F	F	F	F	T
F	T	T	T	F	F	F	T
F	T	F	F	F	F	F	T
F	F	T	F	F	F	F	T
F	F	F	F	F	F	F	T

Identity Properties   If t is a tautology, c is a contradiction, and p is any statement, then

$(p \wedge t) \leftrightarrow p$
$(p \vee c) \leftrightarrow p$

9.5 Properties of Logic

EXAMPLE 9.15 Show that $(p \vee c) \leftrightarrow p$ is a tautology.

Solution

p	c	$p \vee c$	$(p \vee c) \leftrightarrow p$
T	F	T	T
F	F	F	T

Distributive Properties

$[p \wedge (q \vee r)] \leftrightarrow [(p \wedge q) \vee (p \wedge r)]$
$[p \vee (q \wedge r)] \leftrightarrow [(p \vee q) \wedge (p \vee r)]$

EXAMPLE 9.16 Show that the second distributive property is true.

Solution

p	q	r	$q \wedge r$	$p \vee (q \wedge r)$	$p \vee q$	$p \vee r$	$(p \vee q) \wedge (p \vee r)$	$[p \vee (q \wedge r)] \leftrightarrow [(p \vee q) \wedge (p \vee r)]$
T	T	T	T	T	T	T	T	T
T	T	F	F	T	T	T	T	T
T	F	T	F	T	T	T	T	T
T	F	F	F	T	T	T	T	T
F	T	T	T	T	T	T	T	T
F	T	F	F	F	T	F	F	T
F	F	T	F	F	F	T	F	T
F	F	F	F	F	F	F	F	T

Notice that there are two distributive properties of logic, just as there were with sets. The first distributive property will be shown in Problem 12.

Idempotent Properties

$(p \wedge p) \leftrightarrow p$
$(p \vee p) \leftrightarrow p$

Double Negation Property

$[\sim(\sim p)] \leftrightarrow p$

DeMorgan Properties

$(\sim p \wedge \sim q) \leftrightarrow [\sim p \vee q)]$ (NOR)
$(\sim p \vee \sim q) \leftrightarrow [\sim(p \wedge q)]$ (NAND)

Tautology/Contradiction Properties If t is a tautology, c is a contradiction, and p is any statement, then

$(p \vee t) \leftrightarrow t$ $(p \wedge \sim p) \leftrightarrow c$
$(p \wedge c) \leftrightarrow c$ $(p \vee \sim p) \leftrightarrow t$

Problem Set 9.5

In Problems 1–22, use truth tables to verify the property of logic, where t is a tautology and c is a contradiction. Name the property when appropriate.

1. $(p \wedge q) \leftrightarrow (q \wedge p)$
2. $(p \vee q) \leftrightarrow (q \vee p)$
3. $(p \, \bar{\wedge} \, q) \leftrightarrow [\sim(p \wedge q)]$
4. $(p \, \bar{\vee} \, q) \leftrightarrow [\sim(p \vee q)]$
5. $[(p \veebar q)] \leftrightarrow [(\sim p \wedge q) \vee (p \wedge \sim q)]$
6. $(p \rightarrow q) \leftrightarrow [\sim(p \wedge \sim q)]$
7. $(p \leftrightarrow q) \leftrightarrow [(p \wedge q) \vee (\sim p \wedge \sim q)]$
8. $[p \vee (q \vee r)] \leftrightarrow [(p \vee q) \vee r]$
9. $(p \wedge t) \leftrightarrow p$
10. $(p \wedge \sim p) \leftrightarrow c$
11. $(p \vee \sim p) \leftrightarrow t$
12. $[p \wedge (q \vee r)] \leftrightarrow [(p \wedge q) \vee (p \wedge r)]$
13. $(p \wedge p) \leftrightarrow p$
14. $(p \vee p) \leftrightarrow p$
15. $[\sim(\sim p)] \leftrightarrow p$
16. $(\sim p \wedge \sim q) \leftrightarrow [\sim(p \vee q)]$
17. $[(\sim p \vee \sim q)] \leftrightarrow [\sim(p \wedge q)]$
18. $(p \vee t) \leftrightarrow t$
19. $(p \wedge c) \leftrightarrow c$
20. $\sim t \leftrightarrow c$
21. $(c \wedge t) \leftrightarrow c$
22. $(c \vee t) \leftrightarrow t$

23. Is the Exclusive OR operation associative? Show by truth tables.
24. Is the NOR operation associative?
25. Is the NAND operation associative?
26. Is the conditional operation associative?
27. Is $[p \, \bar{\wedge} \, (q \veebar r)] \leftrightarrow [(p \, \bar{\wedge} \, q) \veebar (p \, \bar{\wedge} \, r)]$? In other words, does $\bar{\wedge}$ distribute over $\veebar$?
28. Does $\bar{\vee}$ distribute over $\bar{\wedge}$?
29. Is there an identity for NOR?
30. Is there an identity for NAND?
31. Is Exclusive OR an idempotent operation $[(p \veebar p) \leftrightarrow p]$?
32. Does the biconditional symbol represent an indempotent operation?

All eight operations we have studied can be written in terms of $\wedge$, $\vee$, and $\sim$. In Problems 33–40, find statements involving only $\wedge$, $\vee$, and $\sim$ to yield the truth tables for p and q.

33.

p	q	?
T	T	F
T	F	F
F	T	T
F	F	F

34.

p	q	?
T	T	T
T	F	T
F	T	F
F	F	T

35.

p	q	?
T	T	T
T	F	F
F	T	T
F	F	F

36.

p	q	?
T	T	F
T	F	F
F	T	F
F	F	F

37.

p	q	?
T	T	T
T	F	T
F	T	T
F	F	T

38.

p	q	?
T	T	F
T	F	T
F	T	F
F	F	T

39.

p	q	?
T	T	F
T	F	T
F	T	F
F	F	F

40.

p	q	?
T	T	T
T	F	T
F	T	F
F	F	F

9.6 Arguments

argument
hypothesis
conclusion,
  valid argument
invalid argument

The last topic that we will cover in this chapter is one of the applications of mathematical logic: arguments. An **argument** consists of a set of statements that are assumed to be true, called the **hypothesis**, and one final statement called the **conclusion**. An argument is a **valid argument** iff the truth of all of the statements of the hypothesis *implies* the truth of the conclusion. Otherwise, the argument is an **invalid argument**.

For example, if the hypothesis consists of the statements p, q, and r and the conclusion consists of the statement s, then $(p \wedge q \wedge r) \rightarrow s$ must be a tautology. Since every statement of the hypothesis is assumed true, the only case that must be checked is when p, q, and r are *all* true. Since p, q, and r are often compound (not simple) statements, the easiest way to check validity is to do a complete truth table.

EXAMPLE 9.17 Determine whether the argument is valid or invalid.

a. hypothesis: p
$ \sim q$
conclusion: $p \rightarrow q$

Solution We must determine whether or not $(p \wedge \sim q) \rightarrow (p \rightarrow q)$ is a tautology.

p	q	$\sim q$	$p \wedge \sim q$	$p \rightarrow q$	$(p \wedge \sim q) \rightarrow (p \rightarrow q)$
T	T	F	F	T	T
T	F	T	T	F	F
F	T	F	F	T	T
F	F	T	F	T	T

Since the last column is not always true, this argument is invalid.

b. H: $p \rightarrow q$
$ q \rightarrow r$
C: $p \rightarrow r$

Solution

p	q	r	$p \rightarrow q$	$q \rightarrow r$	$p \rightarrow r$	$(p \rightarrow q) \wedge (q \rightarrow r)$	$[(p \rightarrow q) \wedge (q \rightarrow r)] \rightarrow (p \rightarrow r)$
T	T	T	T	T	T	T	T
T	T	F	T	F	F	F	T
T	F	T	F	T	T	F	T
T	F	F	F	T	F	F	T
F	T	T	T	T	T	T	T
F	T	F	T	F	T	F	T
F	F	T	T	T	T	T	T
F	F	F	T	T	T	T	T

This argument is valid (the last column is always true).

c. H: The savings are lost, or the mortgage payment is due.
If the mortgage payment is not due, then the savings are not lost.
The savings are lost iff the mortgage payment is due.

C: The savings are lost, and the mortgage payment is due.

Solution First, we should use symbols for each simple statement:

p: The savings are lost.
q: The mortgage payment is due.

This notation yields the following symbolic argument:

H: $p \vee q$
$\sim q \rightarrow \sim p$
$p \leftrightarrow \sim q$
C: $p \wedge q$

We must now show whether or not

$$[(p \vee q) \wedge (\sim q \rightarrow \sim p) \wedge (p \leftrightarrow q)] \rightarrow (p \wedge q)$$

For brevity, we will refer to this statement as C in the truth table.

p	q	$\sim p$	$\sim q$	$p \vee q$	$p \wedge q$	$\sim q \rightarrow \sim p$	$p \leftrightarrow q$	$(p \vee q) \wedge (\sim q \rightarrow \sim p) \wedge (p \leftrightarrow q)$	C
T	T	F	F	T	T	T	T	T	T
T	F	F	T	T	F	F	F	F	T
F	T	T	F	T	F	T	F	F	T
F	F	T	T	F	F	T	T	F	T

This argument is valid.

Many statements in logic involve the words *all*, *some*, and *none*. These arguments are more easily solved by Venn diagrams than by truth tables.

Consider the statement "All penguins are smart." This statement means that the set of penguins is a *subset* of the set of smart animals. The corresponding Venn diagram is shown in Figure 9.1, where P = set of penguins and S = set of smart animals.

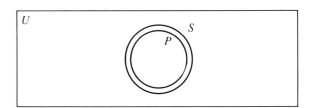

Figure 9.1
All Statement

9.6 Arguments

Now, consider the statement "No penguins are smart." In this case, the set of penguins and the set of smart animals are *disjoint*, as indicated in Figure 9.2.

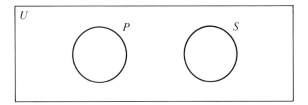

Figure 9.2
None Statement

Finally, consider the statement "Some penguins are smart." In English, this sentence means that at least one penguin is smart and *maybe* all penguins are smart. We will consider *some* to mean *at least one but not all*. So we get *overlapping sets*, as indicated in Figure 9.3.

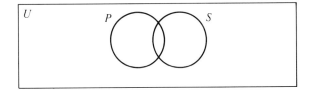

Figure 9.3
Some Statement

In using sets to evaluate arguments, we draw a Venn diagram for all of the statements of the hypothesis. Then we determine whether the conclusion follows from the diagram.

EXAMPLE 9.18 Use a Venn diagram to determine whether the argument is valid or invalid.

   a. H: All bankers are smart.
           All smart people are rich.
        C: All bankers are rich.

Solution Let B = set of bankers, S = set of smart people, and R = set of rich people. The hypothesis gives the Venn diagram shown in Figure 9.4.

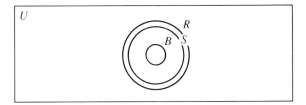

Figure 9.4

From the diagram, $B \subseteq S$ and $S \subseteq R$. It follows that $B \subseteq R$. So all bankers are rich, and the argument is <u>valid</u>.

b.   H: All students work hard.
           <u>No hardworking people fail.</u>
         C: No students fail.

Solution   Let S = set of students, H = set of hardworking people, and F = set of failing people. The hypothesis gives the Venn diagram shown in Figure 9.5.

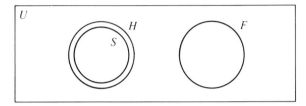

Figure 9.5

From the Venn diagram, we see that the sets S and F are disjoint. So the conclusion "No students fail" is <u>valid</u>.

c.   H: All integers are real.
           <u>Some real numbers are negative.</u>
         C: Some integers are negative.

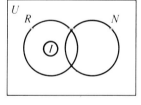 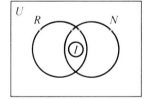 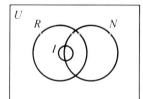

Figure 9.6

Solution   The Venn diagrams are shown in Figure 9.6. In this case, we cannot tell the relationship between the sets N and I. All three Venn diagrams could follow from the hypothesis. The conclusion *could* be true, but it may *not* be true. The conclusion is *not necessarily* true. So the argument is <u>invalid</u>.

■ Problem Set 9.6

In Problems 1–20, use truth tables to determine whether the argument is valid or invalid.

1.   H: p
       $\dfrac{q}{\text{C: } p \wedge q}$

2.   H: p
       $\dfrac{q}{\text{C: } p \vee q}$

3.   H: p
       $\dfrac{p \rightarrow q}{\text{C: } q}$

4.   H: q
       $\dfrac{\sim p}{\text{C: } q \rightarrow \sim p}$

9.6 Arguments

5. H: p
    $\underline{q \lor \sim p}$
    C: q

6. H: q
    $\underline{p \lor \sim q}$
    C: p

7. H: $p \to q$
    $\underline{\sim q \to p}$
    C: $p \leftrightarrow q$

8. H: $\sim p \to \sim q$
    $\underline{\sim q \to \sim p}$
    C: $\sim p \leftrightarrow \sim q$

9. H: $\sim p$
    $\underline{\sim q}$
    C: $\sim(p \lor q)$

10. H: $\sim p$
    $\underline{\sim q}$
    C: $\sim(p \land q)$

11. H: $p \leftrightarrow q$
    $\overline{\text{C: } \sim(p \veebar q)}$

12. H: $p \veebar q$
    $\overline{\text{C: } \sim p \barwedge \sim q}$

13. H: $p \to q$
    $\sim q \to \sim p$
    $\underline{p \land q}$
    C: $p \leftrightarrow q$

14. H: $p \lor q$
    $p \to q$
    $\underline{\sim p \land q}$
    C: $q \to p$

15. H: $p \to q$
    $q \to r$
    $\underline{p \lor r}$
    C: $\sim p \land r$

16. H: $\sim p \land r$
    $r \to q$
    $\underline{q \to p}$
    C: $\sim p \to r$

17. H: Computers use binary numbers or computers use decimals.
    $\underline{\text{Computers don't use binary numbers.}}$
    C: Computers use decimals.

18. H: If the benefits are good, then I will take the job.
    $\underline{\text{I didn't take the job.}}$
    C: The benefits were not good.

19. H: I use seat belts iff I am careful.
    $\underline{\text{If I am careful, then I won't have an accident.}}$
    C: I won't have an accident iff I use seat belts.

20. H: The leaves fall off the trees iff it is autumn.
    $\underline{\text{It is autumn, or the leaves don't fall off the trees.}}$
    C: If it is autumn, then the leaves don't fall off the trees.

For Problems 21–26, use the Venn diagram in Figure 9.7 to determine the relationship between the pairs of sets in terms of *all*, *some*, or *none* (for example, "All A's are B's" and "Some B's are A's").

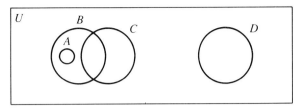

Figure 9.7

21. C and D
22. C and A
23. B and D
24. B and C
25. D and A
26. D and U

In Problems 27–38, use Venn diagrams to determine whether the argument is valid or invalid.

27. H: All A's are B's.
    C: All B's are A's.

28. H: No A's are B's.
    C: No B's are A's.

29. H: Some A's are B's.
    C: Some B's are A's.

30. H: Some B's are A's.
    C: All A's are B's.

31. H: All A's are B's.
    All B's are C's.
    C: All A's are C's.

32. H: No A's are B's.
    No B's are C's.
    C: No A's are C's.

33. H: No B's are C's.
    Some A's are C's.
    C: No A's are C's.

34. H: Some A's are C's.
    All C's are B's.
    C: Some A's are B's.

35. H: Some B's are A's.
    All A's are C's.
    C: Some B's are C's.

36. H: All C's are B's.
    No B's are A's.
    C: Some A's are C's.

37. H: All binary numbers are octal numbers.
    No octal numbers are hexadecimal numbers.
    All even numbers are hexadecimal numbers.
    C: No binary numbers are even numbers.

38. H: Some integers are positive.
    Some positive numbers are rational.
    All rational numbers are integers.
    C: Some integers are rational.

39. Write an argument containing obviously false statements (for instance, "All cats are blue") that leads to an obviously false conclusion which is valid. How can this result happen?

40. Write an argument containing true statements (for instance, "All rational numbers are real") which is invalid. How can this result happen?

CHAPTER SUMMARY

In this chapter, we have studied the basic concepts of mathematical logic. In particular, we have looked at eight operations in logic:

Conjunction	$p \wedge q$	True only when both p and q are true
Disjunction	$p \vee q$	True when at least one of p and q is true
Negation	$\sim p$	True when p is false
NAND	$p \bar{\wedge} q$	True when the conjunction is false
NOR	$p \bar{\vee} q$	True when the disjunction is false
Exclusive OR	$p \veebar q$	True when exactly one of p and q is true
Conditional	$p \rightarrow q$	True except when p is true and q is false
Biconditional	$p \leftrightarrow q$	True when p and q have the same truth value

We also studied eight properties of logic, where t represents a tautology and c represents a contradiction:

Commutative properties	$(p \wedge q) \leftrightarrow (q \wedge p)$
	$(p \vee q) \leftrightarrow (q \vee p)$
Associative properties	$[p \wedge (q \wedge r)] \leftrightarrow [(p \wedge q) \wedge r]$
	$[p \vee (q \vee r)] \leftrightarrow [(p \vee q) \vee r]$
Identity properties	$(p \wedge t) \leftrightarrow p$
	$(p \vee c) \leftrightarrow p$
Distributive properties	$[p \wedge (q \vee r)] \leftrightarrow [(p \wedge q) \vee (p \wedge r)]$
	$[p \vee (q \wedge r)] \leftrightarrow [(p \vee q) \wedge (p \vee r)]$
Idempotent properties	$(p \wedge p) \leftrightarrow p$
	$(p \vee p) \leftrightarrow p$
Double negation property	$[\sim(\sim p)] \leftrightarrow p$
DeMorgan properties	$(\sim p \wedge \sim q) \leftrightarrow [\sim(p \vee q)]$
	$(\sim p \vee \sim q) \leftrightarrow [\sim(p \wedge q)]$
Tautology/contradiction properties	$(p \vee t) \leftrightarrow t$
	$(p \wedge c) \leftrightarrow c$
	$(p \wedge \sim p) \leftrightarrow c$
	$(p \vee \sim p) \leftrightarrow t$

We also looked at an application of logic: arguments. This chapter lays the foundation for a study of how computers use logic, which begins in the next chapter.

REVIEW PROBLEMS

In Problems 1–4, determine whether the sentence is a valid statement. If it is not, state why not.

1. This book is about penguins.
2. How much money did you invest?
3. Sell high!
4. The salesman is obnoxious.

In Problems 5–8, determine whether the statement is simple or compound.

5. The space shuttle always lands in Florida.
6. The interest rate is 5% iff the price is $1200.
7. If complement addition is used, then binary numbers are not used.
8. Real numbers are not rational.

In Problems 9–26, find a truth table for the statement.

9. $\sim(p \land \sim q)$
10. $\sim(\sim p \lor q)$
11. $p \lor (q \land p)$
12. $\sim p \land (p \lor q)$
13. $p \lor (q \to p)$
14. $(p \to q) \to \sim p$
15. $(p \leftrightarrow q) \to (p \land q)$
16. $p \barwedge (p \veebar q)$
17. $(p \to q) \land (q \to \sim p)$
18. $(p \lor q) \leftrightarrow p$
19. $\sim p \veebar (\sim q \lor p)$
20. $\sim p \to \sim p$
21. $(p \to q) \leftrightarrow (q \to p)$
22. $\sim(\sim p \land \sim q)$
23. $(p \lor q) \barwedge r$
24. $\sim p \land (q \veebar r)$
25. $(\sim p \land \sim r) \barvee q$
26. $(p \land \sim q) \leftrightarrow (p \lor r)$

In Problems 27–28, find the converse, inverse, and contrapositive of the statement.

27. $\sim p \to \sim q$
28. If U is the universal set, then A is a subset of U.

In Problems 29–36, determine whether the argument is valid or invalid.

29. H: $p \lor q$
    $\underline{\sim p}$
    C: q

30. H: $\sim p \to q$
    $\underline{q}$
    C: $q \to p$

31. H: $p \lor \sim r$
    $\sim q$
    $\underline{p \to q}$
    C: $r \land p$

32. H: $r \to p$
    $\sim p$
    $\underline{q \land \sim r}$
    C: $\sim p \land r$

33. H: All computers use complement addition.
    Some machines using complement addition break down.
    C: Some computers break down.

34. H: No calculators are computers.
    All computers are inexpensive.
    C: No calculators are inexpensive.

35. H: All X's are Y's.
    No Y's are Z's.
    C: No X's are Z's.

36. H: All word processors are easy to learn.
    Some things that are easy to learn are worthwhile.
    C: Some word processors are worthwhile.

37. What is a tautology?
38. What is a contradiction?

BOOLEAN ALGEBRA

CHAPTER 10

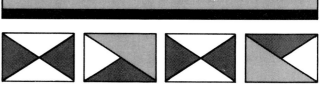

In the 1850s, the English mathematician George Boole took some of the ideas of logic and combined them with algebra to produce what is now called Boolean algebra. It was an exercise of the mind back in the 1850s, but it now forms the foundation for the operation of the modern computer. Little did George Boole realize when he invented Boolean algebra how necessary it would be for the understanding of modern technology over a hundred years later.

In this chapter, we will study Boolean algebra not as an 1850s intellectual topic but as the foundation for circuit design and decision making in computers. This chapter is essential for an understanding of how computers operate.

10.1 Introduction

George Boole was well versed in logic before he developed Boolean algebra. Consider the logic statement p; p is either true or false. There are only two possible outcomes. So our Boolean algebra will consist of only two elements, which we will designate 0 and 1. (Actually, a Boolean algebra can have more than two elements, but we will only consider the case of two elements since that is the case that applies to computers.) We could have used true and false for these two elements, but 0 and 1 fit our needs better.

A Boolean algebra consists of two operations, which are called *addition* and *multiplication*. We will use the symbols $+$ and $\cdot$ to represent these operations, although they are different from addition and multiplication of real numbers. They are defined as illustrated in Figure 10.1.

Figure 10.1
Boolean Addition and Multiplication

+	0	1		·	0	1
0	0	1		0	0	0
1	1	1		1	0	1

Writing these tables in a different way, we get the table shown in Figure 10.2. Comparing the table in Figure 10.2 with the truth tables for $p \vee q$ and $p \wedge q$ from logic shows that the Boolean operations $+$ and $\cdot$ are analogous to the logic operations of $\vee$ and $\wedge$ if 1 is replaced with T and 0 is replaced with F. Tables such as the one in Figure 10.2 will be referred to as **Boolean tables** from now on.

Boolean tables

10.1 Introduction

Figure 10.2
Boolean Table

A	B	A+B	A·B
1	1	1	1
1	0	1	0
0	1	1	0
0	0	0	0

EXAMPLE 10.1 Evaluate the expression by using Figure 10.2.

a. $(0+1) \cdot (1+1)$

Solution $(0+1) \cdot (1+1) = 1 \cdot 1 = \underline{1}$

According to the Boolean table in Figure 10.2,

$0+1 = 1$ and $1+1 = 1$

b. $0 \cdot [(1+0)+1]$

Solution
$0 \cdot [(1+0)+1] = 0 \cdot [1+1]$
$= 0 \cdot 1 = \underline{0}$

c. $[0+(1 \cdot 1)] + [(1 \cdot 0)+(0 \cdot 1)]$

Solution
$[0+(1 \cdot 1)] + [(1 \cdot 0)+(0 \cdot 1)] = (0+1)+(0+0)$
$= 1+0 = \underline{1}$

Any expression involving combinations of 0s, 1s, +, and · can be simplified to either a 0 or a 1, as illustrated in Example 10.1.

We are now ready to define what we mean by a Boolean algebra.

Boolean algebra

Definition 10.1 A **Boolean algebra** is a set consisting of two elements, 0 and 1, and two operations, + and ·, such that the following properties hold.

1. Addition + and multiplication · are commutative:

$A+B = B+A$
$A \cdot B = B \cdot A$

2. Each operation is distributive over the other operation:

$A \cdot (B+C) = (A \cdot B)+(A \cdot C)$
$A+(B \cdot C) = (A+B) \cdot (A+C)$

3. Each operation has an identity:

$A+0 = A$
$A \cdot 1 = A$

4. Each element has a complement, denoted $\bar{A}$, such that

$$A + \bar{A} = 1$$
$$A \cdot \bar{A} = 0$$

Since $+$ and $\cdot$ are analogous to $\vee$ and $\wedge$ from logic, the first three properties should not be too surprising, and they are easy to verify. To verify these properties, we can construct Boolean tables, which are very similar to truth tables. The next example illustrates the process.

EXAMPLE 10.2

Find a Boolean table for $A \cdot (B + C)$.

Solution

A	B	C	B+C	A·(B+C)
1	1	1	1	1
1	1	0	1	1
1	0	1	1	1
1	0	0	0	0
0	1	1	1	0
0	1	0	1	0
0	0	1	1	0
0	0	0	0	0

By finding a Boolean table for $(A \cdot B) + (A \cdot C)$, you can show (see Problem 27) that the result is the same in every case as the table in this example.

What about the complement of 0 and 1? Since our Boolean algebra consists of only the two elements 0 and 1, it seems logical that the complement of 0 be 1 and the complement of 1 be 0. Since $1 + 0 = 1$ and $1 \cdot 0 = 0$, these elements satisfy the fourth property of Definition 10.1. The complement in Boolean algebra is analogous to the negation in logic.

EXAMPLE 10.3

Find a Boolean table for $A + \bar{A}$.

Solution

A	$\bar{A}$	$A + \bar{A}$
1	0	1
0	1	1

Since $A + \bar{A}$ always results in 1, then $A + \bar{A} = 1$.

Problem Set 10.1

In Problems 1–10, evaluate the expression.

1. $(1 + 0) \cdot 1$
2. $(1 \cdot 1) + 1$
3. $(1 \cdot 0) + 1$
4. $1 + [(1 \cdot 0) + 1]$
5. $(1 \cdot 1) + (0 \cdot 1)$
6. $1 + [0 \cdot (1 + 0)]$

10.2 Circuits

7. $1+(1\cdot 1)\cdot(1\cdot 1)$
8. $(1+1)\cdot(0+1)$
9. $1+(0\cdot 1)$
10. $(1+0)\cdot(1+1)$

In Problems 11–20, find a Boolean table for the expression.

11. $A+(A\cdot B)$
12. $(A\cdot B)+B$
13. $\bar{A}+B$
14. $\bar{B}\cdot A$
15. $(A+B)\cdot(A\cdot B)$
16. $(A\cdot B)+(\bar{A}\cdot B)$
17. $(A+B)+C$
18. $(A\cdot B)\cdot C$
19. $A\cdot(A+B)$
20. $\bar{A}\cdot\bar{B}$

In Problems 21–30, verify the Boolean properties by constructing Boolean tables for the expression.

21. $A+B$
22. $B+A$
23. $B\cdot A$
24. $A\cdot B$
25. $A+(B\cdot C)$
26. $(A+B)\cdot(A+C)$
27. $(A\cdot B)+(A\cdot C)$
28. $A\cdot\bar{A}$
29. $A+0$
30. $A\cdot 1$

31. Write a Boolean expression in A and B that gives a 1 when A and B are the same and 0 otherwise.
32. Write a Boolean expression in A and B that gives a 0 when A and B are the same and a 1 otherwise.
33. What operation in logic is analogous to the expression in Problem 31?
34. What operation in logic is analogous to the expression in Problem 32?

10.2 Circuits

In this section, we will relate the concepts of Boolean algebra to computer circuits. A simple circuit may consist of a power source (perhaps a battery) connected by a wire to a destination (perhaps a light bulb) with a switch to control the flow of electricity. This simple circuit is illustrated in Figure 10.3.

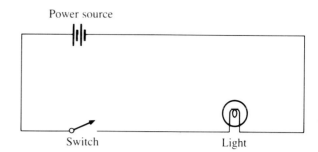

Figure 10.3

When we first studied the binary number system in Chapter 5, we noted how well-suited to computers this number system is, since it uses only two symbols. Consider one switch of a circuit, which is either up, not allowing electricity to flow past it, or down, allowing electricity to pass. These two conditions will be represented by 0 and 1, respectively, and will be drawn as in Figure 10.4. Rather than refer to these conditions as up and down, we will use the terms *open* and *closed*. In general, we will draw circuits and label them with a capital letter, not indicating the condition of the switch.

250 BOOLEAN ALGEBRA

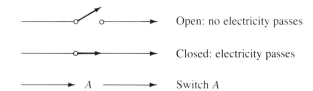

Figure 10.4

network

Any electronic circuit that consists of more than one switch is called a **network**. Networks are constructed by combining two types of circuits, parallel or series. Each type is described in the following subsections.

Parallel Circuit

parallel circuit

A circuit in which the switches are arranged in such a way that electricity passes whenever *at least one* of the switches is closed is called a **parallel circuit**. This circuit is analogous to OR from logic and is described by *Boolean addition* (the result is 1 whenever at least one of A and B is 1). The parallel circuit is drawn as in Figure 10.5. Since the parallel circuit is represented by addition, Figure 10.5 represents $A + B$.

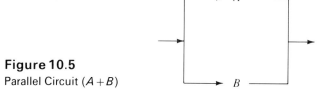

Figure 10.5
Parallel Circuit $(A + B)$

From Figure 10.5, we see that electricity can pass if either (not necessarily both) A or B is closed. Think of switches as drawbridges and the wires as roads. You can travel from left to right if at least one of the drawbridges is down.

■ **EXAMPLE 10.4**   For the parallel circuit of A and B, draw the condition where A is open (0) and B is closed (1). State whether electricity passes ($A + B = 1$) or not ($A + B = 0$).

Solution   The circuit is drawn in Figure 10.6. From the figure, we see that electricity passes ($A + B = 1$).

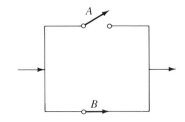

Figure 10.6
$A + B$

10.2 Circuits

Series Circuit

series circuit

A circuit in which the switches are arranged in such a way that electricity passes only when *all* of the switches are closed is called a **series circuit**. This circuit is analogous to AND from logic and is represented by *Boolean multiplication* (the result is 1 only when *both* A and B are 1). The series circuit is drawn as in Figure 10.7. Since the series circuit is represented by multiplication, Figure 10.7 represents $A \cdot B$. So the two basic Boolean operations of addition and multiplication describe the two basic types of circuits: parallel and series circuits.

Figure 10.7
Series Circuit ($A \cdot B$)

———→ A ———→ B ———→

From Figure 10.7, we see that electricity can pass only when both A and B are closed. That is, we can only travel from left to right if both bridges are down.

EXAMPLE 10.5 For the series circuit of A and B, draw the condition where A is open (0) and B is closed (1). State whether electricity passes ($A \cdot B = 1$) or not ($A \cdot B = 0$).

Solution The circuit is drawn in Figure 10.8. From the figure, we see that electricity does not pass ($A \cdot B = 0$).

Figure 10.8
$A \cdot B$

complement

Since switch A has two possible conditions, we can define the **complement** of A to be the switch which is always in the opposite condition from A. The complement will be denoted $\bar{A}$. If A is open (0), then $\bar{A}$ is closed (1). And if A is closed (1), then $\bar{A}$ is open (0).

Problem Set 10.2

In Problems 1–18, draw the circuit with the given conditions, and state whether or not electricity passes.

1. $A + B$, A open and B open
2. $A + B$, A closed and B closed
3. $A + B$, A closed and B open
4. $A \cdot B$, A open and B open
5. $A \cdot B$, A open and B closed
6. $A \cdot B$, A closed and B open
7. $A \cdot B$, A closed and B closed
8. $A + B + C$; A, B open and C closed
9. $A + B + C$; A, B, and C open
10. $A + B + C$; A, B, and C closed
11. $A + B + C$; A, C open and B closed
12. $A + B + C$; B open and A, C closed
13. $A \cdot B \cdot C$; A, B closed and C open
14. $A \cdot B \cdot C$; A, B, and C open
15. $A \cdot B \cdot C$; A, B, and C closed
16. $A \cdot B \cdot C$; A, C open and B closed
17. $\bar{A}$, A closed
18. $\bar{A}$, A open

In Problems 19–30, use Figure 10.9 to determine whether electricity will flow through the circuit under the given conditions.

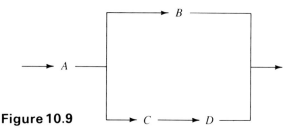

Figure 10.9

19. A open and B, C, and D closed
20. A closed and B, C, and D open
21. A, B, C, and D closed
22. A, B, C, and D open
23. A, B open and C, D closed
24. A, C, and D closed and B open
25. A, B closed and C, D open
26. A, D closed and B, C open
27. Can electricity ever pass if A is open?
28. Will electricity always pass if A is closed? If not, when will it pass?
29. Can electricity ever pass if both C and D are open? If so, when can it pass?
30. Can electricity ever pass if B is open?
31. Draw a series or parallel circuit that will never let electricity pass. (*Hint*: Use $\bar{A}$.)
32. Draw a series or parallel circuit that will always let electricity pass.
33. When will electricity pass through the circuit $A + B + C$?
34. When will electricity pass through the circuit $A \cdot B \cdot C$?

10.3 Combinations of Switches

In this section, we will consider more complicated networks involving numerous switches. As you study this section, remember that parallel circuits represent Boolean addition and series circuits represent Boolean multiplication.

We will first consider the problem of drawing a network to represent a given complicated Boolean expression.

EXAMPLE 10.6 Draw a network to represent the Boolean expression.

a. $A + (B \cdot C)$

Solution This expression is basically the sum of two expressions, A and $(B \cdot C)$, so it is a parallel circuit. One path of the circuit consists only of one switch (A). The other path consists of B and C in series (multiplication). The resulting network is shown in Figure 10.10.

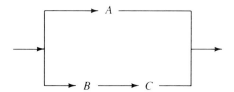

Figure 10.10

10.3 Combinations of Switches

b. $(A+B) \cdot (C+D)$

Solution This expression represents a series circuit (multiplication) consisting of two parallel circuits (addition). The resulting network is shown in Figure 10.11.

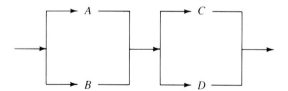

Figure 10.11

c. $(A \cdot B \cdot C) + (\overline{A} + B)$

Solution This expression represents a parallel circuit consisting of a series $(A \cdot B \cdot C)$ circuit and another parallel $(\overline{A} + B)$ circuit. The resulting network is shown in Figure 10.12.

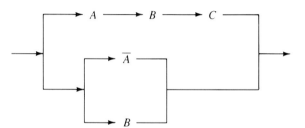

Figure 10.12

d. $A \cdot [B + (C + D)] \cdot E$

Solution This expression is the product of three expressions, so the network consists of a series circuit. The first expression is A, so the first part of the series consists of switch A. The last expression is E, so the third part of the series consists of switch E. The second expression is the sum of B, C, and D, so the second part of the series consists of switches B, C, and D in parallel. The resulting network is shown in Figure 10.13.

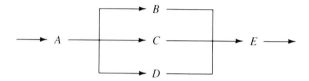

Figure 10.13

Next, we will consider the problem of writing a Boolean expression corresponding to a given network. For this problem, we simply reverse our thinking for the previous procedure.

BOOLEAN ALGEBRA

EXAMPLE 10.7 Write the Boolean expression for the given network.

    a. The network of Figure 10.14

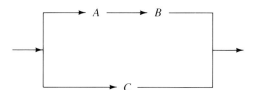

Figure 10.14

Solution This network is a parallel circuit, so the expression is a sum. Since one part of the sum consists of A and B in series, the resulting expression is

$$(A \cdot B) + C$$

    b. The network of Figure 10.15

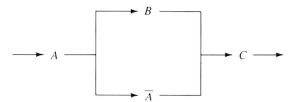

Figure 10.15

Solution This network is a series of three sets of switches, so the expression is a product of three terms. The first term is simply A, and the third term is C. The middle set of switches includes B and $\bar{A}$ in parallel, so the term is $B + \bar{A}$. Thus, the final expression must be

$$A \cdot (B + \bar{A}) \cdot C$$

    c. The network of Figure 10.16

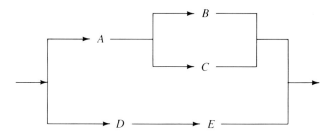

Figure 10.16

Solution This network is basically a parallel circuit, so the expression is a sum:

$$(\quad) + (\quad)$$

10.3 Combinations of Switches

The lower path has D and E in series:

$$(\quad) + (D \cdot E)$$

The upper path has A in series with B and C:

$$[A \cdot (\quad)] + (D \cdot E)$$

Switches B and C are in parallel:

$$[A \cdot (B+C)] + (D \cdot E)$$

To simplify expressions, we will adopt the symbolism of algebra and eliminate writing the $\cdot$ for multiplication when it's not necessary. The answer then becomes

$$\underline{A(B+C) + DE}$$

d. The network of Figure 10.17

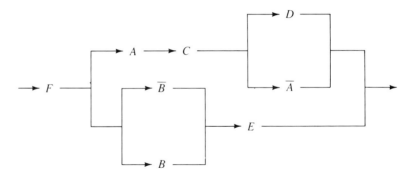

Figure 10.17

Solution Switch F is in series with the rest of the network:

$$F \cdot \{\qquad\}$$

The rest of the network consists of two parallel paths:

$$F \cdot \{[\quad] + [\quad]\}$$

The upper path consists of A and C in series with the parallel circuit of D and $\bar{A}$:

$$F \cdot \{[A \cdot C \cdot (D + \bar{A})] + (\quad)\}$$

The lower path consists of E in series with the parallel circuit of $\bar{B}$ and B:

$$F \cdot \{[A \cdot C \cdot (D + \bar{A})] + [(\bar{B} + B) \cdot E]\}$$
$$\underline{F[AC(D + \bar{A}) + (\bar{B} + B)E]}$$

To find a Boolean table for the expression in Example 10.7d would be a long task since there are six different switches and, therefore, $2^6 = 64$ different possible assignments of values. After discussing the properties of Boolean algebra in the next section, we will attempt to simplify expressions like this one to equivalent expressions involving fewer switches.

Problem Set 10.3

In Problems 1–20, draw a network to represent the Boolean expression.

1. $A+B+C+D$
2. $ABCD$
3. $(A+B)C$
4. $A(B+C)$
5. $AB+AC$
6. $BA+BC$
7. $A+BC$
8. $(A+B)(A+C)$
9. $(A+\bar{A})(B+C)$
10. $(A+A)(\bar{B}+\bar{C})$
11. $A+B(C+D)$
12. $A(B+CD)$
13. $AB(C+\bar{D})$
14. $ABC(D+E)$
15. $A(BC+D\bar{A})$
16. $(A+B)(AB+CD)$
17. $(A+B+C)(D+EF)$
18. $(\bar{A}B\bar{C})+D(A+\bar{E})$
19. $F[(A+CD)+(\bar{E}FG)]$
20. $AB\bar{C}(DAE+\bar{A}C)$

In Problems 21–30, write a Boolean expression for the network.

21. Figure 10.18

Figure 10.18

22. Figure 10.19

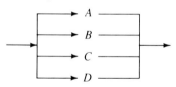

Figure 10.19

23. Figure 10.20

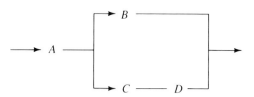

Figure 10.20

24. Figure 10.21

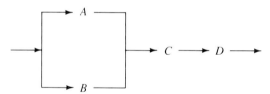

Figure 10.21

25. Figure 10.22

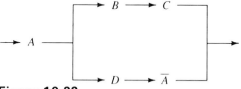

Figure 10.22

26. Figure 10.23

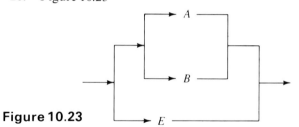

Figure 10.23

10.3 Combinations of Switches

27. Figure 10.24

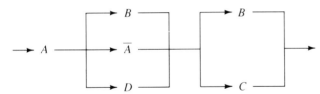

Figure 10.24

28. Figure 10.25

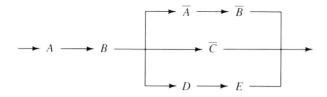

Figure 10.25

29. Figure 10.26

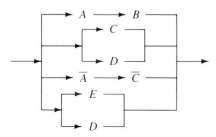

Figure 10.26

30. Figure 10.27

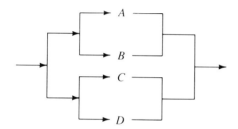

Figure 10.27

In Problems 31–38, draw a network using switches A, B, and C that will allow electricity to pass only if the given conditions hold.

31. At least two switches are closed.
32. At most two switches are closed.
33. Exactly one switch is closed.
34. All switches are closed.
35. No switches are closed. (*Hint:* If A is open, $\bar{A}$ is closed.)
36. Exactly two switches are closed.
37. At least one switch is closed.
38. At most one switch is closed.

In Problems 39–46, A and B are closed, and C and D are open. Determine whether electricity passes through the network. Draw the network with the switches in their designated positions.

39. $(A + B + C)D$
40. $AB + CD$
41. $A(D + \bar{A})$
42. $A + C\bar{D}(B + C)$
43. $A\bar{A}$
44. $A + \bar{A}$
45. $A(\bar{B} + C + D)$
46. $AB(\bar{C} + \bar{D})$

The city council of Dunbar consists of four people. They each cast their vote by pushing a button which closes a switch only if they vote yes. For Problems 47–50, design a network that will allow electricity to pass only if the given condition holds.

47. A majority vote yes.
48. All people vote yes.
49. A tie vote is recorded.
50. Nobody votes yes.

10.4 Properties of Networks

Consider a switch which is *always* closed in a network. Such a switch will always let electricity pass and will, therefore, always have a value of 1. From now on, the number 1 will represent such a switch. Similarly, a switch which is always open and never lets electricity pass will be represented by 0. In reality, the switches 0 and 1 would never be constructed, but they are useful in stating properties and simplifying Boolean expressions.

The switch 1 is analogous to a tautology from logic, which is always true. And switch 0 is analogous to a contradiction, which is always false.

With these ideas in mind, we list the basic properties of networks.

Commutative Properties

$$A + B = B + A \qquad AB = BA$$

Associative Properties

$$A + (B + C) = (A + B) + C$$
$$A(BC) = (AB)C$$

Identity Properties

$$A + 0 = A \qquad A \cdot 1 = A$$

Complement Properties

$$A + \bar{A} = 1 \qquad A \cdot \bar{A} = 0$$

Distributive Properties

$$A(B + C) = AB + AC$$
$$A + BC = (A + B)(A + C)$$

Idempotent Properties

$$A + A = A \qquad AA = A$$

DeMorgan Properties

$$\bar{A} + \bar{B} = \overline{AB} \qquad \bar{A}\bar{B} = \overline{A + B}$$

0 and 1 Properties

$$A \cdot 0 = 0 \qquad A + 1 = 1$$

10.4 Properties of Networks

Double Complement Property

$$\bar{\bar{A}} = A$$

The commutative, identity, complement, and distributive properties are part of the basic definition of a Boolean algebra and were discussed in Section 10.1. The other properties can be verified by finding Boolean tables and drawing circuits. The next example illustrates the process.

EXAMPLE 10.8 Construct Boolean tables for $A+(B+C)$ and $(A+B)+C$, and draw each circuit.

Solution

A	B	C	A+B	B+C	A+(B+C)	(A+B)+C
1	1	1	1	1	1	1
1	1	0	1	1	1	1
1	0	1	1	1	1	1
1	0	0	1	0	1	1
0	1	1	1	1	1	1
0	1	0	1	1	1	1
0	0	1	0	1	1	1
0	0	0	0	0	0	0

Notice that the last two columns are identical, verifying the associative property for addition. The circuits are shown in Figure 10.28.

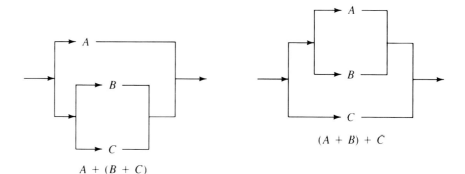

Figure 10.28

EXAMPLE 10.9 Construct a Boolean table for $A \cdot 0$, and draw the circuit.

Solution

A	0	A·0
1	0	0
0	0	0

So $A \cdot 0 = 0$. The circuit is shown in Figure 10.29. From Figure 10.29, we see that electricity can *never* pass through this circuit since 0 is never closed.

Figure 10.29

EXAMPLE 10.10 Verify the first DeMorgan property.

Solution

A	B	$A+B$	$\bar{A}$	$\bar{B}$	$\overline{A+B}$	$\bar{A}\cdot\bar{B}$
1	1	1	0	0	0	0
1	0	1	0	1	0	0
0	1	1	1	0	0	0
0	0	0	1	1	1	1

The last two columns are identical, verifying that $\overline{A+B} = \bar{A}\bar{B}$.

Problem Set 10.4

In Problems 1–16, name the property.

1. $(AB)C = A(BC)$
2. $\overline{AB} = \bar{A} + \bar{B}$
3. $\overline{A+B} = \bar{A}\bar{B}$
4. $(A+B)+C = A+(B+C)$
5. $A(B+C) = AB+AC$
6. $A+0 = A$
7. $A \cdot 1 = A$
8. $A+BC = (A+B)(A+C)$
9. $A+A = A$
10. $AB = BA$
11. $A+B = B+A$
12. $AA = A$
13. $A\bar{A} = 0$
14. $A \cdot 0 = 0$
15. $A+1 = 1$
16. $A+\bar{A} = 1$

In Problems 17–22, construct Boolean tables to verify the property.

17. $A(BC) = (AB)C$
18. $A+A = A$
19. $AA = A$
20. $\overline{AB} = \bar{A}+\bar{B}$
21. $A \cdot 0 = 0$
22. $A+1 = 1$

In Problems 23–36, draw the circuits.

23. $A+B$ and $B+A$
24. AB and BA
25. $A(BC)$ and $(AB)C$
26. $A+0$
27. $A \cdot 1$
28. $A\bar{A}$
29. $A+\bar{A}$
30. $A(B+C)$ and $AB+AC$
31. $A+BC$ and $(A+B)(A+C)$
32. $A+A$
33. AA
34. $\overline{AB}$
35. $\bar{A}+\bar{B}$
36. $A \cdot 1$

37. Compare the two circuits from Problem 31. Which one do you think would be cheaper to produce? Why?
38. Compare the two circuits from Problem 30. Which one do you think would be cheaper to produce? Why?
39. Draw a circuit to represent $\overline{A+B}$ (NOR), using only complements of individual switches.
40. Draw a circuit to represent $\overline{AB}$ (NAND), using only complements of individual switches.
41. Write a Boolean table for $\bar{\bar{A}}$.
42. Write a Boolean table for $\bar{\bar{\bar{A}}}$.
43. What do you think the expression $A+A+A$ can be simplified to?
44. What do you think the expression AAA can be simplified to?
45. If one of the terms in a sum is 1, then what can the sum be simplified to?
46. If one of the factors is 0, then what can the product be simplified to?

10.5 Simplification of Networks

Consider the two networks shown in Figure 10.30 whose Boolean expressions are $A + BC$ and $(A + B)(A + C)$. Both of these networks give the same results (the second distributive property), but $A + BC$ consists of three switches while $(A + B)(A + C)$ consists of four switches. In the design of networks, the fewer switches the better, because fewer switches reduce the complexity and cost of constructing networks. So $A + BC$ is the preferred network here.

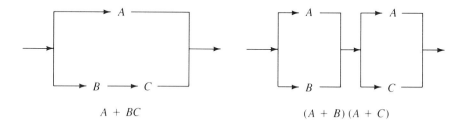

Figure 10.30 $A + BC$ $(A + B)(A + C)$

simplify networks

In this section, we will **simplify networks**. In other words, we will find an equivalent Boolean expression using fewer switches than the original expression. To do so, we must use the properties of Boolean algebra from the previous section. The following example illustrates the procedure.

EXAMPLE 10.11 Simplify the expression $A + AB$.

Solution

$$
\begin{aligned}
A + AB &= A \cdot 1 + AB &&\text{identity property for } \cdot \\
&= A(1 + B) &&\text{first distributive property} \\
&= A \cdot 1 &&\text{property of 1} \\
&= \underline{A} &&\text{identity property for } \cdot
\end{aligned}
$$

We went from an expression involving three switches to an expression involving one switch. Boolean tables will verify that the two expressions give identical results.

How did we know what to do in the first step of Example 10.11? We could have used the second distributive property to get $(A + A)(A + B)$. Simplifying $A + A$ to A would then give us $A(A + B)$. Using the first distributive property at this point would yield $AA + AB$, which simplifies to $A + AB$, which was the original problem.

There are no hard and fast rules for simplifying expressions, but there are some general guidelines. If you can use one of the properties to introduce a 0 or a 1 into the expression, that step will usually simplify the expression. For this reason, the first distributive property was used in the first step of Example 10.11.

Another example should make the procedure clear.

■ **EXAMPLE 10.12** Simplify the expression $A + \bar{A}B$.

Solution

$A + \bar{A}B = (A + \bar{A})(A + B)$	second distributive property
$= 1 \cdot (A + B)$	complement property for $+$
$= \underline{A + B}$	identity property for $\cdot$

Usually, if an expression consists of a switch and its complement, as in Example 10.12, you should try to get them together to simplify the expression. The next example uses this technique, too.

■ **EXAMPLE 10.13** Simplify $(A + B) + \bar{A}$.

Solution

$(A + B) + \bar{A} = (B + A) + \bar{A}$	commutative property for $+$
$= B + (A + \bar{A})$	associative property for $+$
$= B + 1$	complement property for $+$
$= \underline{1}$	property of 1

There is no right or wrong method for simplifying Boolean expressions as long as each step uses one of the Boolean properties. The more simplifications you do, the better you will become at determining which properties will help the simplification. So we will do a few more examples.

■ **EXAMPLE 10.14** Simplify $\bar{A}(A + B)$.

Solution

$\bar{A}(A + B) = \bar{A}A + \bar{A}B$	first distributive property
$= 0 + \bar{A}B$	complement property for $\cdot$
$= \underline{\bar{A}B}$	identity property for $+$

When an expression has been reduced so that each letter appears only once and it contains no 0s or 1s, it is usually reduced as far as possible.

■ **EXAMPLE 10.15** Simplify $\bar{A}(B + C) + AB$.

Solution There is nothing we can do in this situation except use the first distributive property:

$$\bar{A}(B + C) + AB = \bar{A}B + \bar{A}C + AB$$

We would like to get A and $\bar{A}$ together. We can do so by grouping $\bar{A}B$ and AB together and factoring out the B:

10.5 Simplification of Networks

$$\bar{A}B + \bar{A}C + AB = \bar{A}B + AB + \bar{A}C$$
$$= (B\bar{A} + BA) + \bar{A}C$$
$$= B(\bar{A} + A) + \bar{A}C$$
$$= B \cdot 1 + \bar{A}C$$
$$= \underline{B + \bar{A}C}$$

■ EXAMPLE 10.16   Simplify $ABC + A\bar{B}C + \bar{A}\bar{B}C$ (9 switches).

Solution
$$ABC + A\bar{B}C + \bar{A}\bar{B}C = A(BC + \bar{B}C) + \bar{A}\bar{B}C$$
$$= A(CB + C\bar{B}) + \bar{A}\bar{B}C$$
$$= AC(B + \bar{B}) + \bar{A}\bar{B}C$$
$$= AC \cdot 1 + \bar{A}\bar{B}C$$
$$= AC + \bar{A}\bar{B}C$$
$$= CA + C\bar{A}\bar{B}$$
$$= C(A + \bar{A}\bar{B})$$
$$= C(A + \bar{A})(A + \bar{B})$$
$$= C \cdot 1 \cdot (A + \bar{B})$$
$$= \underline{C(A + \bar{B})} \quad \text{(3 switches)}$$

Can you follow and justify each step of the procedure in Example 10.16? The goal is to try to get B and $\bar{B}$ together and then A and $\bar{A}$ together. The complement properties are very important in simplifying networks.

Problem Set 10.5

In Problems 1–12, write the property used in each step of the simplification.

1. $C(A+B) + \bar{A}(\bar{B}+C) + \bar{C}$
    $= CA + CB + \bar{A}(\bar{B}+C) + \bar{C}$
2. $= CA + CB + \bar{A}\bar{B} + \bar{A}C + \bar{C}$
3. $= CA + CB + AB + C\bar{A} + \bar{C}$
4. $= CA + CB + C\bar{A} + \bar{A}\bar{B} + \bar{C}$
5. $= CA + C\bar{A} + CB + \bar{A}\bar{B} + \bar{C}$
6. $= C(A + \bar{A}) + CB + \bar{A}\bar{B} + \bar{C}$
7. $= C \cdot 1 + CB + \bar{A}\bar{B} + \bar{C}$
8. $= C + CB + \bar{A}\bar{B} + \bar{C}$
9. $= C + CB + \bar{C} + \bar{A}\bar{B}$
10. $= C + \bar{C} + CB + \bar{A}\bar{B}$
11. $= 1 + CB + \bar{A}\bar{B}$
12. $= 1$

In Problems 13–38, simplify the expression.

13. $(AA)A$
14. $(A+A) + A$
15. $A + \bar{A}$
16. $A\bar{A}$
17. $(A + \bar{B}) + B$
18. $A(C\bar{A})$
19. $A \cdot B \cdot 0$
20. $A + 1 + B$
21. $A + 1 + B + 0$
22. $A \cdot 1 \cdot B \cdot 0$
23. $BA + BA$
24. $AB + AB$
25. $A + AB + C$
26. $AB + BC$
27. $(A + B)(A + C)$
28. $AB + AC$
29. $AB + A\bar{B}$
30. $A(B + C) + A\bar{B}$
31. $ABCD + AD$
32. $(ABC)(ABC)$
33. $A + B + C + \bar{A}\bar{C}$
34. $A + \bar{A}B + \bar{B}$
35. $ABC + \bar{A}B\bar{C} + BC$
36. $BC + B(\bar{C} + \bar{B})$

37. $(A + BC)(B + AC)$
38. $(A + B)(A + C)(A + B)$
39. Find a Boolean expression for the Exclusive OR operation and simplify it as far as possible.
40. Find a Boolean expression for iff and simplify it as far as possible.
41. Can you write a Boolean expression involving A, B, and C and no 0s or 1s that contains each letter exactly once and can be simplified to fewer letters? If so, write such an expression.
42. Can you write a Boolean expression involving A, B, C, and 1 that cannot be simplified such that the 1 is eliminated? If so, write such an expression.
43. Can you write a Boolean expression involving A, B, C, and 0 that cannot be simplified such that the 0 is eliminated? If so, write such an expression.
44. Write a Boolean expression involving both 0 and 1 that simplifies to 1.
45. Write a Boolean expression involving both 0 and 1 that simplifies to 0.
46. Write a Boolean expression involving both 0 and 1 that simplifies to something other than 0 or 1.

10.6 Logic Circuits

logic circuits
logic gates

We have studied switching networks up to this point. We will now consider **logic circuits**, circuits made up of logic gates. You can think of a **logic gate** as a device which accepts input (0s and/or 1s) from one or more locations and produces output (a 0 or a 1) at one location. Logic gates can be combined to form logic circuits.

We will first consider the three basic logic gates, which correspond to the three basic operations of logic. Then we will combine them to form logic circuits.

The AND Gate

The AND gate, which is shown in Figure 10.31, receives input from A and B and produces output at C by the following rule:

C will be 1 only if both A and B are 1.
Otherwise, C will be 0.

Figure 10.31

This rule is simply the *multiplication* operation of Boolean algebra ($A \cdot B$).

The OR Gate

The OR gate, which is shown in Figure 10.32, produces a 1 at C if at least one of A and B is a 1. This rule is the *addition* operation of Boolean algebra ($A + B$).

10.6 Logic Circuits

Figure 10.32

The NOT Gate

inverter

The NOT gate, also called an **inverter**, is shown in Figure 10.33. It produces a 1 at C if A is a 0 and a 0 at C if A is a 1. This rule is the *complement* operation of Boolean algebra.

Figure 10.33

A logic gate can only process one set of data at a time, so if a series of bits is to be processed, it is processed one bit at a time.

EXAMPLE 10.17 Find the output produced by the specified logic gate.

Figure 10.34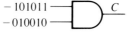

a. The gate of Figure 10.34

Solution The first pair of values to be processed is $^1_0$, so the result is 0. Continuing the procedure, $^1_1$ yields 1, $^0_0$ yields 0, and so on, giving a final output of 000010.

Figure 10.35

b. The gate of Figure 10.35

Solution Only $^0_0$ results in 0, so the output is 111011.

Figure 10.36

c. The gate of Figure 10.36.

Solution This gate yields the 1's complement: 010100.

Circuits

Logic gates can be combined to form more complicated logic circuits corresponding to more complicated Boolean expressions. We will look at some examples to investigate the process.

EXAMPLE 10.18 Write a logic circuit to represent the NAND operation (a NAND gate).

Solution The NAND operation is $\overline{AB}$. The complement is taken *after* the multiplication. The NAND gate is drawn in Figure 10.37.

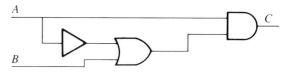

Figure 10.37

EXAMPLE 10.19 Write a logic circuit for $A(\overline{A} + B)$.

Solution This circuit is basically a multiplication problem (AND), whose first factor is A. See Figure 10.38.

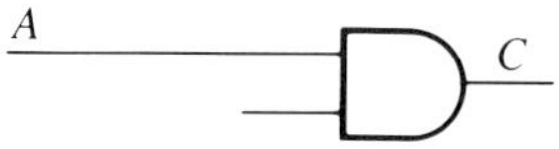

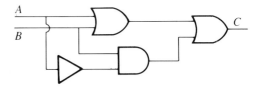

Figure 10.38

The second factor is the complement (NOT) of A with B. The circuit is drawn in Figure 10.39.

Figure 10.39

EXAMPLE 10.20 Find a Boolean expression for the logic circuit shown in Figure 10.40.

Figure 10.40

Solution The last operation is OR:

$$(\quad) + (\quad)$$

The top path leading to this OR is A OR B:

$$(A + B) + (\quad)$$

The lower path leading to this OR is B AND (NOT A):

$$(A + B) + (B \cdot \overline{A})$$

10.6 Logic Circuits

half-adder

We now consider a special logic circuit called a **half-adder**, which will accept two bits and return the sum of those bits in the form of a "sum" bit and a "carry" bit. The half-adder circuit is shown in Figure 10.41.

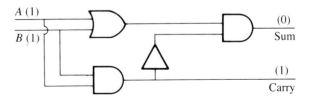

Figure 10.41

Figure 10.41 indicates that if A and B are both 1, the sum is 0 and the carry is 1, corresponding to $1_2 + 1_2 = 10_2$. When 1_2 and 1_2 are added, the result is 10_2, with the 1_2 being carried into the next column. Similarly, if A is 1 and B is 0, the result $(1_2 + 0_2)$ is 1_2, with 0_2 carried over to the next column.

Logic circuits similar to the one in Figure 10.41 are combined in computers to perform more complicated operations than simply the sum of two bits. Recall that addition is the basic operation of computers. The problem set will further explore this circuit.

Problem Set 10.6

In Problems 1–6, find the output from the pairs of numbers passing through an AND gate.

1. 1001
   1101
2. 00110
   11001
3. 1111
   0000
4. 111000
   010101
5. 11001001
   10000111
6. 10010111
   01101110

In Problems 7–12, find the output from the pairs of numbers passing through an OR gate.

7. 1001
   1101
8. 00110
   11001
9. 1111
   0000
10. 111000
    010101
11. 11001001
    10000111
12. 10010111
    01101110

In Problems 13–18, find the output from the numbers passing through a NOT gate.

13. 1011
14. 1001
15. 100011
16. 10111
17. 11011101
18. 1001011

In Problems 19–30, write a logic circuit to represent the expression.

19. The NOR operation
20. The Exclusive OR operation
21. The conditional
22. The biconditional
23. $\bar{A}B$
24. $A + \bar{B}$
25. $\bar{A}(A + B)$
26. $A + \bar{A}B$
27. $(A + B)(\bar{A} + \bar{B})$
28. $\bar{A}\bar{B} + AB$
29. $A + B + C$
30. ABC

In Problems 31–34, find Boolean expressions for the logic circuit.

31. Figure 10.42

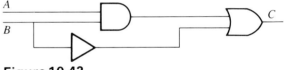

Figure 10.42

268 BOOLEAN ALGEBRA

32. Figure 10.43

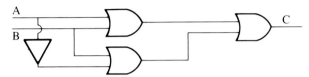

Figure 10.43

33. Figure 10.44

Figure 10.44

34. Figure 10.45

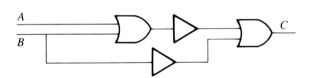

Figure 10.45

In Problems 35–36, find the output from the half-adder circuit for the given conditions.

35. $A = 0, B = 1$
36. $A = 0, B = 0$

Consider the circuit of Figure 10.46. In Problems 37–44, find the output from this logic circuit under the given conditions.

37. $A = 0, B = 1, C = 1$
38. $A = 0, B = 0, C = 0$
39. $A = 1, B = 1, C = 1$
40. $A = 0, B = 0, C = 1$
41. $A = 1, B = 0, C = 0$
42. $A = 1, B = 1, C = 0$
43. $A = 1, B = 0, C = 1$
44. $A = 0, B = 1, C = 0$

45. Find the Boolean expression for the logic circuit of Figure 10.46.
46. What does the logic circuit of Figure 10.46 do?
47. Change the half-adder circuit so that the sum bit is correct but the carry bit is always wrong.
48. Change the half-adder circuit so that the sum bit is always wrong but the carry bit is correct.
49. Write the expression from the half-adder circuit that yields the sum bit.
50. Write the expression from the half-adder circuit that yields the carry bit.

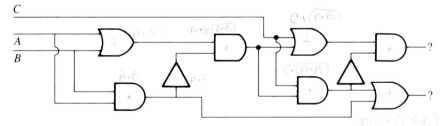

Figure 10.46

CHAPTER SUMMARY

In this chapter, we studied the basics of Boolean algebra and how it relates to circuits in a computer. A Boolean algebra is a set consisting of two elements (0 and 1), corresponding to a switch being open or closed, and two operations ($+$ and $\cdot$), corresponding to parallel or series circuits.

The properties of a Boolean algebra include the following:

Commutative properties	$A + B = B + A$, $AB = BA$
Associative properties	$A + (B + C) = (A + B) + C$
	$A(BC) = (AB)C$
Identity properties	$A + 0 = A$, $A \cdot 1 = A$
Complement properties	$A + \bar{A} = 1$, $A\bar{A} = 0$
Distributive properties	$A(B + C) = AB + AC$
	$A + BC = (A + B)(A + C)$
Idempotent properties	$A + A = A$, $AA = A$
DeMorgan properties	$\overline{A + B} = \bar{A}\bar{B}$, $\overline{AB} = \bar{A} + \bar{B}$
0 and 1 properties	$A \cdot 0 = 0$, $A + 1 = 1$
Double complement property	$\bar{\bar{A}} = A$

These properties were used to simplify Boolean expressions and the corresponding networks. Finally, we considered logic gates and networks and how they can be used inside computers to perform simple arithmetic.

REVIEW PROBLEMS

In Problems 1–4, find a Boolean table for the expression.

1. $A(B + \bar{A})$
2. $AB + A\bar{B}$
3. $AB + B\bar{C}$
4. $A(\bar{B} + \bar{C})B$

For Problems 5–8, use the circuit of Figure 10.47. Determine whether electricity will pass through the circuit under the given conditions.

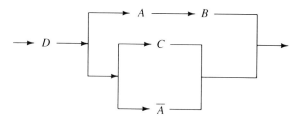

Figure 10.47

5. A, B, and C open and D closed
6. A, C closed and B, D open
7. A, B, C, and D closed
8. A open and B, C, and D closed

In Problems 9–12, draw a network to represent the Boolean expression.

9. $A(B + C) + D$
10. $ABC + D + ED$
11. $ABC(D + \bar{A}F)$
12. $A + BC + D\bar{E}\bar{A}$

In Problems 13–16, write a Boolean expression for the network.

13. Figure 10.48

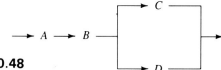

Figure 10.48

14. Figure 10.49

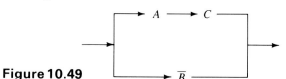

Figure 10.49

15. Figure 10.50

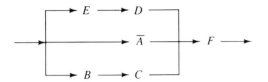

Figure 10.50

16. Figure 10.51

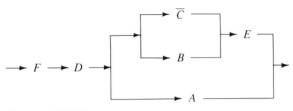

Figure 10.51

17. Draw a network involving A, B, C, and D that will allow electricity to pass only if A and B are closed and C and D are open.
18. Draw a network involving A, B, C, and D that will *not* let electricity pass when A, B, C, and D are all closed.

In Problems 19–22, name the property.

19. $\overline{A+B} = \overline{A}\overline{B}$
20. $A + BC = (A+B)(A+C)$
21. $A(BC) = (AB)C$
22. $A + \overline{A} = 1$

In Problems 23–28, simplify the expression.

23. $A(\overline{A} + B + C)$
24. $A + ABC$
25. $ABC + AC + B\overline{C}$
26. $A(BA + C + B)$
27. $AB + A(B + C)$
28. $ABC + BCA + ACB$

In Problems 29–34, write a logic circuit to represent the expression.

29. $A + \overline{B}$
30. $A(B + \overline{C})$
31. $A + AB + B$
32. $AB + B\overline{A}$
33. $A\overline{B}C + B$
34. $AB + \overline{C}(A + B)$

In Problems 35–38, find a Boolean expression for the logic circuit.

35. Figure 10.52

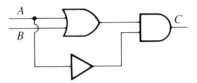

Figure 10.52

36. Figure 10.53

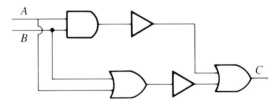

Figure 10.53

37. Figure 10.54

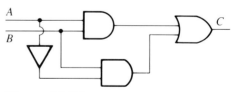

Figure 10.54

38. Figure 10.55

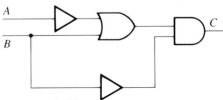

Figure 10.55

Review Problems

39. Write a logic circuit that will take two bits, A and B, and return the product (AB).

40. Write a logic circuit that will take two bits, A and B, and return the complement of the product ($\overline{AB}$).

COMPUTER LOGIC AND STRUCTURED PROGRAMMING

CHAPTER **11**

The purpose of this chapter is to introduce you to the logical thinking necessary to solve problems and to design and write programs that a computer can understand. We will discuss algorithms, which are step-by-step lists of how a problem can be solved, and flowcharts, which are graphical representations of algorithms.

We will cover the basic structures that are used in all programming languages and give numerous examples involving algorithms and flowcharts of these structures. While knowledge of a programming language is not necessary to understand this chapter, students of BASIC should be able to change the flowcharts into BASIC programs. A knowledge of this chapter will put you on the road toward becoming an organized, logical programmer.

▬ 11.1 Algorithms

algorithm

An **algorithm** is simply an ordered list of instructions to solve a problem. For example, an algorithm for starting a car might look like this:

1. Get into car.
2. Put on seat belt.
3. Lock door.
4. Make sure car is in park.
5. Put key in ignition.
6. Turn key.
7. Put car in drive or reverse.
8. Drive.

You are probably more familiar with actually doing things than with describing how to do something. Programmers don't solve problems. They write programs, which are merely lists of instructions for the computer to solve problems. The first step in writing a program is to design an algorithm to solve the problem.

If you had mistakenly left out step 5 in the previous algorithm, any person following the algorithm to start her or his car would probably have figured out that this step should be there. However, a computer doesn't think. It merely follows the instructions step by step. If a computer were following the algorithm and step 5 was missing, the car would not start.

Computers do not tolerate mistakes. We must be precise in our instructions and think of every possible step that might occur in our problem.

11.1 Algorithms

Let's examine a mathematical example.

EXAMPLE 11.1 Write an algorithm to take three numbers and find their average.

Solution
1. Get the three numbers.
2. Find the sum of the numbers.
3. Divide the sum by 3.
4. Print the results.

inputting

outputting

The computer does not *know* the values of numbers until you tell the computer what the values are. That is why step 1 is necessary. This step involves the process of **inputting** values into the program. Steps 2 and 3 can easily be combined into one step, if desired. Step 4 is needed because the computer does not automatically print out the results of a program. You must tell the computer what values you want printed out. This step involves the process of **outputting** values from the program. Without step 4, the computer will still perform the calculations, but it won't reveal the results.

EXAMPLE 11.2 Write an algorithm to take the length and width of a rectangle and find the area and perimeter of the rectangle.

Solution
1. Input the length and width of the rectangle.
2. Find the area of the rectangle.
3. Find the perimeter of the rectangle.
4. Print the area and perimeter of the rectangle.

Ideally, each step of the algorithm should correspond to one command to the computer. In the BASIC programming language, input can be achieved by use of an INPUT statement, and output is produced by a PRINT statement. Since you usually don't know what the values are when the program is written (or even if you do), make the program general and use variables to represent the values.

As we have done in previous chapters, we will usually use single letters to represent numeric variables. If more than one variable is to be inputted or printed, the variables will be separated by commas. Any calculations that are performed by a computer are usually performed in an assignment statement. Thus, a BASIC program for the algorithm in Example 11.2 may look like this:

```
10   INPUT L,W
20   A = L * W
30   P = (2 * L) + (2 * W)
40   PRINT A,P
```

Note that BASIC statements are ordered by statement numbers preceding each statement.

EXAMPLE 11.3 Write a BASIC program to carry out the algorithm in Example 11.1.

Solution
```
10   INPUT X,Y,Z
20   S = X+Y+Z
30   A = S/3
40   PRINT A
```

EXAMPLE 11.4 Write an algorithm and BASIC program to input the number of quarters, dimes, nickels, and pennies in a purse and to find the total amount.

Solution
1. Input number of quarters, number of dimes, number of nickels, and number of pennies.
2. Calculate total amount.
3. Output total amount.

```
10   INPUT Q,D,N,P
20   T = (25*Q)+(10*D)+(5*N)+P
30   PRINT T
```

In calculating T, we must multiply the number of each type of coin by its value. So 3 quarters yields 25 * 3 cents, and so on.

Any time you follow a recipe or a list of instructions to put something together, you are using an algorithm. If one step is missing, it could have a dramatic effect on the final result.

Problem Set 11.1

In Problems 1–14, write an algorithm to carry out the procedure.

1. Stop at a gas station and put gas in your car.
2. Write a letter and mail it.
3. Take the three coefficients a, b, and c from a quadratic equation and find the two roots by using the quadratic formula (assume that there are two real roots).
4. Take the coordinates of two points and find the distance between them.
5. Input the number of feet and inches and convert to total inches.
6. Input the number of inches and convert to feet.
7. Input the number of field goals (2 points) and free throws (1 point) that a basketball team made and find how many points they scored.
8. Input the number of touchdowns (6 points), extra points (1 point), field goals (3 points), and safeties (2 points) scored by a football team and find how many points it scored.
9. Take the base and height of a triangle and find the area of the triangle (area = one-half the base times the height).
10. Take a Fahrenheit temperature and change it to the corresponding Celsius temperature $[C = (\frac{5}{9})(F-32)]$.

11.2 Flowcharts

11. Take a binary number and convert it to a decimal number.
12. Take a decimal integer and convert it to a binary number.
13. Take an octal number and convert it to a binary number.
14. Take a hexadecimal number and convert it to an octal number.

In Problems 15–22, write a BASIC program for the algorithm in the indicated problem.

15. Problem 3
16. Problem 4
17. Problem 5
18. Problem 6
19. Problem 7
20. Problem 8
21. Problem 9
22. Problem 10

In Problems 23–26, determine what is wrong (if anything) with the algorithm.

23. (1) Find the sum of A and B.
    (2) Input A,B.
    (3) Output the sum.
24. (1) Input A, B, and C.
    (2) Find $A + B + C$.
    (3) Find A squared.
25. (1) Input number of centimeters.
    (2) Output number of inches.
26. (1) Get the length of a side of a square.
    (2) Find the area of the square.
    (3) Find the perimeter of the square.
    (4) Output the area and perimeter of the square.

In Problems 27–30, write a BASIC program for the corrected algorithm of the indicated problem.

27. Problem 23
28. Problem 24
29. Problem 25
30. Problem 26

11.2 Flowcharts

flowcharts

Algorithms are step-by-step procedures for solving problems. **Flowcharts** are graphical representations of algorithms. We will use graphic symbols to represent different types of steps in algorithms. The first three flowchart symbols that we will use are shown in Figure 11.1. These three symbols can be used to construct flowcharts similar to the algorithms of the previous section.

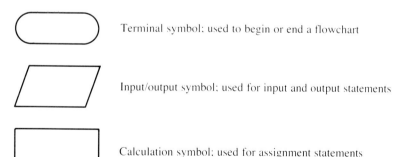

Figure 11.1

The algorithms of the previous section involved inputting some information, performing some calculations, and outputting the results. These steps are an example of a sequence structure, which is the first of three important structures that we will consider in this chapter.

278 COMPUTER LOGIC AND STRUCTURED PROGRAMMING

sequence structure

A **sequence structure** for a program proceeds from top to bottom, step by step, with the statements performed in the order that they are written. There are no steps that instruct the computer to skip over some statements, to make a decision about which statement to perform next, or to repeat some statements. These other types of structures will be discussed in the next two sections.

Now, let's look at some examples of flowcharts and the sequence structure.

EXAMPLE 11.5 Construct a flowchart for a procedure to input the price of an item and find the new price if a 10% discount is offered and the tax rate is 5%. Tax is based on the actual selling price, not on the original list price.

Solution The flowchart is illustrated in Figure 11.2.

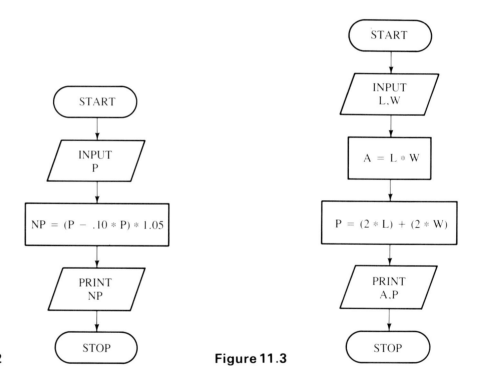

Figure 11.2

Figure 11.3

EXAMPLE 11.6 Write a flowchart for a procedure to take the length and width of a rectangle and find and print the area and perimeter of the rectangle (see Example 11.2).

Solution The flowchart appears in Figure 11.3.

EXAMPLE 11.7 Write a flowchart for a procedure to input the number of quarters, dimes, nickels, and pennies in a purse and find the total amount (see Example 11.4).

Solution The flowchart appears in Figure 11.4.

11.2 Flowcharts

EXAMPLE 11.8 Write a flowchart for a procedure to read in a person's age and bank balance. Add 1 to the person's age, calculate her interest (5.5% of bank balance) and new bank balance, and print all the information.

Solution The flowchart appears in Figure 11.5.

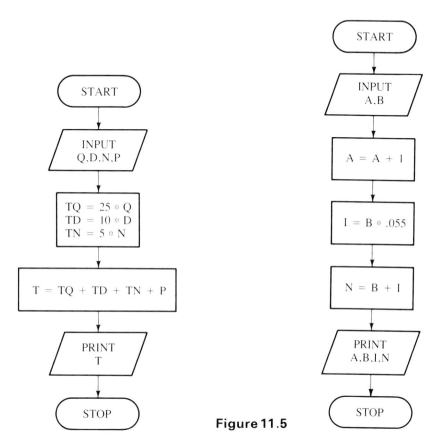

Figure 11.4

Figure 11.5

Problem Set 11.2

1. Write an algorithm for the flowchart in Example 11.5.
2. Write an algorithm for the flowchart in Example 11.8.
3. Write a BASIC program for the flowchart in Example 11.5.
4. Write a BASIC program for the flowchart in Example 11.8.

In Problems 5–10, state which flowchart symbol is used to represent the step in an algorithm.

5. Start an algorithm.
6. Print out two test scores.
7. Input a person's name.
8. End an algorithm.
9. Calculate an average.
10. Add 1 to a person's age.

In Problems 11–16, write a flowchart for the procedure.

11. Input two numbers, and find and print their sum and difference.
12. Input two numbers, find their product, and print both numbers and the product.
13. Input the coefficients (a, b, and c) of a quadratic function, and find and print the two values of x given by the quadratic formula. Assume that division by zero will not occur.
14. Input the coefficients of two points (x_1, y_1) and (x_2, y_2), and find and print the distance between the points and the slope of the line between the points.
15. Input the digits of a four-digit binary number (for example, if the number is 1101, then 1, 1, 0, and 1 are inputted), and convert the number to a decimal number. Print out the decimal number.
16. Input a distance in terms of yards, feet, and inches (for example, 4 yards, 2 feet, and 7 inches), and convert the distance to total inches. Print out the distance in total inches.

In Problems 17–22, write an algorithm for the flowchart in the indicated problem.

17. Problem 11   18. Problem 12
19. Problem 13   20. Problem 14
21. Problem 15   22. Problem 16

In Problems 23–28, write a BASIC program for the flowchart in the indicated problem.

23. Problem 11   24. Problem 12
25. Problem 13   26. Problem 14
27. Problem 15   28. Problem 16

29. In all of the examples that we have looked at so far, the INPUT statements come before the calculations. Why must they?
30. In all of the examples that we have looked at so far, the assignment statements come before the PRINT statements. Why must they?

11.3 The Decision Structure

logical decisions

As mentioned in the previous section, computers can do more than perform mathematical calculations. They can also make **logical decisions**, decisions which have exactly two possible outcomes: true and false. These decisions are represented by a diamond symbol and are usually written by using an IF statement in programming.

For example, suppose that a student will make the dean's list if his grade point average (GPA) is greater than 3.5. Otherwise, he won't make the dean's list. The decision GPA > 3.5 is either true or false. Figure 11.6 illustrates the flowchart for this **decision structure**. Note that there are two possible paths. A good programming practice is to bring the two paths together as soon as possible. A small circle is used as a **connector symbol** when two branches of a flowchart come together.

decision structure

connector symbol

There are no standard positions for the true and false paths in a flowchart, but we will always put the true path on the right side, for consistency.

Only one branch of a decision will actually be performed in a program. When writing an algorithm, flowchart, or computer program, the programmer does not usually know which branch will be performed; so each branch must be completely thought out.

11.3 The Decision Structure

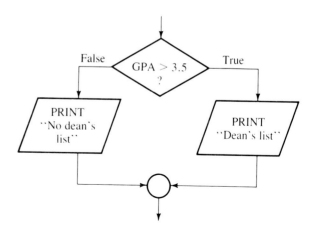

Figure 11.6

EXAMPLE 11.9 Write an algorithm to carry out the decision of Figure 11.6.

Solution To parallel what actual program statements look like, we will use an IF in our decision algorithms, followed by the decision, and THEN and ELSE branches indicating what happens when the decision is true (THEN) and false (ELSE). The step will look like this:

1. IF GPA > 3.5 THEN Print "Dean's list"
   ELSE Print "No dean's list"

When decision algorithms are written as in Example 11.9, they closely resemble the actual programming statements. In fact, the step in Example 11.9 *is* the BASIC statement for this decision. The closer that algorithms match actual statements, the easier it is to change algorithms to programs. Thus, many programmers are forgoing flowcharts in favor of **pseudocode**, which consists of algorithms written in a form which roughly corresponds to actual programming statements.

pseudocode

Many times in programming (and in real life, for that matter), a decision consists of an "empty" branch. No action will be taken unless a certain condition is met. The next example illustrates this situation.

EXAMPLE 11.10 Write a flowchart for a decision to add 6% sales tax to the price of a mail-order item only if the customer is from Michigan.

Solution The flowchart is shown in Figure 11.7. Since there is nothing to do if the decision is false, there are no statements on the false branch.

EXAMPLE 11.11 Write an algorithm for the decision given in Example 11.10.

Solution
1. IF Customer is from Michigan
   THEN P = P + (.06 * P)

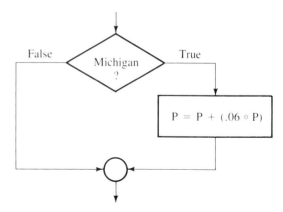

Figure 11.7

Since there are no statements to be performed if the customer is *not* from Michigan, there is no need for an ELSE.

A logical decision will always consist of exactly two paths. If a decision involving more than two possible outcomes is required, it must be broken into several logical decisions. Note, however, that FORTRAN has an *arithmetic IF* which consists of *three* paths, depending on whether the variable in question is negative, zero, or positive. Inside the computer, this decision must be broken down into a series of logical decisions, though. Consider Example 11.12.

■ **EXAMPLE 11.12** Write an algorithm and a flowchart to input a real number and assign −1, 0, or 1 to another variable, depending on whether the real number is negative, zero, or positive.

Solution In this case, we must choose between *three* possible outcomes. This choice cannot be made by using one logical decision, so we must use at least two logical decisions. If the number is not positive, then it is either zero or negative. In other words, there are two possible outcomes if the number is not positive. These outcomes can be dealt with by using another IF. The algorithm looks like this:

1. INPUT N
2. IF N>0 THEN A = 1
       ELSE IF N = 0 THEN A = 0
           ELSE A = −1

There is no need for a third IF statement. The only way the program will ever get to the ELSE of the second IF is if N is not positive and not zero. In that case, N must be negative. The flowchart for this algorithm appears in Figure 11.8.

Another way to write the algorithm for this problem is to use three separate IF statements (see Problem 1):

1. INPUT N
2. IF N>0 THEN A = 1

11.3 The Decision Structure

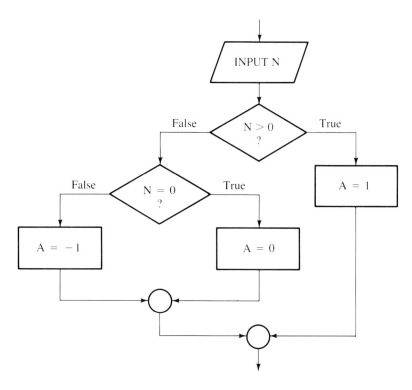

Figure 11.8

3. IF N = 0 THEN A = 0
4. IF N < 0 THEN A = −1

If more then three possible outcomes exist, then more combinations of logical decisions are needed.

compound decisions

We use Boolean connectives to make **compound decisions** similar to compound statements in logic. For example, if you must decide whether a number is between 80 and 90, in mathematics you can simply write $80 < X < 90$. But this statement is not acceptable in programming. Each simple logical decision in programming must consist of only *one* comparison. So we have to write a compound decision like this: IF X > 80 AND X < 90. The Boolean connectives AND and OR mean the same thing here as they did in Boolean algebra. The Exclusive OR, NOR, and NAND connectives are also used in programming, but not as frequently as AND and OR.

■ EXAMPLE 11.13 Write compound logical decisions for the problem.

a. Add $1000 to a student's bank account if the student has a GPA above 3.4 and more than $10,000 in the bank.

Solution   IF GPA > 3.4 AND A > 10000
THEN A = A + 1000

b. Print "Accepted" if a person is in her twenties or is single.

Solution We need a variable S which is set to 1 if she is single and 0 otherwise:

IF (A > 20 AND A < 30) OR (S = 1)
THEN PRINT "Accepted"

In this example, we need the parentheses to be sure that the AND is performed first.

Perhaps just as important as being able to write algorithms and flowcharts is being able to follow someone else's algorithm and/or flowchart and discover what it does. Consider the next example.

■ EXAMPLE 11.14 Follow the flowchart in Figure 11.9, and find the output if 7 and 9 are the values inputted for X and Y, respectively.

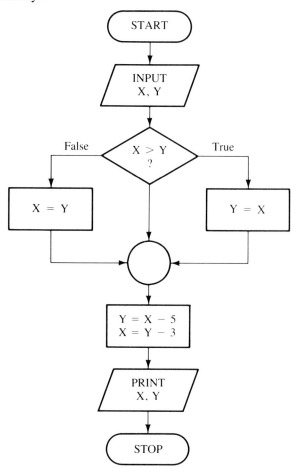

Figure 11.9

Solution We can make a list of each variable and the output and follow each step of the flowchart. You should verify that the following results are obtained:

X	Y	Output
7	9	
9	4(9 − 5)	
6		6 and 4

Problem Set 11.3

1. Write a flowchart for the second algorithm in Example 11.12.
2. Write a flowchart for the procedure in Example 11.13a.
3. Write an algorithm for the flowchart in Example 11.14.
4. Verify the results of Example 11.14.

In Problems 5–16, write an algorithm for the procedure.

5. Input a person's age and salary. If the age equals 65, then print the age.
6. Input a number and print out the number only if it is larger than 10.
7. Input a number and print out the number only if it is smaller than 0.
8. Input the amount of rainfall last month. If the rainfall was larger than 7 inches, print "Wet". Otherwise, print "Dry".
9. Read in three numbers and find their sum. Print "Positive", "Negative", or "Zero", depending on whether or not the sum is positive, negative, or zero.
10. Read in three numbers and print out the smallest number.
11. Read in three numbers and print out the largest number.
12. Input a whole number between 0 and 6 and decide whether it is an even number (2 or 4) or an odd number (1, 3, or 5).
13. Input a person's age and salary. If the age is over 32, then add 1000 to the salary. Otherwise, subtract 500 from the salary. Finally, print out the age and new salary.
14. Input the number of hours that an employee worked last week and his hourly wage. Calculate and print his pay if he gets double pay for each hour over 40 that he worked.
15. Input the length and width of a carpet and the price per square foot to clean the carpet. Calculate and print the area of the carpet (in square feet) and the cost of cleaning the carpet if the cost is double for every square foot past the first 100.
16. Input three test scores and find their average. Then print out the grade based on the following scale:

100–90	A
89–80	B
79–70	C
69–60	D
59–0	F

In Problems 17–28, write a flowchart for the algorithm in the indicated problem.

17. Problem 5
18. Problem 6
19. Problem 7
20. Problem 8
21. Problem 9
22. Problem 10
23. Problem 11
24. Problem 12
25. Problem 13
26. Problem 14
27. Problem 15
28. Problem 16

In Problems 29–32, write compound decisions for the Boolean expression, using only AND and OR as connectives.

29. $p \vee q$
30. $p \wedge q$
31. $(p \vee q) \wedge (p \vee r)$
32. $p \vee [q \wedge (p \vee r)]$

In Problems 33–38, follow each step of the flowchart in Figure 11.10, using the given values for X and Y, respectively, and find the output.

33. $X = 4, Y = 7$
34. $X = 7, Y = 9$
35. $X = 6, Y = 6$
36. $X = -2, Y = -2$
37. $X = 7, Y = -3$
38. $X = 12, Y = 8$

39. Write an algorithm for the flowchart in Figure 11.10.
40. Write a flowchart to input three numbers and print the numbers in increasing order.

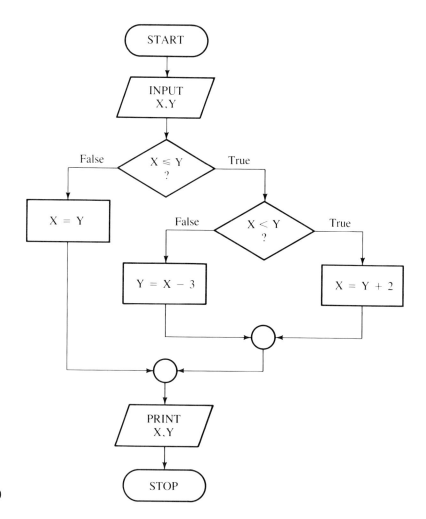

Figure 11.10

11.4 The Repetition Structure

repetition structure, loop

So far in this chapter, we have written algorithms and flowcharts for procedures which go from top to bottom without repeating any statements. The power of the computer is that it can *repeat* steps over and over again, as many times as necessary. This ability leads to the **repetition structure**, or **loop**, which allows a sequence or decision structure to be repeated.

One of the hardest decisions in programming a loop is which statements go inside the loop and which statements go outside the loop. Here are the general guidelines: Any statement that is to be repeated many times should go inside the loop. Any statement that is to be performed only once goes outside the loop.

DO–WHILE

There are two types of loops commonly used in programming. The first type of loop that we will consider is the **DO–WHILE**. In this case, the logical decision of whether or not to process the loop is made at the *beginning* of the loop, and the loop is processed whenever the logical decision is true. Thus, it is possible that a DO–WHILE loop will *never* be performed if the logical decision is false the first time the program gets to the loop.

counter

We will use a **counter** (C) to count how many times we go through a loop. This counter should be set equal to zero *before* the loop starts (C = 0). Each time the loop is performed, 1 should be added to the counter (C = C + 1). This statement should be *inside* the loop so that 1 is added each time the loop is performed. We will exit the loop when C has reached the desired value.

To illustrate a DO–WHILE with a counter, let's change the flowchart of Figure 11.3 to an algorithm that will work for 43 different rectangles.

■ EXAMPLE 11.15 Write an algorithm for a procedure to take the length and width of 43 different rectangles and find and print the area and perimeter of the rectangles.

Solution The algorithm follows.

1.	C = 0	:	Set the counter to 0
2.	DO–WHILE (C < 43)	:	Start the DO–WHILE loop
	a. INPUT L,W	:	Input the length and width
	b. A = L * W	:	Find the area of the rectangle
	c. P = (2 * L) + (2 * W)	:	Find the perimeter of the rectangle
	d. C = C + 1	:	Add 1 to the counter
	e. PRINT A,P	:	Print the area and perimeter of the rectangle
3.	END DO–WHILE	:	End the DO–WHILE loop

REPEAT–UNTIL

The second type of loop that we will consider is the **REPEAT–UNTIL** loop. This loop is similar to the DO–WHILE except for two main differences. First, the logical decision whether or not to process the loop is made at the *end* of the loop. Second, the loop is processed *until* the logical decision is true. Unlike a

DO–WHILE, a REPEAT–UNTIL loop will always be processed at least once, since the test is not made until the end of the loop.

Example 11.15 illustrated how a sequence structure can be converted into a loop by using a DO–WHILE. We will now do the same for a decision structure, using a REPEAT–UNTIL.

EXAMPLE 11.16 Rewrite the flowchart of Example 11.14 (Figure 11.9) so that it repeats for 121 different pairs of numbers.

Solution The flowchart appears in Figure 11.11.

Another way of writing the flowchart of Example 11.16 is to use the decision $C < 121$ and to repeat the loop as long as the condition is true. This algorithm combines the two types of loops discussed in this section (see Problem 1), and it is commonly used in some programming languages.

The first two examples of this section included a counter. Many languages include statements which will automatically increment the counter and process the loop until the desired value of the counter is exceeded. For example, with the use of the FOR and NEXT statements from BASIC, the loop from Example 11.15 is programmed as follows:

```
100    FOR C = 1 TO 43
110        INPUT L,W
120        A = L * W
130        P = (2 * L) + (2 * W)
140        PRINT A,P
150    NEXT C
```

Each time through the loop, the value of C is increased by 1. When 43 is surpassed, the program goes to the statement following 150.

Many times, a programmer will not know how many times a loop should be processed. In this case, a counter is of no benefit. Many programming languages have a built-in indicator for the end of an input file. If input is attempted after the last item has been read in, the end-of-file indicator, which we will call EOF, is set to TRUE. If our loop starts with DO–WHILE (NOT EOF), for example, then the loop will be processed as long as there is information to read in, since NOT EOF will be TRUE. When input is attempted past the last item, EOF becomes TRUE, so NOT EOF becomes FALSE; and the program exits the loop. The next example illustrates.

EXAMPLE 11.17 Write an algorithm using a DO–WHILE structure for a procedure to calculate 5% sales tax on a number of items in a department store given the prices of the items. Print out the prices and tax on all items.

11.4 The Repetition Structure

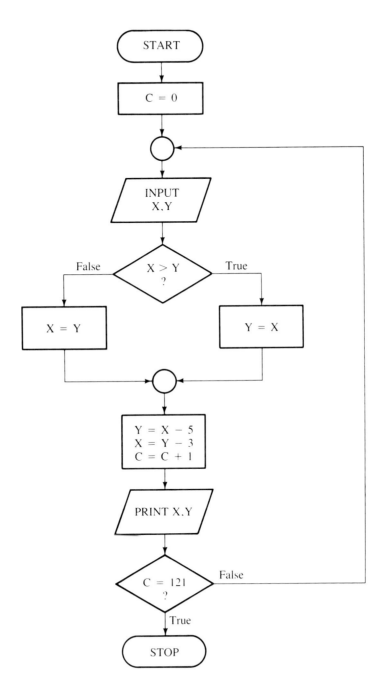

Figure 11.11

Solution	1. INPUT P	: Get price of first item
	2. DO–WHILE (NOT EOF)	: Start DO–WHILE loop
	a. T = P * .05	: Calculate sales tax
	b. PRINT P,T	: Print price and sales tax
	c. INPUT P	: Get price of next item
	3. END DO–WHILE	: End DO–WHILE loop

Finding an average is a common activity in many programming applications. To find an average, we first find the sum of all numbers involved. We can find this sum by using a loop. We keep a running sum of the numbers inside the loop as the loop processes one number at a time. If we use the assignment statement $S = S + N$, where S represents the sum and N represents the current number, each time the program goes through the loop, the sum will be updated by adding the current number to it.

Understanding why we use the same variable (S) on both sides of the assignment statement is very important. We want to take the value for the sum so far (S), add the current number to it (N), and assign the result to the sum (S). In this manner, the sum is updated each time the loop is performed. The value of S should be set to zero before the loop begins.

When the loop is finished, the value of S will be the final (total) sum of all of the numbers. This sum can then be divided by the number of numbers that were added, to give the average. This last step should be performed *after* the loop is finished since it is only the *final* average that is usually of interest. The next example illustrates the procedure for finding an average.

EXAMPLE 11.18   Write an algorithm for a procedure to read in salaries for an unspecified number of employees and find and print the average salary.

Solution   In this case, we need a counter even though it will not be used to determine when to exit the loop. We need to know how many salaries have been inputted in order to find the average. The algorithm follows.

1. C = 0   : Set counter to zero
2. S = 0   : Set sum to zero
3. INPUT W   : Read in first salary
4. DO–WHILE (NOT EOF)   : Start DO–WHILE loop
   a. S = S + W   : Add current salary to sum
   b. C = C + 1   : Add 1 to counter
   c. INPUT W   : Read in next salary
5. END DO–WHILE   : End DO–WHILE loop
6. A = S/C   : Calculate final average
7. PRINT A   : Print average

EXAMPLE 11.19   Write a flowchart and an algorithm for a procedure to enter students' test scores and find the highest and lowest scores. Use a REPEAT–UNTIL structure.

Solution   The flowchart appears in Figure 11.12.

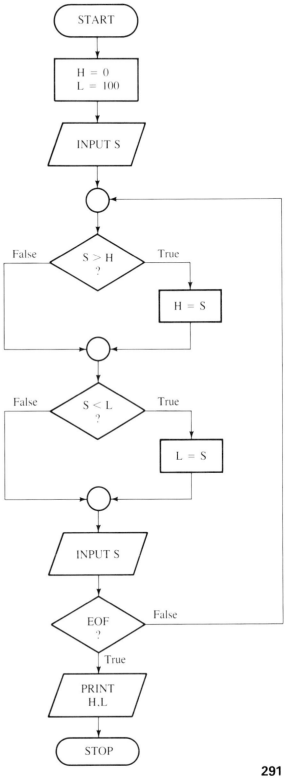

Figure 11.12

The algorithm follows.

1.	H = 0	:	Set high score to 0
2.	L = 100	:	Set low score to 100
3.	INPUT S	:	Input first test score
4.	REPEAT	:	Start REPEAT–UNTIL loop
	a. IF S > H THEN H = S	:	If score is larger than the high, it becomes new high
	b. IF S < L THEN L = S	:	If score is smaller than the low, it becomes new low
	c. INPUT S	:	Input next test score
5.	UNTIL (EOF)	:	End REPEAT–UNTIL loop
6.	PRINT H,L	:	Print high and low score

Setting the high score to a small number and the low score to a high number at the beginning of the problem may seem strange, but this step is necessary to force H and L to change to their real values in the program. If H were set to 100 at the beginning of the procedure and no score was as high as 100, the value of H would be 100 at the end of the procedure instead of the highest test score. This result will be demonstrated in the problem set (Problems 3 and 4).

One of the more interesting repetition structures is a loop inside a loop. Consider, for example, a class of 76 students, with each student having 10 test scores. To find each student's average, we use a loop which repeats 10 times to process that student's test scores. Then this loop is inside another loop which repeats 76 times, once for each student in the class. The inner loop is performed repeatedly (10 times) every time the outside loop is performed once. The procedure is illustrated in the next example.

■ **EXAMPLE 11.20** Write a flowchart for a procedure to find the average test score for each of 76 students in a class. Each student has 10 test scores.

Solution The flowchart appears in Figure 11.13.

■ Problem Set 11.4

1. Write a flowchart for the procedure in Example 11.16 in which the loop is repeated if the decision is true.
2. Write an algorithm for the procedure in Example 11.16.
3. Follow the flowchart for Example 11.19, and find the final values for H and L for the following set of test scores: 87, 45, 98, 34, and 77.
4. Change the first two steps of the algorithm for Example 11.19 to H = 100 and L = 0. Follow the algorithm, and find the final values for H and L for the following set of test scores: 87, 45, 98, 34, and 77.

11.4 The Repetition Structure

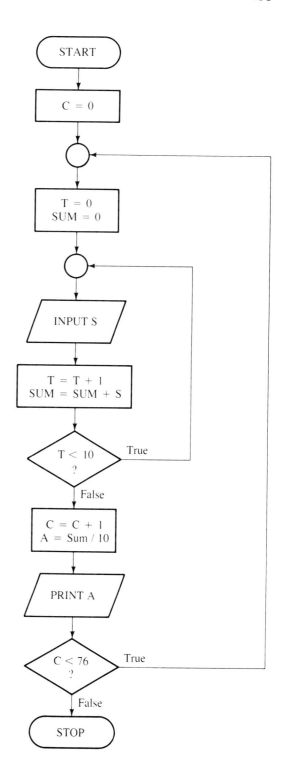

Figure 11.13

5. Write an algorithm for the procedure in Example 11.20, using REPEAT–UNTIL structures.
6. How many total times will the inner loop from Example 11.20 be performed throughout the entire procedure?

In Problems 7–12, use DO–WHILE structures to write an algorithm for the procedure.

7. Input the number of honor points and credits for 587 students, and calculate and print each student's grade point average (honor points divided by credits).
8. Input the height in inches of an unspecified number of people, and change each person's height to feet and inches. Print out each changed height.
9. Convert all whole number Fahrenheit temperatures from 32 to 212 inclusive to Celsius temperatures $[C = (\frac{5}{9})(F - 32)]$. Print out each Fahrenheit and corresponding Celsius temperature. (*Hint:* No input is needed.)
10. Convert all whole number Celsius temperatures from 0 to 100 inclusive to Fahrenheit temperatures $[F = (\frac{9}{5})C + 32]$. Print out each Celsius and Fahrenheit temperature.
11. Input the number of hours that each employee of the Hard-Work Company worked last week and his or her hourly wage. Calculate and print each worker's pay if the worker gets double pay for each hour over 40 that he or she works. Print out each employee's pay.
12. Input the salaries for workers of the Double-Loop Company, and calculate and print the average salary.

In Problems 13–18, use REPEAT–UNTIL structures to write an algorithm for the procedure.

13. Find and print the average of three test scores to be inputted for each of 77 students.
14. Input the amount of rainfall for each month in the past ten years. For each month, print "Wet" if the rainfall is greater than 7 inches and print "Dry" otherwise.
15. Find the total sales for each salesperson of the Superman Company who each worked 52 weeks. Each salesperson's weekly sales are inputted. Print out the total sales for each salesperson. (*Hint:* Use a loop for the salespeople and a loop for the weeks—one loop inside the other loop.)
16. Calculate Y from the formula $Y = 2A^2 - 3B + 7$ for all cases where A and B can be any whole number value from 0 to 34. (*Hint:* Use a loop inside a loop.)
17. Input the lengths of the three sides of a triangle and calculate the perimeter of the triangle. Continue the calculations until there are no more triangles to be inputted. Print out the perimeter for each triangle.
18. Input the grade point average of a student. If the grade point average is under 2.00, print "Probation". Otherwise, print "No probation". Continue this process as long as there are grade point averages to be processed.
19. Write a BASIC program by using FOR and NEXT statements for the flowchart in Example 11.16.
20. Write a BASIC program for the procedure in Problem 10.

CHAPTER SUMMARY

In this chapter, we studied algorithms and flowcharts, which are aids to writing computer programs. An algorithm is a list of the steps necessary to solve a problem. A flowchart is a graphical representation of an algorithm.

Three fundamental structures in programming were considered in this chapter:

- Sequence structure
- Decision structure
- Repetition structure

A sequence structure proceeds from step to step in orderly sequence, with no loops or decisions. A decision structure involves a logical decision, a decision that is either true or false. Both branches of a logical decision must be written. A repetition structure, or loop, allows the computer to repeat a series of steps as many times as is necessary to solve a problem.

Two types of loops are commonly used in programming:

- DO–WHILE, which repeats when a logical condition is true; the test is performed at the beginning of the loop
- REPEAT–UNTIL, which repeats until a logical condition is true; the test is performed at the end of the loop

All programs that are written must use combinations of the three basic structures.

REVIEW PROBLEMS

In Problems 1–8, write an algorithm for the procedure.

1. Input three numbers, and find and print their sum and product.
2. Input a person's salary, and find and print the new salary, assuming a $1300 raise.
3. Input the number of 13¢ stamps and 17¢ stamps that were purchased, and calculate and print the total money spent on the stamps.
4. Input the number of field goals (2 points) and free throws (1 point) that a basketball team scored in a game, and calculate and print the total points scored by the team.
5. Input a person's salary and age. If the person is older than 35, then give him a $4000 raise. Otherwise, give him a pay cut of $2300. Print out the new salary.
6. Input two numbers, and add 3 to the smaller number. Print out both numbers.
7. Input three numbers, and print out only the middle number (that is, not the largest or smallest number).
8. Input a student's number of credits, and print out the student's class by the following rule:

0–29 credits	Freshman
30–59 credits	Sophomore
60–89 credits	Junior
90–120 credits	Senior

In Problems 9–14, write a flowchart for the procedure given in the indicated problem.

9. Problem 1
10. Problem 2
11. Problem 3
12. Problem 4
13. Problem 5
14. Problem 6

In Problems 15–18, use DO–WHILE structures to write an algorithm for the procedure.

15. Use a loop to input the names and amounts of money in savings accounts for 560 people. Print out each name and amount of savings and the total sum of all accounts.

16. Use a loop to take all inches from 0 to 120 and convert them to centimeters, using the formula C = 2.54 * I. Print out all values for inches and centimeters.

17. Rewrite Problem 5 so that it will work for an unspecified number of people.

18. Rewrite Problem 8 so that it will work for an unspecified number of students.

In Problems 19–22, rewrite the algorithm for the indicated problem by using a REPEAT–UNTIL structure.

19. Problem 15
20. Problem 16
21. Problem 17
22. Problem 18

ARRAYS AND MATRICES

CHAPTER 12

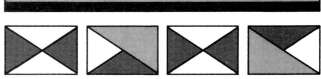

Many applications involving mathematics and computers involve organizing large amounts of related pieces of information. For example, a teacher may want to keep a grade book of all students and their test results, or a bank may keep records on all of its customers.

One way of handling large amounts of data is by using arrays. In this chapter, we will study such arrays. In particular, we will study one type of array, matrices, in detail. We will emphasize the connection to computers whenever possible and use computer notation as much as possible.

spreadsheet

One of the most useful applications of computers is the electronic **spreadsheet**, which is simply the concept of matrices applied to real-world problems. Matrices can also be used to simplify the process of finding solutions to systems of equations. In this chapter, we will investigate these and other applications of arrays.

12.1 Arrays

Consider the test scores for a class of six students. If each position corresponds to a student, this list is an ordered list of numbers. The numbers can be stored in a **one-dimensional array**, also called a **vector**, as follows:

one-dimensional array, vector

$$A = [76 \quad 84 \quad 32 \quad 71 \quad 93 \quad 81]$$

row vector
column vector

If the elements of the vector A are listed horizontally, as here, A is a **row vector**. If the elements are listed vertically, A is a **column vector**. The vector A that follows is a column vector:

$$A = \begin{bmatrix} 76 \\ 84 \\ 32 \\ 71 \\ 93 \\ 81 \end{bmatrix}$$

The elements of an array are referred to by the use of *subscripts*. The first element of A is a_1, the second element a_2, and so on. This mathematical notation is not possible in computers, however, because all symbols must be written on the same line, and no subscripts are possible. For this reason, programmers have adopted the notation $A(1)$ to represent a_1, $A(2)$ to represent a_2, and so forth. The

12.1 Arrays

index

number in parentheses is the subscript, or **index**. We will usually use this computer notation in this book. To refer to a particular element in an array, we use the name of the array and the subscript, as in A(2).

Arrays are ideally suited to computer calculation. To process a vector, for example, we use a loop to examine each element. The index for the array is the variable that controls the loop. The flowchart of such an operation might look like the one shown in Figure 12.1.

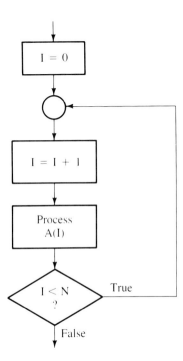

Figure 12.1

To read elements into a 20-element array A, for example, we can use the following segment of a BASIC program:

```
FOR I = 1 to 20
          READ A(I)
NEXT I
```

two-dimensional array, matrix, elements

A **two-dimensional array**, called a **matrix**, is a rectangular arrangement of numbers, or **elements**, as we will call them from now on. A matrix consists of horizontal rows and vertical columns. So *two* subscripts are now necessary to reference a particular element. The first subscript gives the row, and the second subscript gives the column of the element.

Consider this matrix:

$$A = \begin{bmatrix} 2 & 6 & 0 \\ -1 & 3 & 25 \end{bmatrix}$$

dimensions   Matrix A has two rows and three columns, so the **dimensions** of A are said to be 2×3 (2 by 3). Note that a 2×3 matrix is *not* the same as a 3×2 matrix. Element $A(1, 2)$ is the number 6, since 6 appears in the first row and second column. Note that the subscript representing the row always is written first, and the column is always designated second. So $A(2, 1)$ is -1. As mentioned earlier, mathematicians refer to $A(2, 1)$ as a_{21} and to an arbitrary element of A as a_{ij}, the element in the ith row and jth column of A. We will refer to this arbitrary element as $A(I, J)$.

equal arrays   **Definition 12.1**   Two arrays A and B are **equal arrays** iff (if and only if) A and B have the same dimensions and all corresponding elements are equal.

■ **EXAMPLE 12.1**   Are arrays A and B equal?

Solution
$$A = \begin{bmatrix} 2 & 6 & 0 \\ -1 & 3 & 25 \end{bmatrix} \qquad B = \begin{bmatrix} \sqrt{4} & 6 & 0 \\ -1 & \frac{6}{2} & 5^2 \end{bmatrix}$$

Both A and B are 2×3 arrays. Since $2 = \sqrt{4}$, $\frac{6}{2} = 3$, and $25 = 5^2$, they are equal.

To process matrices by computer requires a loop inside a loop: one loop for rows and one loop for columns. For instance, as I goes from 1 to 4 and J goes from 1 to 5, $A(I, J)$ will process all elements of a 4×5 matrix, as illustrated in the flowchart of Figure 12.2.

Whether you are inputting, outputting, adding, or performing any other operation on a matrix which involves every element, a loop inside a loop is the standard method for processing the matrix. Furthermore, note that all arrays are stored as single lists in computer memory.

three-dimensional array   We can also consider a **three-dimensional array** consisting of rows, columns, *levels*   and **levels** (see Figure 12.3). (Whether levels go *down* or *back* depends on your own interpretation.) For such an array, $A(4, 3, 7)$ references the element in the 4th row, 3rd column, and 7th level of A. To process a three-dimensional array requires a loop inside a loop inside a loop: one loop for each dimension.

A three-dimensional array can be used to store students' test scores in several classes. For example, a teacher may have 5 classes (levels), each class consisting of 30 students (rows), who each take 4 tests (columns). A $30 \times 4 \times 5$ array can be used to hold all these test scores. Element $A(17, 3, 4)$, for example, holds the 3rd test score for the 17th student in class 4.

We could continue to discuss arrays with even more dimensions. But we will concentrate on two-dimensional arrays in this chapter, because they are easy to

12.1 Arrays

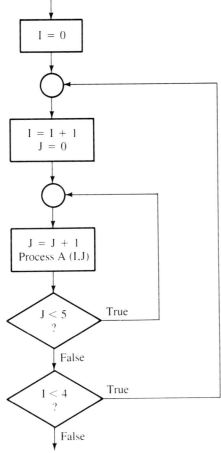

Figure 12.2

Figure 12.3
Three-Dimensional Array

understand and they follow the same rules as arrays of higher dimensions.

dimension (DIM) statement

In BASIC, by use of a **dimension (DIM) statement**, you tell the computer at the beginning of the program the dimensions of the arrays you will be using. For example, DIM A(7), B(2,3) tells the computer that you will be using an array A of dimension 7 and an array B of dimension 2×3.

Problem Set 12.1

In Problems 1–22, use the following arrays A and B. Find the specified information or element (if possible).

$$A = \begin{bmatrix} 3 & 2 & -4 & 1 & 5 \end{bmatrix} \quad B = \begin{bmatrix} 1 & 0 & 5 & -4 \\ 2 & -1 & 7 & 6 \\ 4 & 3 & 9 & -2 \end{bmatrix}$$

1. The dimensions of A
2. The dimensions of B
3. a_4
4. b_{12}
5. b_{31}
6. a_2
7. $A(5)$
8. $B(3, 2)$
9. $B(2, 1)$
10. $A(3)$
11. $B(3, 4)$
12. $B(1, 4)$
13. $A(?) = -4$
14. $A(?) = 3$
15. $B(?, ?) = 7$
16. $B(?, ?) = 6$

17. $A(6)$
18. $B(4, 2)$
19. The number of rows in A
20. The number of columns in A
21. The number of columns of B
22. The number of rows of B
23. Can a one-dimensional array ever equal a two-dimensional array?
24. Can a 3×4 matrix ever equal a 4×3 matrix?
25. Does $[1 \quad 2 \quad 3] = [3 \quad 2 \quad 1]$?
26. Can a one-dimensional array ever have more elements than a two-dimensional array? If so, give an example.
27. Five employees have salaries of \$17,075, \$18,091, \$15,043, \$20,261, and \$24,000 a year. Write these salaries as a row vector.
28. A student received test scores of 88, 62, 73, 64, and 91. Write these test scores as a column vector.
29. Five players had the following scores on the first four holes of a golf tournament:

Player	Hole			
	1	2	3	4
Joe	3	4	5	4
Sam	2	5	7	4
Bo	3	3	2	2
Bill	5	4	1	6
Jan	3	3	3	3

Write these scores as a matrix.
30. Bob received quiz scores of 5, 6, and 10; Nancy had scores of 10, 10, and 10; and Jason had scores of 4, 5, and 6. Write these scores as a matrix.

12.2 Fundamental Matrix Operations

There are several fundamental operations for matrices, as there are for real numbers, but matrix operations are more restrictive. Addition of matrices, for example, is defined only between matrices that have the same dimensions.

sum

Definition 12.2 If A and B are both $m \times n$ matrices, then the **sum** of A and B, designated $A + B = C$, is defined as

$$C(I, J) = A(I, J) + B(I, J)$$

In other words, corresponding elements are added.

EXAMPLE 12.2 Find the sum of A and B:

$$A = \begin{bmatrix} 1 & 0 & 2 & 4 \\ 3 & 1 & 0 & -1 \\ 2 & 5 & 7 & 3 \end{bmatrix} \quad B = \begin{bmatrix} 3 & 2 & 5 & -3 \\ -3 & 2 & 1 & 0 \\ 0 & 1 & -3 & 4 \end{bmatrix}$$

Solution

$$\begin{bmatrix} 1 & 0 & 2 & 4 \\ 3 & 1 & 0 & -1 \\ 2 & 5 & 7 & 3 \end{bmatrix} + \begin{bmatrix} 3 & 2 & 5 & -3 \\ -3 & 2 & 1 & 0 \\ 0 & 1 & -3 & 4 \end{bmatrix} = \begin{bmatrix} 4 & 2 & 7 & 1 \\ 0 & 3 & 1 & -1 \\ 2 & 6 & 4 & 7 \end{bmatrix}$$

12.2 Fundamental Matrix Operations

If two matrices are *not* the same size, they cannot be added. So addition of matrices is not possible on many occasions.

Let's consider another example illustrating an application.

EXAMPLE 12.3 Three students received the following grades on four different tests:

Student	Test 1	Test 2	Test 3	Test 4
1	70	65	52	43
2	21	16	0	17
3	55	34	18	51

Since the scores were not very good, some extra-credit problems were assigned, with the following results:

Student	Test 1	Test 2	Test 3	Test 4
1	10	5	6	0
2	9	0	7	1
3	7	4	9	0

Find the new test scores after the extra credit has been added.

Solution By finding the sum of the associated matrices, we can find the new text scores:

$$\begin{bmatrix} 70 & 65 & 52 & 43 \\ 21 & 16 & 0 & 17 \\ 55 & 34 & 18 & 51 \end{bmatrix} + \begin{bmatrix} 10 & 5 & 6 & 0 \\ 9 & 0 & 7 & 1 \\ 7 & 4 & 9 & 0 \end{bmatrix} = \begin{bmatrix} 80 & 70 & 58 & 43 \\ 30 & 16 & 7 & 18 \\ 62 & 38 & 27 & 51 \end{bmatrix}$$

Subtraction for matrices can also be defined.

difference

Definition 12.3 If A and B are both $m \times n$ matrices, then the **difference** of A and B, designated $A - B = C$, is defined as

$$C(I, J) = A(I, J) - B(I, J)$$

Thus, corresponding elements are subtracted.

EXAMPLE 12.4 Find $A - B$:

$$A = \begin{bmatrix} 1 & 0 & 2 & 4 \\ 3 & 1 & 0 & -1 \\ 2 & 5 & 7 & 3 \end{bmatrix} \quad B = \begin{bmatrix} 3 & 2 & 5 & -3 \\ -3 & 2 & 1 & 0 \\ 0 & 1 & -3 & 4 \end{bmatrix}$$

Solution

$$\begin{bmatrix} 1 & 0 & 2 & 4 \\ 3 & 1 & 0 & -1 \\ 2 & 5 & 7 & 3 \end{bmatrix} - \begin{bmatrix} 3 & 2 & 5 & -3 \\ -3 & 2 & 1 & 0 \\ 0 & 1 & -3 & 4 \end{bmatrix} = \begin{bmatrix} -2 & -2 & -3 & 7 \\ 6 & -1 & -1 & -1 \\ 2 & 4 & 10 & -1 \end{bmatrix}$$

Addition and subtraction of matrices is straightforward. Multiplication is a bit complex. We will look at two different types of multiplication. **Scalar multiplication** involves multiplication of a matrix by a real number, called a **scalar**. In scalar multiplication, every element of the matrix is multiplied by the scalar. The product is designated by $n \cdot M$, or simply nM, where n is a real number and M is a matrix. The next example illustrates the process.

scalar multiplication
scalar

■ **EXAMPLE 12.5** Find the scalar product $5A$, using matrix A of Example 12.4.

Solution

$$5 \cdot \begin{bmatrix} 1 & 0 & 2 & 4 \\ 3 & 1 & 0 & -1 \\ 2 & 5 & 7 & 3 \end{bmatrix} = \begin{bmatrix} 5 & 0 & 10 & 20 \\ 15 & 5 & 0 & -5 \\ 10 & 25 & 35 & 15 \end{bmatrix}$$

Multiplication between two matrices will be covered in the next section.
One more operation on matrices should be mentioned at this time: transposition.

transpose

Definition 12.4 The **transpose** of a matrix A, designated A^T, is the matrix obtained by interchanging the rows and columns of A.

■ **EXAMPLE 12.6** Find A^T for A:

$$A = \begin{bmatrix} 2 & 4 & 0 & -1 \\ 3 & 7 & -2 & 5 \\ 1 & -3 & 0 & -4 \end{bmatrix}$$

Solution

$$A^T = \begin{bmatrix} 2 & 3 & 1 \\ 4 & 7 & -3 \\ 0 & -2 & 0 \\ -1 & 5 & -4 \end{bmatrix}$$

Problem Set 12.2

In Problems 1–30, perform the indicated matrix operations (if possible) for the following matrices:

$$A = \begin{bmatrix} -4 & 0 & 2 \\ 3 & 1 & 5 \\ 1 & 2 & 1 \end{bmatrix} \quad B = \begin{bmatrix} 0 & 1 & -1 \\ 2 & 4 & 2 \\ 3 & 5 & 1 \end{bmatrix}$$

$$C = \begin{bmatrix} 1 & 0 & 0 \\ 0 & 1 & 0 \\ 0 & 0 & 1 \end{bmatrix} \quad D = \begin{bmatrix} 3 & 0 \\ 2 & -1 \\ 1 & -2 \end{bmatrix}$$

1. $A + B$
2. $A + C$
3. $A + D$
4. $B + D$
5. $B + C$
6. $C + B$
7. $A + (B + C)$
8. $C + A$
9. $(A + B) + C$
10. $C + (B + D)$
11. $C + B$
12. $(C + B) + D$
13. $A - B$
14. $C - B$
15. $C - D$
16. $D - A$
17. $7A$
18. $-3D$
19. $-2B$
20. $5C$
21. A^T
22. C^T
23. D^T
24. B^T
25. $3A + 2C$
26. $4C - 3B$
27. $3(A - D)$
28. $-7(C - B)$
29. $-3A^T + 2B$
30. $C^T - B^T$

In Problems 31–36, give an example to support each answer.

31. Is matrix addition commutative?
32. Is matrix subtraction commutative?
33. Is matrix subtraction associative?
34. Is matrix addition associative?
35. Is there an identity matrix for addition?
36. Does every matrix have an inverse for addition?

37. Two football teams gain the following number of yards in four quarters:

Team	Quarter			
	1	2	3	4
1	80	73	50	26
2	16	43	82	104

    They also incur the following penalties (in yards):

Team	Quarter			
	1	2	3	4
1	10	0	25	30
2	5	5	0	5

    Use matrix subtraction to find the net yards for each team in each quarter.

38. The Emperor Penguin Company produced three types of stuffed penguins in January at three different locations, as follows:

Type	Location		
	Big Rapids	Clio	Elsie
Small	220	100	45
Medium	160	80	120
Large	50	150	190

    Projections call for all plants to triple their output in February. Use scalar multiplication to show the projected output for February.

39. Find a matrix A such that $A = A^T$.
40. When is $A + A^T$ possible?
41. Does $(A^T)^T$ always equal A?
42. Does $n(A + B)$ always equal $nA + nB$ for matrices A and B and a real number n?

12.3 Matrix Multiplication

Matrix multiplication involves multiplication of two matrices. We will first consider multiplication of two vectors. Suppose a salesperson has four different items and sells 10, 16, 4, and 8 of the items, respectively. We can construct a row vector to hold these values:

$$A = \begin{bmatrix} 10 & 16 & 4 & 8 \end{bmatrix}$$

Now, suppose each of these items sells for different prices: $5, $7, $4, and $2.50, respectively. We can construct a column vector to hold these values:

$$B = \begin{bmatrix} 5 \\ 7 \\ 4 \\ 2.5 \end{bmatrix}$$

To find the total sales for the salesperson, we multiply the number of each item sold by its price and add the products:

$$\begin{aligned} \text{total sales} &= (10 \cdot 5) + (16 \cdot 7) + (4 \cdot 4) + (8 \cdot 2.50) \\ &= 50 + 112 + 16 + 20 \\ &= 198 \text{ dollars} \end{aligned}$$

In vector form, this corresponds to multiplying each number in the row vector by the corresponding number in the column vector and adding the products. The result is the **vector product** of the two vectors.

vector product

$$A \cdot B = \begin{bmatrix} 10 & 16 & 4 & 8 \end{bmatrix} \cdot \begin{bmatrix} 5 \\ 7 \\ 4 \\ 2.5 \end{bmatrix} = \begin{bmatrix} 198 \end{bmatrix}$$

Now suppose that there are three salespeople who each sell four different items in the following amounts:

Salesperson	Screwdrivers	Hammers	Saws	Wrenches
1	10	16	4	8
2	4	3	7	10
3	12	5	4	2

A screwdriver costs $5, a hammer $7, a saw $4, and a wrench $2.50. We wish to find the total sales for *each* salesperson.

12.3 Matrix Multiplication

We first construct a sales matrix,

$$\begin{bmatrix} 10 & 16 & 4 & 8 \\ 4 & 3 & 7 & 10 \\ 12 & 5 & 4 & 2 \end{bmatrix}$$

and a cost matrix,

$$\begin{bmatrix} 5 \\ 7 \\ 4 \\ 2.5 \end{bmatrix}$$

The vector product of each row of the sales matrix and the cost matrix will yield a matrix containing the total sales for each salesperson:

$$\begin{bmatrix} 10 & 16 & 4 & 8 \\ 4 & 3 & 7 & 10 \\ 12 & 5 & 4 & 2 \end{bmatrix} \cdot \begin{bmatrix} 5 \\ 7 \\ 4 \\ 2.5 \end{bmatrix} = \begin{bmatrix} (10\cdot 5)+(16\cdot 7)+(4\cdot 4)+(8\cdot 2.5) \\ (4\cdot 5)+(3\cdot 7)+(7\cdot 4)+(10\cdot 2.5) \\ (12\cdot 5)+(5\cdot 7)+(4\cdot 4)+(2\cdot 2.5) \end{bmatrix}$$

$$= \begin{bmatrix} 50+112+16+20 \\ 20+21+28+25 \\ 60+35+16+5 \end{bmatrix} = \begin{bmatrix} 198 \\ 94 \\ 116 \end{bmatrix}$$

The first salesperson had total sales of $198; the second had sales of $94; and the third had sales of $116.

The example illustrates the principle behind matrix multiplication. To multiply two matrices A and B, we multiply each row of matrix A by each column of matrix B, as we did in vector multiplication. Thus, the number of elements in a row of A must be the same as the number of elements in a column of B. Otherwise, matrix multiplication is not possible.

product

Definition 12.5 If A is an $m \times n$ matrix and B is an $n \times p$ matrix, then the **product** of A and B, designated $A \cdot B = C$ or simply $AB = C$, is the $m \times p$ matrix obtained by performing vector multiplication between each row of A and each column of B. The product $C(I, J)$ is the vector product of the ith row of A and the jth column of B.

EXAMPLE 12.7 Find AB, if multiplication is possible, for these matrices:

$$A = \begin{bmatrix} 0 & 4 & 3 \\ 1 & 2 & 1 \\ 0 & 3 & 7 \\ 7 & 4 & 2 \end{bmatrix} \quad B = \begin{bmatrix} 4 & 1 \\ 2 & 3 \\ 0 & 4 \end{bmatrix}$$

Solution Matrix A is 4×3, and B is 3×2, so multiplication is possible. The product AB will be a 4×2 matrix. If the dimensions of the two matrices are written side by side, multiplication will be possible if the *inner* numbers are the same. The *outer* numbers give the dimensions of the product. For AB, we have the following dimensions:

$$\underbrace{(4 \times 3) \quad \overbrace{(3 \times 2)}^{\text{same}}}_{\text{dimensions}}$$

Now, we multiply:

$$\begin{bmatrix} 0 & 4 & 3 \\ 1 & 2 & 1 \\ 0 & 3 & 7 \\ 7 & 4 & 2 \end{bmatrix} \begin{bmatrix} 4 & 1 \\ 2 & 3 \\ 0 & 4 \end{bmatrix} = \begin{bmatrix} (0 \cdot 4)+(4 \cdot 2)+(3 \cdot 0) & (0 \cdot 1)+(4 \cdot 3)+(3 \cdot 4) \\ (1 \cdot 4)+(2 \cdot 2)+(1 \cdot 0) & (1 \cdot 1)+(2 \cdot 3)+(1 \cdot 4) \\ (0 \cdot 4)+(3 \cdot 2)+(7 \cdot 0) & (0 \cdot 1)+(3 \cdot 3)+(7 \cdot 4) \\ (7 \cdot 4)+(4 \cdot 2)+(2 \cdot 0) & (7 \cdot 1)+(4 \cdot 3)+(2 \cdot 4) \end{bmatrix}$$

$$= \begin{bmatrix} 0+8+0 & 0+12+12 \\ 4+4+0 & 1+6+4 \\ 0+6+0 & 0+9+28 \\ 28+8+0 & 7+12+8 \end{bmatrix} = \begin{bmatrix} 8 & 24 \\ 8 & 11 \\ 6 & 37 \\ 36 & 27 \end{bmatrix}$$

■ **EXAMPLE 12.8** Find AC, if multiplication is possible, for matrices A and C:

$$A = \begin{bmatrix} 1 & 0 & 2 \\ 2 & 3 & -1 \\ 6 & 4 & 5 \end{bmatrix} \quad C = \begin{bmatrix} 1 & 3 & 0 & -1 \\ 0 & 4 & 1 & 1 \\ 2 & 5 & 0 & 2 \end{bmatrix}$$

Solution

$$AC = \begin{bmatrix} 1 & 0 & 2 \\ 2 & 3 & -1 \\ 6 & 4 & 5 \end{bmatrix} \cdot \begin{bmatrix} 1 & 3 & 0 & -1 \\ 0 & 4 & 1 & 1 \\ 2 & 5 & 0 & 2 \end{bmatrix} = \begin{bmatrix} 5 & 13 & 0 & 3 \\ 0 & 13 & 3 & -1 \\ 16 & 59 & 4 & 8 \end{bmatrix}$$

■ **EXAMPLE 12.9** Find CB if possible:

$$C = \begin{bmatrix} 1 & 2 \\ 0 & 3 \\ 7 & 1 \end{bmatrix} \quad B = \begin{bmatrix} 2 & 5 \\ 3 & 0 \\ -1 & -6 \end{bmatrix}$$

Solution Multiplication is *not* possible since C has two columns and B has three rows.

12.3 Matrix Multiplication

Problem Set 12.3

In Problems 1–40, perform the matrix operations, if possible, for the following matrices:

$$A = \begin{bmatrix} 3 & 1 & 4 \\ 2 & 0 & -1 \end{bmatrix} \quad B = \begin{bmatrix} 3 & 2 & 3 \\ 0 & 1 & -1 \\ 5 & 1 & 0 \end{bmatrix} \quad C = \begin{bmatrix} 7 & -2 \\ 4 & 1 \\ 2 & 3 \end{bmatrix}$$

$$D = \begin{bmatrix} 1 & 0 & 0 \\ 0 & 1 & 0 \\ 0 & 0 & 1 \end{bmatrix} \quad E = \begin{bmatrix} 1 & 3 & 5 \\ 2 & 4 & 6 \\ 0 & 0 & 0 \end{bmatrix} \quad F = \begin{bmatrix} 1 & 2 \\ 0 & -1 \end{bmatrix}$$

1. AB
2. AC
3. AD
4. AE
5. AF
6. FA
7. EA
8. DA
9. CA
10. BA
11. BC
12. BD
13. BE
14. BF
15. FB
16. EB
17. DB
18. CB
19. CD
20. CE
21. CF
22. FC
23. EC
24. DC
25. DE
26. DF
27. FD
28. ED
29. EF
30. FE
31. $A(B+E)$
32. B^2
33. $A(BF)$
34. $(AB)F$
35. $C(4E)$
36. $D(B-F)$
37. A^2
38. $(F-D)C$
39. $(B+D)C$
40. $(6A)(4D)$
41. $AB + AE$
42. $DB - DF$

43. Is matrix multiplication commutative? Use examples from Problems 1–40 to support your answer.

44. Is matrix multiplication associative? Use an example from Problems 1–40 to support your answer.

45. What can you conclude from Problems 31 and 41?

46. What can you conclude about the product of any 3×3 matrix with matrix D?

47. Consider this system of equations:

$$2x + 3y = 4$$
$$-x + 4y = -2$$

Form the 2×2 "coefficient" matrix, consisting of the coefficients of x and y; form a 2×1 "variable" matrix, consisting of x and y; and form a 2×1 "constant" matrix, consisting of 4 and -2. Write an equation relating these three matrices.

48. Consider this system of equations:

$$x + 2y + 4z = 5$$
$$3x - y + 3z = 1$$
$$2x - 4y + z = 0$$

Form the three matrices as in Problem 47, and find an equation relating the matrices.

For Problems 49–52, suppose the Emperor Penguin Company produces three kinds of stuffed penguins, small, medium, and large, at three different locations. In Big Rapids, it produces 150, 50, and 200, respectively, of each type of penguin; Clio produces 300, 100, and 100, respectively; Elsie produces 80, 150, and 500, respectively. Small penguins cost $4 each to produce and $2 each to transport; medium penguins cost $7 each to produce and $3 each to transport; large penguins cost $15 each to produce and $4 each to transport.

49. Form the 3×3 production matrix.

50. Form the 3×2 cost matrix.

51. Find the product of the two matrices in Problems 49 and 50.

52. What does the matrix in Problem 51 tell you about the Emperor Penguin Company?

12.4 Identity and Inverse Matrices for Multiplication

square matrices
identity matrix

We will only consider **square matrices**, matrices with the same number of rows and columns, when discussing multiplicative identities and inverses. The **identity matrix** for an $n \times n$ matrix is the matrix consisting of 1s on the *main diagonal*—the diagonal consisting of all entries whose subscripts are the same, such as $A(1, 1)$, $A(2, 2), \ldots, A(N, N)$—and 0s elsewhere. The 2×2 and 3×3 identity matrices are

$$I_2 = \begin{bmatrix} 1 & 0 \\ 0 & 1 \end{bmatrix} \quad I_3 = \begin{bmatrix} 1 & 0 & 0 \\ 0 & 1 & 0 \\ 0 & 0 & 1 \end{bmatrix}$$

The $n \times n$ identity matrix will be referred to as I_n, or sometimes simply as I.

■ **EXAMPLE 12.10** Illustrate that I_2 and I_3 are identity matrices.

Solution We will multiply some arbitrarily chosen matrices by I_2 and I_3, showing that this multiplication does not change the matrix. Thus,

$$\begin{bmatrix} 3 & 2 \\ -4 & 1 \end{bmatrix} \cdot \begin{bmatrix} 1 & 0 \\ 0 & 1 \end{bmatrix} = \begin{bmatrix} (3 \cdot 1)+(2 \cdot 0) & (3 \cdot 0)+(2 \cdot 1) \\ (-4 \cdot 1)+(1 \cdot 0) & (-4 \cdot 0)+(1 \cdot 1) \end{bmatrix}$$

$$= \begin{bmatrix} 3+0 & 0+2 \\ -4+0 & 0+1 \end{bmatrix} = \begin{bmatrix} 3 & 2 \\ -4 & 1 \end{bmatrix}$$

and

$$\begin{bmatrix} 2 & 0 & 1 \\ 3 & -1 & 4 \\ 1 & 0 & 2 \end{bmatrix} \cdot \begin{bmatrix} 1 & 0 & 0 \\ 0 & 1 & 0 \\ 0 & 0 & 1 \end{bmatrix} = \begin{bmatrix} 2 & 0 & 1 \\ 3 & -1 & 4 \\ 1 & 0 & 2 \end{bmatrix}$$

The matrices in Example 12.10 were not changed by multiplying by I_2 or I_3. Now, consider the following example.

■ **EXAMPLE 12.11** Perform the multiplication:

$$\begin{bmatrix} 2 & 1 \\ 0 & 3 \end{bmatrix} \cdot \begin{bmatrix} \frac{1}{2} & -\frac{1}{6} \\ 0 & \frac{1}{3} \end{bmatrix}$$

Solution
$$\begin{bmatrix} 2 & 1 \\ 0 & 3 \end{bmatrix} \cdot \begin{bmatrix} \frac{1}{2} & -\frac{1}{6} \\ 0 & \frac{1}{3} \end{bmatrix} = \begin{bmatrix} 1+0 & -\frac{2}{6}+\frac{1}{3} \\ 0+0 & 0+1 \end{bmatrix}$$
$$= \begin{bmatrix} 1 & 0 \\ 0 & 1 \end{bmatrix}$$

12.4 Identity and Inverse Matrices for Multiplication

multiplicative inverses

The two matrices in Example 12.11 are **multiplicative inverses** of each other because their product is I_2. Multiplication will always verify whether or not two matrices are inverses of each other. The inverse of A is denoted A^{-1}.

Given a square matrix, how do you find an inverse? While the procedure is straightforward, a short discussion is necessary to help you understand why the steps we will be taking are valid.

First, consider the following system of equations:

$$2x + 3y = 1$$
$$x - 2y = 4$$

We can form a matrix of coefficients for this system:

$$\begin{bmatrix} 2 & 3 & 1 \\ 1 & -2 & 4 \end{bmatrix}$$

Each row of this matrix represents one equation. This representation is a classic application of matrices. In finding inverses, we will be using three rules, called row transformations. These rules are based on the matrix representation of systems of equations such as the matrix of coefficients just presented. The rules are given next.

Rules for Row Transformations

1. Two rows may be interchanged. In other words, rearranging the order of the equations does not change the system.
2. A row may be multiplied by a nonzero constant. That is, multiplying both sides of an equation by a nonzero constant does not change the solution of an equation.
3. A multiple of one row may be added to another row. For example, we used this rule in eliminating a variable in the multiplication-addition method of solving systems of equations.

We will now use these rules to find the inverse of a matrix. To find the inverse of a matrix A, we will write the matrix side by side with I. We will then use the three rules to change A to I while performing the same steps on I. Once A has been changed to I, the original I will have been changed to A^{-1}.

Why does this method work? It works because when we are performing one of the row transformations, we are really multiplying by a matrix. For example, to interchange the two rows of a 2×2 matrix, we multiply the matrix by the matrix formed from interchanging the two rows of I_2:

$$\begin{bmatrix} 0 & 1 \\ 1 & 0 \end{bmatrix} \cdot \begin{bmatrix} 1 & 2 \\ 3 & 4 \end{bmatrix} = \begin{bmatrix} 3 & 4 \\ 1 & 2 \end{bmatrix}$$

In the row transformation, we start with the equation $AA^{-1} = I$. We successively multiply both sides of this equation by the matrices needed to perform the necessary row transformations to A. We are left with IA^{-1} on the left side of the equation, which equals A^{-1} on the right side of the equation. The procedure is illustrated in the next example.

■ **EXAMPLE 12.12** Find A^{-1} for

$$A = \begin{bmatrix} 1 & -2 \\ 2 & 3 \end{bmatrix}$$

Solution We first write A side by side with I_2:

$$\begin{bmatrix} 1 & -2 & | & 1 & 0 \\ 2 & 3 & | & 0 & 1 \end{bmatrix}$$

We then perform the following steps:

$$\begin{bmatrix} 1 & -2 & | & 1 & 0 \\ 0 & 7 & | & -2 & 1 \end{bmatrix} \quad \text{add } -2 \text{ times row 1 to row 2}$$

$$\begin{bmatrix} 1 & -2 & | & 1 & 0 \\ 0 & 1 & | & -\frac{2}{7} & \frac{1}{7} \end{bmatrix} \quad \text{multiply row 2 by } \frac{1}{7}$$

$$\begin{bmatrix} 1 & 0 & | & \frac{3}{7} & \frac{2}{7} \\ 0 & 1 & | & -\frac{2}{7} & \frac{1}{7} \end{bmatrix} \quad \text{add 2 times row 2 to row 1}$$

You can verify that

$$A^{-1} = \begin{bmatrix} \frac{3}{7} & \frac{2}{7} \\ -\frac{2}{7} & \frac{1}{7} \end{bmatrix}$$

by multiplying AA^{-1} and $A^{-1}A$ to get I_2 (see Problems 9 and 10).

When we find the inverse of a 3×3 matrix, the procedure is the same. However, we should proceed one column at a time, starting with column 1, in changing A to I to avoid complications.

■ **EXAMPLE 12.13** Find A^{-1} if

$$A = \begin{bmatrix} 0 & 2 & 1 \\ -3 & -10 & 2 \\ 1 & 4 & 0 \end{bmatrix}$$

12.4 Identity and Inverse Matrices for Multiplication

Solution
$$\left[\begin{array}{rrr|rrr} 0 & 2 & 1 & 1 & 0 & 0 \\ -3 & -10 & 2 & 0 & 1 & 0 \\ 1 & 4 & 0 & 0 & 0 & 1 \end{array}\right]$$

Since $A(1,1)$ is 0, we would be better off switching rows 1 and 3 to start the procedure:

$$\left[\begin{array}{rrr|rrr} 1 & 4 & 0 & 0 & 0 & 1 \\ -3 & -10 & 2 & 0 & 1 & 0 \\ 0 & 2 & 1 & 1 & 0 & 0 \end{array}\right] \text{ interchange rows 1 and 3}$$

$$\left[\begin{array}{rrr|rrr} 1 & 4 & 0 & 0 & 0 & 1 \\ 0 & 2 & 2 & 0 & 1 & 3 \\ 0 & 2 & 1 & 1 & 0 & 0 \end{array}\right] \begin{array}{l}\text{add 3 times row 1 to row 2 (column 1} \\ \text{complete)}\end{array}$$

$$\left[\begin{array}{rrr|rrr} 1 & 4 & 0 & 0 & 0 & 1 \\ 0 & 1 & 1 & 0 & \frac{1}{2} & \frac{3}{2} \\ 0 & 2 & 1 & 1 & 0 & 0 \end{array}\right] \text{ multiply row 2 by } \tfrac{1}{2}$$

$$\left[\begin{array}{rrr|rrr} 1 & 0 & -4 & 0 & -2 & -5 \\ 0 & 1 & 1 & 0 & \frac{1}{2} & \frac{3}{2} \\ 0 & 2 & 1 & 1 & 0 & 0 \end{array}\right] \text{ add } -4 \text{ times row 2 to row 1}$$

$$\left[\begin{array}{rrr|rrr} 1 & 0 & -4 & 0 & -2 & -5 \\ 0 & 1 & 1 & 0 & \frac{1}{2} & \frac{3}{2} \\ 0 & 0 & -1 & 1 & -1 & -3 \end{array}\right] \begin{array}{l}\text{add } -2 \text{ times row 2 to row 3} \\ \text{(column 2 complete)}\end{array}$$

$$\left[\begin{array}{rrr|rrr} 1 & 0 & -4 & 0 & -2 & -5 \\ 0 & 1 & 1 & 0 & \frac{1}{2} & \frac{3}{2} \\ 0 & 0 & 1 & -1 & 1 & 3 \end{array}\right] \text{ multiply row 3 by } -1$$

$$\left[\begin{array}{rrr|rrr} 1 & 0 & 0 & -4 & -2 & 7 \\ 0 & 1 & 1 & 0 & \frac{1}{2} & \frac{3}{2} \\ 0 & 0 & 1 & -1 & 1 & 3 \end{array}\right] \text{ add 4 times row 3 to row 1}$$

$$\left[\begin{array}{rrr|rrr} 1 & 0 & 0 & -4 & 2 & 7 \\ 0 & 1 & 0 & 1 & -\frac{1}{2} & -\frac{3}{2} \\ 0 & 0 & 1 & -1 & 1 & 3 \end{array}\right] \begin{array}{l}\text{add } -1 \text{ times row 3 to row 2} \\ \text{(column 3 complete)}\end{array}$$

$$A^{-1} = \begin{bmatrix} -4 & 2 & 7 \\ 1 & -\frac{1}{2} & -\frac{3}{2} \\ -1 & 1 & 3 \end{bmatrix}$$

EXAMPLE 12.14   Find A^{-1} for

$$A = \begin{bmatrix} 1 & 0 \\ 2 & 0 \end{bmatrix}$$

Solution

$$\left[\begin{array}{cc|cc} 1 & 0 & 1 & 0 \\ 2 & 0 & 0 & 1 \end{array}\right]$$

$$\left[\begin{array}{cc|cc} 1 & 0 & 1 & 0 \\ 0 & 0 & -2 & 1 \end{array}\right] \quad \text{add } -2 \text{ times row 1 to row 2 (column 1 complete)}$$

At this point we are stuck. We cannot get $A(2, 2)$ to equal 1. Thus, A does not have an inverse.

The result of Example 12.14 leads to a definition.

singular matrix,
nonsingular matrix

Definition 12.6   A square matrix that does not have an inverse is called a **singular matrix**. A matrix that does have an inverse is a **nonsingular matrix**.

In the next section, we will find a simple test to determine whether or not a matrix has an inverse. Several applications of inverse matrices will be discussed in the problem set and the next few sections.

Problem Set 12.4

1. What is I_4?
2. What is I_1?

In Problems 3–8, verify that $AI = A$ for the matrix.

3. $A = \begin{bmatrix} 2 & 1 \\ 3 & 4 \end{bmatrix}$

4. $A = \begin{bmatrix} -1 & 0 \\ 1 & 2 \end{bmatrix}$

5. $A = \begin{bmatrix} 1 & 1 \\ 1 & 1 \end{bmatrix}$

6. $A = \begin{bmatrix} 0 & 1 \\ 1 & 0 \end{bmatrix}$

7. $A = \begin{bmatrix} 1 & 2 & 4 \\ 0 & 3 & -1 \\ 1 & 0 & 1 \end{bmatrix}$

8. $A = \begin{bmatrix} 1 & 1 & 1 \\ 2 & 2 & 2 \\ 3 & 3 & 3 \end{bmatrix}$

9. Find AA^{-1} for Example 12.12.
10. Find $A^{-1}A$ for Example 12.12.
11. Find $A^{-1}A$ for Example 12.13.
12. Find AA^{-1} for Example 12.13.

In Problems 13–18, determine whether or not A and B are inverses.

13. $A = \begin{bmatrix} 1 & 3 \\ 2 & 0 \end{bmatrix}, B = \begin{bmatrix} 1 & 2 \\ 3 & 0 \end{bmatrix}$

14. $A = \begin{bmatrix} 2 & 0 \\ 0 & 2 \end{bmatrix}, B = \begin{bmatrix} \frac{1}{2} & 0 \\ 0 & \frac{1}{2} \end{bmatrix}$

15. $A = \begin{bmatrix} 0 & 1 \\ 1 & 0 \end{bmatrix}, B = \begin{bmatrix} 0 & 1 \\ 1 & 0 \end{bmatrix}$

16. $A = \begin{bmatrix} 2 & 1 \\ 1 & 0 \end{bmatrix}, B = \begin{bmatrix} 0 & 1 \\ 1 & -2 \end{bmatrix}$

12.4 Identity and Inverse Matrices for Multiplication

17. $A = \begin{bmatrix} 0 & 0 & 3 \\ 0 & 3 & 0 \\ 3 & 0 & 0 \end{bmatrix}, B = \begin{bmatrix} 0 & 0 & \frac{1}{3} \\ 0 & \frac{1}{3} & 0 \\ \frac{1}{3} & 0 & 0 \end{bmatrix}$

18. $A = \begin{bmatrix} 1 & 0 & 4 \\ 0 & 1 & 2 \\ 1 & 0 & 0 \end{bmatrix}, B = \begin{bmatrix} 1 & 0 & 1 \\ 0 & 1 & 0 \\ 4 & 2 & 0 \end{bmatrix}$

In Problems 19–36, find A^{-1} if possible.

19. $A = [6]$
20. $A = [-4]$
21. $A = \begin{bmatrix} 1 & 0 \\ 0 & 4 \end{bmatrix}$
22. $A = \begin{bmatrix} 0 & 1 \\ 1 & 0 \end{bmatrix}$
23. $A = \begin{bmatrix} 1 & 1 \\ 1 & 1 \end{bmatrix}$
24. $A = \begin{bmatrix} 0 & 3 \\ 3 & 1 \end{bmatrix}$
25. $A = \begin{bmatrix} 5 & 0 \\ 0 & 5 \end{bmatrix}$
26. $A = \begin{bmatrix} 1 & 2 \\ -2 & -4 \end{bmatrix}$
27. $A = \begin{bmatrix} 1 & 2 \\ 3 & 4 \end{bmatrix}$
28. $A = \begin{bmatrix} -1 & 1 \\ 2 & 5 \end{bmatrix}$
29. $A = \begin{bmatrix} 0 & 0 & -2 \\ 0 & -2 & 0 \\ -2 & 0 & 0 \end{bmatrix}$
30. $A = \begin{bmatrix} 4 & 0 & 0 \\ 0 & 4 & 0 \\ 0 & 0 & 4 \end{bmatrix}$
31. $A = \begin{bmatrix} 1 & 0 & 1 \\ 0 & 2 & 3 \\ 5 & 1 & 0 \end{bmatrix}$
32. $A = \begin{bmatrix} 1 & 2 & 5 \\ 0 & 1 & 4 \\ 3 & 2 & 6 \end{bmatrix}$
33. $A = \begin{bmatrix} 2 & 6 & -4 \\ 0 & 1 & 3 \\ -1 & -3 & 2 \end{bmatrix}$
34. $A = \begin{bmatrix} 2 & 7 & 7 \\ 0 & 0 & 0 \\ -1 & 3 & 5 \end{bmatrix}$
35. $A = \begin{bmatrix} 1 & 2 & 0 & 1 \\ 3 & 4 & 0 & 3 \\ 1 & 0 & 2 & 1 \\ 0 & 1 & 2 & 1 \end{bmatrix}$
36. $A = \begin{bmatrix} 5 & 0 & 0 & 5 \\ 0 & 1 & 2 & 3 \\ 4 & 0 & 2 & -2 \\ 1 & 0 & 3 & 1 \end{bmatrix}$

In economics, the matrix $(I - A)^{-1}$ is important. For Problems 37–40, find $(I - A)^{-1}$ for matrix A of the indicated problem.

37. Problem 5
38. Problem 6
39. Problem 7
40. Problem 8
41. Can a matrix that contains a row of 0s have an inverse? If so, give an example.
42. Can a matrix that contains a column of 0s have an inverse? If so, give an example.
43. Can a matrix that contains two identical columns have an inverse? Make up an example and try it.
44. Can a matrix that contains two identical rows have an inverse? Make up an example and try it.
45. There are four 2×2 matrices that are their own inverse ($AA = I_2$). Find them.
46. There are several idempotent 2×2 matrices ($AA = A$). Find two of them.

Division of matrices can be defined as multiplication by the inverse ($A/B = AB^{-1}$, if B^{-1} exists). In Problems 47–50, find A/B for the matrices.

47. $A = \begin{bmatrix} 1 & 3 \\ 2 & 0 \end{bmatrix}, B = \begin{bmatrix} 1 & 2 \\ 3 & 4 \end{bmatrix}$
48. $A = \begin{bmatrix} 1 & 2 \\ 0 & 1 \end{bmatrix}, B = \begin{bmatrix} -1 & 1 \\ 2 & 5 \end{bmatrix}$
49. $A = \begin{bmatrix} 0 & 2 & 1 \\ 4 & 3 & 0 \\ 1 & 2 & 1 \end{bmatrix}, B = \begin{bmatrix} 1 & 0 & 1 \\ 0 & 2 & 3 \\ 5 & 1 & 0 \end{bmatrix}$
50. $A = \begin{bmatrix} 3 & -1 & 0 \\ 1 & 4 & 2 \\ 3 & -2 & 0 \end{bmatrix}, B = \begin{bmatrix} 1 & 2 & 5 \\ 0 & 1 & 4 \\ 3 & 2 & 6 \end{bmatrix}$

12.5 Determinants

Every square matrix has a number associated with it called a determinant, which is useful in many applications. For example, determinants will be used to find the multiplicative inverses of matrices and to solve systems of linear equations. We will discuss how to find determinants in this section. We will give some reasons behind the procedures when we get to a specific application in the next section.

determinant

The **determinant**, denoted $|A|$, of a 2×2 matrix A is the product of the two elements on the main diagonal minus the product of the other two elements. The next example illustrates.

■ **EXAMPLE 12.15** Find the determinant of the matrix.

a. $A = \begin{bmatrix} 3 & 4 \\ 1 & 2 \end{bmatrix}$

Solution $|A| = (3 \cdot 2) - (4 \cdot 1) = 6 - 4 = \underline{2}$

b. $B = \begin{bmatrix} 1 & -3 \\ -2 & 6 \end{bmatrix}$

Solution $|B| = (1 \cdot 6) - (-3 \cdot -2) = 6 - 6 = \underline{0}$

If the determinant of a matrix is 0, *the matrix does not have an inverse.* Otherwise, the matrix does have an inverse. In Example 12.15, A has an inverse ($|A| \neq 0$), but B does not have an inverse ($|B| = 0$). So finding the determinant of a matrix provides a quick test about whether or not a matrix is nonsingular.

To find the determinant of a 3×3 matrix is more complicated. But we first need to talk about what we mean by the minor and the cofactor of an element.

Consider *any* element in a 3×3 matrix. If we cross out the row and column that contain this element, we are left with a 2×2 matrix. The determinant of this 2×2 matrix is the **minor** of the element, denoted m_{ij}.

minor

■ **EXAMPLE 12.16** Find the minor of each element of A:

$$A = \begin{bmatrix} 1 & 0 & 2 \\ 3 & 1 & -1 \\ 2 & 1 & 4 \end{bmatrix}$$

Solution To find m_{11}, we cross out row 1 and column 1 of A, leaving

$$\begin{bmatrix} 1 & -1 \\ 1 & 4 \end{bmatrix}$$

12.5 Determinants

So

$$m_{11} = \begin{vmatrix} 1 & -1 \\ 1 & 4 \end{vmatrix} = (1 \cdot 4) - (-1 \cdot 1) = 4 - (-1) = \underline{5}$$

Similarly, we get the following minors:

$$m_{12} = \begin{vmatrix} 3 & -1 \\ 2 & 4 \end{vmatrix} = 12 - (-2) = \underline{14}$$

$$m_{13} = \begin{vmatrix} 3 & 1 \\ 2 & 1 \end{vmatrix} = 3 - 2 = \underline{1}$$

$$m_{21} = \begin{vmatrix} 0 & 2 \\ 1 & 4 \end{vmatrix} = 0 - 2 = \underline{-2}$$

$$m_{22} = \begin{vmatrix} 1 & 2 \\ 2 & 4 \end{vmatrix} = 4 - 4 = \underline{0}$$

$$m_{23} = \begin{vmatrix} 1 & 0 \\ 2 & 1 \end{vmatrix} = 1 - 0 = \underline{1}$$

$$m_{31} = \begin{vmatrix} 0 & 2 \\ 1 & -1 \end{vmatrix} = 0 - 2 = \underline{-2}$$

$$m_{32} = \begin{vmatrix} 1 & 2 \\ 3 & -1 \end{vmatrix} = -1 - 6 = \underline{-7}$$

$$m_{33} = \begin{vmatrix} 1 & 0 \\ 3 & 1 \end{vmatrix} = 1 - 0 = \underline{1}$$

cofactor

The **cofactor** of an element, denoted c_{ij}, is the minor times 1 or -1, depending on the following rule:

$$c_{ij} = m_{ij} \text{ if } i+j \text{ is even}$$
$$c_{ij} = -m_{ij} \text{ if } i+j \text{ is odd}$$

■ **EXAMPLE 12.17** Find the cofactor of each element of A in Example 12.16.

Solution
$c_{11} = \underline{5}$ $(1+1$ is even$)$
$c_{12} = \underline{-14}$ $(1+2$ is odd$)$

$$c_{13} = \underline{1}$$
$$c_{21} = -(-2) = \underline{2}$$
$$c_{22} = \underline{0}$$
$$c_{23} = \underline{-1}$$

$$c_{31} = \underline{-2}$$
$$c_{32} = \underline{7}$$
$$c_{33} = \underline{1}$$

To find the determinant of a 3×3 matrix, we can take *any row or column* and find the sum of each element in the row or column times its cofactor.

■ **EXAMPLE 12.18** Find $|A|$ for matrix A of Example 12.16.

Solution Let's pick row 1. Then

$$\begin{aligned} |A| &= (1 \cdot c_{11}) + (0 \cdot c_{12}) + (2 \cdot c_{13}) \\ &= (1 \cdot 5) + (0 \cdot -14) + (2 \cdot 1) \\ &= 5 + 0 + 2 \\ &= \underline{7} \end{aligned}$$

As a check of our answer, we will now find $|A|$ by using column 3:

$$\begin{aligned} |A| &= (2 \cdot c_{13}) + (-1 \cdot c_{23}) + (4 \cdot c_{33}) \\ &= (2 \cdot 1) + (-1 \cdot -1) + (4 \cdot 1) \\ &= 2 + 1 + 4 = \underline{7} \end{aligned}$$

Since you can choose any row or column to find the determinant, you should choose the one that contains the simplest numbers (0s and 1s). In Example 12.18, perhaps the "easiest" choice would have been column 2, which consists of a 0 and two 1s. In this case, $|A| = 0 + c_{22} + c_{32}$, which is $0 + 0 + 7$, or $\underline{7}$.

The method of cofactors is easily adaptable to larger matrices.

There is another method for finding the determinant of a 3×3 matrix, which may be easier in some cases than the first method. But it is not adaptable to larger matrices. We will describe this method by using an example.

Consider the matrix A of Example 12.16. Write the first two columns of A to the right of A as follows:

$$\begin{bmatrix} 1 & 0 & 2 & | & 1 & 0 \\ 3 & 1 & -1 & | & 3 & 1 \\ 2 & 1 & 4 & | & 2 & 1 \end{bmatrix}$$

Starting with the main diagonal, there are three 3-element diagonals from upper left to lower right and three 3-element diagonals from lower left to upper right, as follows:

12.5 Determinants

$$\begin{bmatrix} 1 & 0 & 2 & 1 & 0 \\ 3 & 1 & -1 & 3 & 1 \\ 2 & 1 & 4 & 2 & 1 \end{bmatrix}$$

The determinant $|A|$ is simply the sum of the products of the first three diagonals minus the sum of the products of the other three diagonals:

$$\begin{aligned} |A| &= [(1\cdot 1\cdot 4)+(0\cdot -1\cdot 2)+(2\cdot 3\cdot 1)] \\ &\quad -[(2\cdot 1\cdot 2)+(1\cdot -1\cdot 1)+(4\cdot 3\cdot 0)] \\ &= (4+0+6)-(4+-1+0) \\ &= 10-3 = \underline{7} \end{aligned}$$

Determinants and cofactors can be used in an alternative method for finding inverses of matrices. The steps of the procedure are as follows:

Step 1 Find the determinant of the matrix. If the determinant is zero, there is no inverse.

Step 2 Form the *matrix of cofactors* by replacing each element of the matrix with its cofactor.

Step 3 Transpose the matrix of cofactors.

Step 4 Multiply the resulting matrix by the reciprocal of the determinant.

This method is illustrated in the next example.

■ **EXAMPLE 12.19** Find A^{-1} if possible.

a. $A = \begin{bmatrix} 1 & 3 \\ 2 & 6 \end{bmatrix}$

Solution $|A| = (1\cdot 6)-(3\cdot 2) = 6-6 = 0$

Therefore, <u>no inverse</u> exists.

b. $A = \begin{bmatrix} 3 & 4 \\ 1 & 2 \end{bmatrix}$

Solution **Step 1** $|A| = (3\cdot 2)-(4\cdot 1) = 6-4 = 2$

Step 2 The cofactor of an element in a 2×2 matrix is plus or minus the only element left when the row and column are crossed out:

$$\begin{bmatrix} 2 & -1 \\ -4 & 3 \end{bmatrix} \quad \text{(1 and 4 have a negative sign because } i+j \text{ is odd)}$$

Step 3 $\begin{bmatrix} 2 & -4 \\ -1 & 3 \end{bmatrix}$ (interchange rows and columns)

Step 4 $A^{-1} = \frac{1}{2} \cdot \begin{bmatrix} 2 & -4 \\ -1 & 3 \end{bmatrix} = \begin{bmatrix} 1 & -2 \\ -\frac{1}{2} & \frac{3}{2} \end{bmatrix}$

c. $A = \begin{bmatrix} 1 & 0 & 2 \\ 3 & 1 & -1 \\ 2 & 1 & 4 \end{bmatrix}$

Solution Step 1 $|A| = 7$ (from Example 12.18)

Step 2 $\begin{bmatrix} 5 & -14 & 1 \\ 2 & 0 & -1 \\ -2 & 7 & 1 \end{bmatrix}$ (matrix of cofactors)

Step 3 $\begin{bmatrix} 5 & 2 & -2 \\ -14 & 0 & 7 \\ 1 & -1 & 1 \end{bmatrix}$ (transpose)

Step 4 $A^{-1} = \frac{1}{7} \cdot \begin{bmatrix} 5 & 2 & -2 \\ -14 & 0 & 7 \\ 1 & -1 & 1 \end{bmatrix} = \begin{bmatrix} \frac{5}{7} & \frac{2}{7} & -\frac{2}{7} \\ -2 & 0 & 1 \\ \frac{1}{7} & -\frac{1}{7} & \frac{1}{7} \end{bmatrix}$

Problem Set 12.5

In Problems 1–10, use the following matrix A to find the specified minors and cofactors.

$A = \begin{bmatrix} 7 & 3 & 1 \\ 3 & 0 & -2 \\ 0 & -1 & 4 \end{bmatrix}$

1. m_{12}
2. m_{31}
3. m_{22}
4. m_{12}
5. c_{11}
6. c_{33}
7. c_{21}
8. c_{13}
9. c_{23}
10. c_{22}

In Problems 11–20, find the determinants of the matrix.

11. $\begin{bmatrix} 1 & 0 \\ 2 & 4 \end{bmatrix}$
12. $\begin{bmatrix} -1 & 2 \\ 6 & 3 \end{bmatrix}$
13. $\begin{bmatrix} 3 & 2 \\ 4 & -1 \end{bmatrix}$
14. $\begin{bmatrix} 1 & 0 \\ 0 & 1 \end{bmatrix}$
15. $\begin{bmatrix} 3 & 4 \\ 2 & -1 \end{bmatrix}$
16. $\begin{bmatrix} 3 & -5 \\ 2 & 7 \end{bmatrix}$
17. $\begin{bmatrix} 0 & 1 \\ 1 & 0 \end{bmatrix}$
18. $\begin{bmatrix} -3 & -2 \\ -1 & 4 \end{bmatrix}$
19. $\begin{bmatrix} 2 & -8 \\ 3 & -6 \end{bmatrix}$
20. $\begin{bmatrix} 3 & -3 \\ 3 & 3 \end{bmatrix}$

12.6 Systems of Equations: By Matrices

In Problems 21–30, find the determinant of the matrix by use of cofactors.

21. $\begin{bmatrix} 1 & 0 & 0 \\ 0 & 1 & 0 \\ 0 & 0 & 1 \end{bmatrix}$

22. $\begin{bmatrix} 0 & 0 & 1 \\ 0 & 1 & 0 \\ 1 & 0 & 0 \end{bmatrix}$

23. $\begin{bmatrix} 2 & 0 & 1 \\ 3 & 0 & -2 \\ 2 & 0 & 4 \end{bmatrix}$

24. $\begin{bmatrix} 1 & 3 & -5 \\ 2 & 0 & 4 \\ 0 & 0 & 0 \end{bmatrix}$

25. $\begin{bmatrix} 1 & 1 & 1 \\ 1 & 1 & 1 \\ 1 & 1 & 1 \end{bmatrix}$

26. $\begin{bmatrix} 2 & 0 & 1 \\ 0 & 3 & 4 \\ 2 & 1 & 0 \end{bmatrix}$

27. $\begin{bmatrix} 3 & 2 & -1 \\ 2 & 4 & 3 \\ 1 & 1 & 2 \end{bmatrix}$

28. $\begin{bmatrix} 2 & -2 & 2 \\ 1 & 4 & 3 \\ 5 & 1 & 2 \end{bmatrix}$

29. $\begin{bmatrix} 0 & 7 & 0 \\ 7 & 0 & 7 \\ 0 & 7 & 0 \end{bmatrix}$

30. $\begin{bmatrix} 2 & 1 & 5 \\ -2 & 4 & 1 \\ 2 & 3 & 2 \end{bmatrix}$

For Problems 31–40, find the determinant of the matrix in the indicated problem by using the method of diagonals.

31. Problem 21
32. Problem 22
33. Problem 23
34. Problem 24
35. Problem 25
36. Problem 26
37. Problem 27
38. Problem 28
39. Problem 29
40. Problem 30

In Problems 41–60, use cofactors to find the inverse of the matrix in the indicated problem.

41. Problem 11
42. Problem 12
43. Problem 13
44. Problem 14
45. Problem 15
46. Problem 16
47. Problem 17
48. Problem 18
49. Problem 19
50. Problem 20
51. Problem 21
52. Problem 22
53. Problem 23
54. Problem 24
55. Problem 25
56. Problem 26
57. Problem 27
58. Problem 28
59. Problem 29
60. Problem 30

61. Does $|A| = |A^T|$ for 2×2 matrices? (*Hint:* Consider Problems 13 and 15.)
62. Does $|A| = |A^T|$ for 3×3 matrices?
63. Write an assignment statement in BASIC to find $|A|$ for a 2×2 matrix.
64. Write four assignment statements in BASIC to find the matrix of cofactors for a 2×2 matrix.

12.6 Systems of Equations: By Matrices

Gaussian elimination

In this section, we are going to again consider solving systems of equations. This time we will use matrices. The procedure that we will follow is called **Gaussian elimination**, and it follows the principles of the multiplication-addition method.

Consider the following system of equations:

$$3x + y = 5$$
$$6x + y = 17$$

We can eliminate the x term from the second equation by adding -2 times the first equation to the second equation:

$$3x + y = 5$$
$$-y = 7$$

We can then eliminate the y term from the first equation by adding the second equation to the first equation:

$$3x = 12$$
$$-y = 7$$

Each equation can now be solved for the designated variable:

$$x = 4$$
$$y = -7$$

The solution is $(4, -7)$.

We are going to use this method with matrices in the next example.

■ **EXAMPLE 12.20** Solve this system by using matrices.

$$3x + y = 5$$
$$6x + y = 17$$

Solution First, we form the matrix connected with the system as follows:

$$\begin{bmatrix} 3 & 1 & | & 5 \\ 6 & 1 & | & 17 \end{bmatrix}$$

augmented matrix

This matrix is called the **augmented matrix**.

Now, we use the procedure for finding an inverse to transform the first two columns to I_2. The solution can then be found in the last column:

$$\begin{bmatrix} 3 & 1 & | & 5 \\ 0 & -1 & | & 7 \end{bmatrix} \quad \text{add } -2 \text{ times row 1 to row 2}$$

$$\begin{bmatrix} 3 & 0 & | & 12 \\ 0 & -1 & | & 7 \end{bmatrix} \quad \text{add row 2 to row 1}$$

$$\begin{bmatrix} 1 & 0 & | & 4 \\ 0 & -1 & | & 7 \end{bmatrix} \quad \text{multiply row 1 by } \tfrac{1}{3}$$

$$\begin{bmatrix} 1 & 0 & | & 4 \\ 0 & 1 & | & -7 \end{bmatrix} \quad \text{multiply row 2 by } -1$$

The solution is $(4, -7)$.

12.6 Systems of Equations: By Matrices

The procedure for solving a system of three equations in three variables is similar. The first three columns are transformed to I_3, and the fourth column contains the solution.

EXAMPLE 12.21 Solve this system of equations:

$$2x + y - 3z = 0$$
$$x + 2y = 3$$
$$4x + 2z = -4$$

Solution First, form the augmented matrix:

$$\begin{bmatrix} 2 & 1 & -3 & | & 0 \\ 1 & 2 & 0 & | & 3 \\ 4 & 0 & 2 & | & -4 \end{bmatrix}$$

Now, reduce the first three columns to I_3:

$$\begin{bmatrix} 1 & \frac{1}{2} & -\frac{3}{2} & | & 0 \\ 1 & 2 & 0 & | & 3 \\ 4 & 0 & 2 & | & -4 \end{bmatrix} \quad \text{multiply row 1 by } \tfrac{1}{2}$$

$$\begin{bmatrix} 1 & \frac{1}{2} & -\frac{3}{2} & | & 0 \\ 0 & \frac{3}{2} & \frac{3}{2} & | & 3 \\ 4 & 0 & 2 & | & -4 \end{bmatrix} \quad \text{add } -1 \text{ times row 1 to row 2}$$

$$\begin{bmatrix} 1 & \frac{1}{2} & -\frac{3}{2} & | & 0 \\ 0 & \frac{3}{2} & \frac{3}{2} & | & 3 \\ 0 & -2 & 8 & | & -4 \end{bmatrix} \quad \text{add } -4 \text{ times row 1 to row 3 (column 1 complete)}$$

$$\begin{bmatrix} 1 & \frac{1}{2} & -\frac{3}{2} & | & 0 \\ 0 & 1 & 1 & | & 2 \\ 0 & -2 & 8 & | & -4 \end{bmatrix} \quad \text{multiply row 2 by } \tfrac{2}{3}$$

$$\begin{bmatrix} 1 & 0 & -2 & | & -1 \\ 0 & 1 & 1 & | & 2 \\ 0 & -2 & 8 & | & -4 \end{bmatrix} \quad \text{add } -\tfrac{1}{2} \text{ times row 2 to row 1}$$

$$\begin{bmatrix} 1 & 0 & -2 & | & -1 \\ 0 & 1 & 1 & | & 2 \\ 0 & 0 & 10 & | & 0 \end{bmatrix} \quad \text{add 2 times row 2 to row 3 (column 2 complete)}$$

$$\begin{bmatrix} 1 & 0 & -2 & | & -1 \\ 0 & 1 & 1 & | & 2 \\ 0 & 0 & 1 & | & 0 \end{bmatrix} \quad \text{multiply row 3 by } \tfrac{1}{10}$$

$$\begin{bmatrix} 1 & 0 & 0 & | & -1 \\ 0 & 1 & 1 & | & 2 \\ 0 & 0 & 1 & | & 0 \end{bmatrix} \quad \text{add 2 times row 3 to row 1}$$

$$\begin{bmatrix} 1 & 0 & 0 & | & -1 \\ 0 & 1 & 0 & | & 2 \\ 0 & 0 & 1 & | & 0 \end{bmatrix} \quad \text{add } -1 \text{ times row 3 to row 2 (column 3 complete)}$$

The solution is $(-1, 2, 0)$.

Not all systems of equations have unique solutions. When we try to solve an inconsistent or dependent system, a row of 0s will appear to the left of the dotted line, making it impossible to obtain I. If the number to the right of the dotted line in that row is zero ($0 = 0$), the system is dependent, having many solutions. Otherwise, the system is inconsistent.

■ **EXAMPLE 12.22** Solve the system of equations, if possible.

a. $2x - 3y = 4$
$4x - 6y = 10$

Solution

$$\begin{bmatrix} 2 & -3 & | & 4 \\ 4 & -6 & | & 10 \end{bmatrix}$$

You can verify (Problem 9) that this matrix reduces to:

$$\begin{bmatrix} 1 & -\tfrac{3}{2} & | & 2 \\ 0 & 0 & | & 2 \end{bmatrix}$$

The last row corresponds to the equation $0 = 2$. The system is inconsistent and has no solution.

b. $x - 2y + z = 3$
$2x - 4y + 2z = 6$
$x + y + z = 3$

Solution

$$\begin{bmatrix} 1 & -2 & 1 & | & 3 \\ 2 & -4 & 2 & | & 6 \\ 1 & 1 & 1 & | & 3 \end{bmatrix}$$

12.6 Systems of Equations: By Matrices

You can verify (Problem 10) that this matrix reduces to

$$\begin{bmatrix} 1 & 0 & 1 & | & 3 \\ 0 & 1 & 0 & | & 0 \\ 0 & 0 & 0 & | & 0 \end{bmatrix}$$

The last row corresponds to $0 = 0$. The system is dependent and has an infinite number of solutions.

This matrix method is adaptable to systems of more equations in more variables.

Problem Set 12.6

In Problems 1–6, use Gaussian elimination without matrices to solve the system of equations.

1. $2x + y = 4$
   $x - y = 2$

2. $3x - y = 7$
   $x + 2y = 0$

3. $x + 3y = 2$
   $2x + 4y = 2$

4. $x + 5y = 7$
   $-2x - 10y = 4$

5. $-x + 3y = 4$
   $3x - 9y = -12$

6. $5x + 3y = 6$
   $2x - 4y = 7$

7. Show that $(4, -7)$ is the solution to the system in Example 12.20.

8. Show that $(-1, 2, 0)$ is the solution to the system in Example 12.21.

9. Show that the matrix reduction in Example 12.22a is correct.

10. Show that the matrix reduction in Example 12.22b is correct.

In Problems 11–32, use matrices to solve the system of equations, if possible.

11. $x + y = 7$
    $y = 3$

12. $x = 4$
    $x + y = -2$

13. $x - 3y = 4$
    $2x - y = 4$

14. $x + 2y = -2$
    $3x - y = 15$

15. $2x + 3y = 10$
    $4x - y = 6$

16. $3x - 4y = 12$
    $4x - 3y = 12$

17. $2x + 3y = 4$
    $-4x - 6y = 2$

18. $8x - 2y = 10$
    $4x - y = 5$

19. $2x + 5y = -6$
    $-2x - 5y = 6$

20. $8x - 6y = 6$
    $4x - 3y = 0$

21. $x + y = 5$
    $y = 2$
    $z = 0$

22. $x = 5$
    $x + y = 7$
    $x + y + z = 10$

23. $x + 2y = 5$
    $x - 2y + z = 0$
    $y - z = -1$

24. $y - 2z = 2$
    $3x + 3z = 3$
    $2x + 3y = 0$

25. $2x + y + z = 2$
    $x - y + z = 3$
    $3x + 2y + z = 1$

26. $2x + 3y - z = 0$
    $x + 4y + 5z = 1$
    $-4x - 6y + 2z = 0$

27. $x - 3y + 2z = 2$
    $3x + 9y - 6z = 8$
    $2x - y - z = 9$

28. $2x + 2y - z = 0$
    $4x + 7y - 3z = 0$
    $x - 11y + 10z = 0$

29. $3x - y + z = 7$
    $2x + y - z = 3$
    $-8x - 4y + 4z = -12$

30. $2x - 6y + 2z = 0$
    $x + y + z = 3$
    $-4x + 12y - 4z = 10$

31. $\quad x+2y+z-w = 6$
$\quad\quad 2x+y+w = -1$
$\quad\quad 3x+y-z+2w = -5$
$\quad\quad -x+5y+2z+w = 0$
32. $\quad 2x-y+z-w = 10$
$\quad\quad y+3z-4w = -5$
$\quad\quad 4x+2y-3z = -4$
$\quad\quad x+w = 3$

In Problems 33–36, use matrices to solve the application.

33. The difference between two positive numbers is four. If the larger number is three times the smaller number, then find the numbers.
34. Find the lengths of the three sides of an isosceles triangle if the two equal sides are each 16 inches longer than the third side and the perimeter is 230 inches.
35. Four cans of orange juice and six cans of grapefruit juice cost $7.20. Five cans of orange juice and three cans of grapefruit juice cost $6.30. Find the cost per can for both items.
36. A company wants to mix two kinds of oil together, both containing the secret chemical JX–40. The first type of oil contains 10% JX–40, and the second type of oil contains 7% JX–40. How much of each type should be added together to achieve a 10-quart solution which is 8% JX–40?

In Problems 37–42, consider the following system of equations:

$$2x+y-z = 1$$
$$x-2y+z = 0$$
$$3x-y+2z = 7$$

37. Form the 3×3 coefficient matrix consisting of the coefficients of x, y, and z.
38. Form the 3×1 variable matrix consisting of x, y, and z.
39. Form the 3×1 constant matrix consisting of 1, 0, and 7.
40. Find an equation relating the matrices of Problems 37–39.
41. Find the inverse of the matrix in Problem 37.
42. Multiply both sides of the equation of Problem 40 by the inverse found in Problem 41. What do you have?

■ 12.7 Systems of Equations: By Cramer's Rule

Cramer's rule

In this section, we will consider one final method for solving systems of equations. This method involves neither matrices nor algebra, although the derivation is pure algebra. The method is called **Cramer's rule** after the Swiss mathematician who developed it.

Consider the general system of two equations in two variables:

$$a_1 x + b_1 y = c_1$$
$$a_2 x + b_2 y = c_2$$

If Gaussian elimination is used on this general system (Problems 5 and 6), the result is

12.7 Systems of Equations: By Cramer's Rule

$$x = \frac{b_2 c_1 - b_1 c_2}{a_1 b_2 - a_2 b_1}$$

$$y = \frac{a_1 c_2 - a_2 c_1}{a_1 b_2 - a_2 b_1}$$

The numerator and denominator of each fraction is the evaluation of a 2 × 2 determinant. The two denominators are identical and are simply the determinant of the coefficient matrix, which we will call D:

$$D = \begin{vmatrix} a_1 & b_1 \\ a_2 & b_2 \end{vmatrix}$$

The two numerators are the determinants of the matrices obtained from the coefficient matrix by replacing the column representing the variable desired by the constants c_1 and c_2:

$$D_x = \begin{vmatrix} c_1 & b_1 \\ c_2 & b_2 \end{vmatrix} \qquad D_y = \begin{vmatrix} a_1 & c_1 \\ a_2 & c_2 \end{vmatrix}$$

The solution then becomes

$$x = \frac{D_x}{D} \quad \text{and} \quad y = \frac{D_y}{D}$$

■ **EXAMPLE 12.23** Solve the system of equations.

a. $2x - 3y = 7$
    $3x + 4y = 2$

Solution

$$D = \begin{vmatrix} 2 & -3 \\ 3 & 4 \end{vmatrix} = 8 - (-9) = 17$$

$$D_x = \begin{vmatrix} 7 & -3 \\ 2 & 4 \end{vmatrix} = 28 - (-6) = 34$$

$$D_y = \begin{vmatrix} 2 & 7 \\ 3 & 2 \end{vmatrix} = 4 - 21 = -17$$

$$x = \frac{D_x}{D} = \frac{34}{17} = 2$$

$$y = \frac{D_y}{D} = \frac{-17}{17} = -1$$

The solution is $(2, -1)$.

b. $2x = 3 + 4y$
$y = -3x + 2$

Solution   Before we can use Cramer's rule, we must rearrange these equations:

$$2x - 4y = 3$$
$$3x + y = 2$$

Then we obtain

$$D = \begin{vmatrix} 2 & -4 \\ 3 & 1 \end{vmatrix} = 2 - (-12) = 14$$

$$D_x = \begin{vmatrix} 3 & -4 \\ 2 & 1 \end{vmatrix} = 3 - (-8) = 11$$

$$D_y = \begin{vmatrix} 2 & 3 \\ 3 & 2 \end{vmatrix} = 4 - 9 = -5$$

$$x = \frac{11}{14} \quad \text{and} \quad y = \frac{-5}{14}$$

The solution is $(\frac{11}{14}, -\frac{5}{14})$.

Cramer's rule can be easily extended by using square matrices of any size. The next example illustrates its use with a 3×3 system.

■ **EXAMPLE 12.24**   Solve the system of equations:

$$x + 3z = 2$$
$$2x - y + z = 4$$
$$3x + y - z = 1$$

Solution

$$D = \begin{vmatrix} 1 & 0 & 3 \\ 2 & -1 & 1 \\ 3 & 1 & -1 \end{vmatrix} = 15 \qquad D_y = \begin{vmatrix} 1 & 2 & 3 \\ 2 & 4 & 1 \\ 3 & 1 & -1 \end{vmatrix} = -25$$

$$D_x = \begin{vmatrix} 2 & 0 & 3 \\ 4 & -1 & 1 \\ 1 & 1 & -1 \end{vmatrix} = 15 \qquad D_z = \begin{vmatrix} 1 & 0 & 2 \\ 2 & -1 & 4 \\ 3 & 1 & 1 \end{vmatrix} = 5$$

$$x = \frac{D_x}{D} = \frac{15}{15} = 1 \qquad y = \frac{D_y}{D} = \frac{-25}{15} = \frac{-5}{3} \qquad z = \frac{D_z}{D} = \frac{5}{15} = \frac{1}{3}$$

The solution is $(1, -\frac{5}{3}, \frac{1}{3})$.

If the coefficient matrix D is zero, the system of equations is either inconsistent or dependent and cannot be solved by using Cramer's rule. Another method must be used.

Problem Set 12.7

1. Verify that $(2, -1)$ is the solution to the system in Example 12.23a.
2. Verify that $(\frac{11}{14}, -\frac{5}{14})$ is the solution to the system in Example 12.23b.
3. Verify the calculations of D, D_x, D_y, and D_z in Example 12.24.
4. Verify that $(1, -\frac{5}{3}, \frac{1}{3})$ is the solution to the system in Example 12.24.
5. Use Gaussian elimination without matrices to verify the formula for x for the general system of two equations in two variables.
6. Use Gaussian elimination without matrices to verify the formula for y for the general system of two equations in two variables.

In Problems 7–28, use Cramer's rule to solve the system of equations, if possible.

7. $2x = 4$
   $y = 3$
8. $2y = 6$
   $x - 3y = 8$
9. $x + 3y = 1$
   $2x - y = 7$
10. $2x - 5y = 0$
    $x + y = 7$
11. $4x - 3y = 2$
    $-3x + 4y = 2$
12. $6x - 2y = 8$
    $x - 7y = 8$
13. $2x - 3y = 1$
    $-6x + 9y = -3$
14. $3x - 9y = 3$
    $2x + 8y = 4$
15. $-3x + 5y = -7$
    $10x - 3y = 9$
16. $x - 7y = 6$
    $-x + 7y = 4$
17. $x = 3$
    $2x - y = 4$
    $3x + y + z = 8$
18. $y + 3z = 7$
    $2x - z = 2$
    $x - y = 1$
19. $2x - y + 2z = -4$
    $x + 4y + 10z = 16$
    $3x - y - 4z = 4$
20. $2x - 3y + 7z = 0$
    $3x + y - 6z = 0$
    $x + y - 4z = 0$
21. $5x + 3y + 3z = 7$
    $2x - 2y + 10z = -2$
    $6x + 4y - 2z = 0$
22. $2x + y - 3z = 2$
    $-7x + y - z = -7$
    $3x + 2y + z = 3$
23. $2x + z = 0$
    $4x + y + 2z = 7$
    $4x + 2z = 7$
24. $2x - y + z = 1$
    $y = 4x - 7$
    $4x - 4y - 1 = z$
25. $2x = -z$
    $4x + 6y - 2z = -12$
    $x - 3y + z = -3$
26. $x + 2y - z = 1$
    $2x - y + 3z = 2$
    $3x + 6y - 3z = 3$

27. $\begin{aligned} x+2y-w &= 0 \\ 3x-y+z+w &= 1 \\ 2x+3y-4z+w &= -4 \\ x-7y+5z-w &= 5 \end{aligned}$

28. $\begin{aligned} 3x+2y-z+4w &= 4 \\ y+3z+2w &= 0 \\ x+2y-z+w &= 6 \\ 2x+3w &= -1 \end{aligned}$

In Problems 29–32, solve the application by using Cramer's rule, if possible.

29. The length of a rectangle is 6 inches more than its width. The perimeter is 84 inches. Find the dimensions of the rectangle.

30. Ten pounds of beans and 4 pounds of cauliflower cost $4.18. Seven pounds of beans and 11 pounds of cauliflower cost $5.14. Find the cost per pound of each item.

31. A customer bought 33 stamps. They were all 22¢ stamps or 14¢ stamps. If she spent $6.22 for stamps, how many of each type did she get?

32. A student paid $ for tuition and room and board. If tuition was $300 less than three times room and board, how much did the student spend for tuition?

CHAPTER SUMMARY

In this chapter we studied arrays, which are very important in computer applications. In particular, we studied two-dimensional arrays, or matrices, in detail. We defined addition, subtraction, scalar multiplication, and matrix multiplication for matrices.

Determinants were also discussed, and several applications of determinants were covered. The determinant of a 2×2 matrix is simply the difference of the products of the elements on the two diagonals. Finding determinants of larger matrices requires the use of cofactors.

Some matrices have multiplicative inverses such that $AA^{-1} = I$, where I is the identity matrix for multiplication. Several techniques for finding inverses were discussed.

Systems of equations were solved by using matrices and row transformations. They were also solved by using Cramer's rule, which involves finding the coefficient matrix and, if that is not zero, proceeding to find corresponding determinants for each variable. These techniques are adaptable to any size of square matrix.

REVIEW PROBLEMS

In Problems 1–28, find the expression for the following arrays A, B, C, D, E, and F:

$A = \begin{bmatrix} 2 & 3 & 4 & 0 \end{bmatrix}$ $\qquad$ $B = \begin{bmatrix} 1 & 0 & -4 & 5 \end{bmatrix}$ $\qquad$ $C = \begin{bmatrix} 1 & -2 \\ 3 & 7 \end{bmatrix}$ $\qquad$ $D = \begin{bmatrix} -2 & 0 \\ 5 & 3 \end{bmatrix}$

Review Problems

$$E = \begin{bmatrix} 2 & 3 & 5 \\ 1 & -1 & 2 \\ 4 & 6 & -3 \end{bmatrix} \quad F = \begin{bmatrix} 4 & 0 & -1 \\ 2 & 1 & 3 \\ -1 & 5 & 0 \end{bmatrix}$$

1. a_2
2. b_3
3. c_{21}
4. e_{32}
5. $D(1, 2)$
6. $F(2, 3)$
7. $A + B$
8. $C + D$
9. $E + F$
10. $B - A$
11. $D - C$
12. $F - E$
13. $4E$
14. $-3D$
15. C^T
16. E^T
17. FE
18. CD
19. $3D - 4C$
20. $3E - 2F$
21. C^{-1}
22. D^{-1}
23. E^{-1}
24. F^{-1}
25. $|C|$
26. $|D|$
27. $|E|$
28. $|F|$

In Problems 29–32, solve the system of equations by matrices, if possible.

29. $2x + 3y = 4$
    $-3x + y = -6$
30. $4x + 7y - 4 = 0$
    $2x = y - 5$
31. $2x + y - 3z = 0$
    $-4x - 2y + 6z = 10$
    $3x - y - z = 6$
32. $x - 5y + 2z = 4$
    $3x + 4y - 5z = 6$
    $-2x + y + 2z = 10$

In Problems 33–36, solve the system of equations by Gaussian elimination.

33. $3x - y = 7$
    $2x + 5y = -1$
34. $-2x + 3y = 3$
    $8x - 12y = 12$
35. $x - y + 2z = -11$
    $3x + 2y - z = 3$
    $4x - 3y - 2z = -11$
36. $3x - y + 4z = 0$
    $x - y - 2z = 0$
    $2x + 3y + z = 2$

37. The sum of two numbers is 16. The larger number is 13 more than the smaller number. Find the numbers.

38. A basketball team scored eight more field goals (2 points) than they scored free throws (1 point). The total points scored were 67. How many free throws were made?

39. Write an algorithm to find the sum of two $m \times n$ matrices.

40. Write a BASIC decision to decide whether or not two matrices can be multiplied.

LINEAR PROGRAMMING

CHAPTER **13**

linear programming

One of the most frequently used applications of mathematics and computers is the problem of trying to maximize or minimize some quantity (like profit or cost, for example), subject to certain restrictions on the variables. In many cases, the quantity and restrictions can be described by linear equations and inequalities. The study of such problems is called **linear programming**.

In this chapter, we will solve linear programming problems by graphing and by a general procedure called the simplex method. The simplex method can be easily programmed on a computer, since there is a direct relationship between this method and solving a system of linear equations. Linear programming usually involves solving a system of linear inequalities.

13.1 Introduction to Linear Programming

objective function

constraints

A linear programming problem is one in which the quantity to be maximized or minimized and all of the restrictions are linear. The quantity to be maximized or minimized is called the **objective function** and will be represented by the letter z. The variables that are used to describe the objective function will be referred to as x_1, x_2, x_3, and so on. The restrictions that these variables must satisfy are referred to as the **constraints** of the problem.

Therefore, a linear programming problem consists of one objective function and one or more constraints. The goal of a linear programming problem is to find the *optimal solution*, the values of all variables which give the *best* value for the objective function (either a maximum or a minimum, whichever is desired), subject to the given constraints.

One of the hardest steps in solving a linear programming problem is to take the statement of the problem and change it into an objective function and a list of constraints. This task is the subject of the examples in this section. There are no rules to follow in translating words into mathematical equations and inequalities. However, the more of these problems you do, the better you will become at it.

The following four examples show how to set up the linear programming problems. We will solve problems like these in the following sections of this chapter.

EXAMPLE 13.1

Set up this problem: Two negative numbers are restricted such that their sum never exceeds -2. Find the smallest possible difference between the numbers.

Solution

Since we want the smallest possible difference, our objective function is $x_1 - x_2$, which we would like to minimize. Both x_1 and x_2 must be negative, and their sum cannot exceed -2. The problem looks like this:

13.1 Introduction to Linear Programming

$$\begin{aligned}\text{minimize:} \quad & z = x_1 - x_2 \\ \text{constraints:} \quad & x_1 < 0 \qquad x_2 < 0 \\ & x_1 + x_2 \le -2\end{aligned}$$

EXAMPLE 13.2 Set up this problem: The Spirit of Christmas Company sells three kinds of trees: plastic, aluminum, and real. Plastic trees sell for $10 each, aluminum trees sell for $20 each, and real trees sell for $15 each. The company wants to have no more than 100 total trees in stock and no more than 50 of any kind of tree. Maximize income under these conditions.

Solution If x_1, x_2, and x_3 represent the number of plastic, aluminum, and real trees, respectively, then the income is given by $10x_1 + 20x_2 + 15x_3$: the sum of the price of each type of tree times the number of each type of tree sold. Although not specifically mentioned as a constraint, we see from the statement of the problem that all variables must be nonnegative. Nonnegative variables are the case in *most* linear programming applications. Thus, our problem looks like this:

$$\begin{aligned}\text{maximize:} \quad & z = 10x_1 + 20x_2 + 15x_3 \\ \text{constraints:} \quad & x_1 \ge 0 \qquad x_2 \ge 0 \qquad x_3 \ge 0 \\ & x_1 + x_2 + x_3 \le 100 \\ & x_1 \le 50 \qquad x_2 \le 50 \qquad x_3 \le 50\end{aligned}$$

Since, usually, all variables will be nonnegative, rather than list each variable separately as in Example 13.2, we will simply list the constraint $x_i \ge 0$ whenever it applies in the future.

EXAMPLE 13.3 Set up this problem: The Cutting Edge Company produces two kinds of knives: hunting knives and kitchen knives. It can only spend up to $500 a week; making a hunting knife costs $7, and making a kitchen knife costs $5. The company can make no more than 80 knives a week and no more than 60 of either kind of knife. Hunting knives sell for $11 each, and kitchen knives sell for $8 each. Maximize the weekly profit.

Solution Profit is income minus cost, so the profit for a hunting knife is $4 ($11 − $7) and for a kitchen knife $3 ($8 − $5). Again, the variables are all nonnegative by the nature of the problem. Cost is also restricted to be no more than $500. The problem setup follows.

$$\begin{aligned}\text{maximize:} \quad & z = 4x_1 + 3x_2 \qquad \text{(profit)} \\ \text{constraints:} \quad & x_i \ge 0 \\ & 7x_1 + 5x_2 \le 500 \qquad \text{(cost)} \\ & x_1 + x_2 \le 80 \qquad x_1 \le 60 \qquad x_2 \le 60\end{aligned}$$

EXAMPLE 13.4 Set up this problem: The Picture Perfect Company produces slides, pictures, and posters at two stores with the following daily output:

Store	Slides	Pictures	Posters
1	100	200	150
2	200	300	100

The daily cost of operating the two stores is $100 and $150, respectively. The company must produce at least 3000 slides, 6000 pictures, and 1500 posters during the month of January. Find the minimum cost needed to meet demand. Also, find the number of days each store should be open if neither store can be open more than 20 days this month and the sum of the days both stores can be open is limited to 35.

Solution

minimize: $z = 100x_1 + 150x_2$ (cost)

constraints: $x_i \geq 0$

$$\left.\begin{array}{l} 100x_1 + 200x_2 \geq 3000 \\ 200x_1 + 300x_2 \geq 6000 \\ 150x_1 + 100x_2 \geq 1500 \end{array}\right\} \text{(to meet demand)}$$

$x_1 \leq 20 \quad x_2 \leq 20 \quad x_1 + x_2 \leq 35$

Problem Set 13.1

In Problems 1–6, determine whether the objective function is linear.

1. $z = 2x_1 - x_2$
2. $z = 15x_1 + 72x_2$
3. $z = x_1^2 - 4x_2 + 3$
4. $z = x_1 + 3x_2$
5. $z = 2^x + 3^x$
6. $z = \log_{10}(x_1 + x_2)$

In Problems 7–10, determine whether the inequality is linear.

7. $2x_1 + 4x_2 \geq 7$
8. $x_1^2 + x_2 \leq 4$
9. $5x_1 - 4 < x_2^2$
10. $-3x_1 + 30 > 17x_2$

In Problems 11–20, find the objective function and all constraints for the linear programming problem.

11. Two nonnegative numbers are restricted so that twice the first number plus three times the second number is never larger than 12. Find the smallest sum of these numbers under these conditions.

12. Two nonnegative numbers are restricted so that their sum is never larger than 10. What values for these numbers will make the sum of five times the first number and three times the second number as large as possible?

13. Two nonpositive numbers are restricted so that neither number is smaller than -10 and their sum is at least as large as -15. Find the values of the numbers that will make the difference of the first number and twice the second number as small as possible.

14. Two nonnegative numbers are restricted so that their sum cannot be larger than 4 and their difference cannot be larger than 2. Find the values of these numbers that will make the sum of three times the larger number and ten times the smaller number as large as possible.

15. The Penguin Puppet Company makes two kinds of puppets: large and small. Working at peak efficiency, it can make no more than 100

large puppets a day and no more than 200 small puppets a day. The total number of puppets produced cannot exceed 150. If small puppets sell for $5 each and large puppets sell for $7 each, find the maximum income possible.

16. The Space Puzzle Company makes two kinds of puzzles: beginner and advanced. It can make no more than 80 beginner puzzles and 60 advanced puzzles per week, but it is limited to making no more than 100 total puzzles per week. If beginner puzzles sell for $6 each and advanced puzzles sell for $7 each, find the number of each type of puzzle that produces the maximum income.

17. The Silver Bullet Company produces two kinds of guns: muzzle loader rifles and flintlock rifles. It has only $400 to spend; making a muzzle loader costs $40, and making a flintlock costs $50. The company only wants to make a maximum of nine guns per week, and it can only get materials for seven muzzle loaders at most and six flintlocks at most per week. The company wants to sell the muzzle loaders for $140 each and the flintlocks for $200 each. At these prices, the company will have no trouble selling all it makes. Find the maximum profit possible per week.

18. Dinos Unlimited produces two kinds of stuffed dinosaurs: brontosaurus and tyrannosaurus. It can only spend $300 a day; making a brontosaurus costs $5, and making a tyrannosaurus costs $4. The company doesn't want to make more than 60 dinosaurs a day and can only get materials for 50 brontosaurus and 45 tyrannosaurus a day. A brontosaurus sells for $9, and a tyrannosaurus sells for $7. Find the maximum profit possible per day.

19. The Printing Press produces hardcover books, paperbacks, and magazines at three locations, with the following monthly output:

Location	Hardcovers	Paperbacks	Magazines
Newton	100	200	300
Fayette	100	50	0
LaFollet	0	150	100

The company can only afford to open one location each month. It needs at least 1000 of each type of document yearly, and to operate each location costs $1000, $500, and $400 a month, respectively. Find the minimum yearly cost to meet demand.

20. The Magnetic Attraction produces large and small magnets at two factories, with the following weekly output:

Location	Large	Small
Port Huron	800	1000
Boca Raton	1250	900

Maximize the total number of magnets that can be produced in a ten-week period if neither factory can be open more than seven weeks, the number of large magnets cannot exceed 7500, and the number of small magnets cannot exceed 8500. Assume that both factories cannot be operated in the same week.

13.2 Graphing Linear Inequalities

The first method that we will use to solve linear programming problems is the method of graphing all of the constraints on one graph. Since the constraints are usually inequalities, in this section, we will discuss how to graph linear inequalities.

Consider the linear equation $2x+3y=6$. The graph of this equation is, of course, a line. For every point (x, y) on this line, $2x+3y$ and 6 are equal. For every point (x, y) *not* on this line, $2x+3y$ is either larger than 6 or smaller than 6. So the line $2x+3y=6$ divides the graph into two portions: points to the right of (or above) the line and points to the left of (or below) the line. These two portions correspond to $2x+3y>6$ and $2x+3y<6$. Hence, all we need to determine for a linear inequality is which side of the line yields the correct inequality.

Graphing linear inequalities is a three-step procedure:

1. Graph the corresponding linear equation.
2. Choose a point that is *not* on the line.
3. Substitute the chosen point into the given inequality. If the resulting inequality is true, shade in that side of the line. Otherwise, shade in the other side of the line.

EXAMPLE 13.5 Graph the linear inequality.

a. $2x+3y<6$

Solution The graph of the line $2x+3y=6$ is shown in Figure 13.1. Since we can choose *any* point not on $2x+3y=6$, we might as well pick the easiest point to work with, namely $(0,0)$. Substituting this point into the linear inequality yields $(2 \cdot 0) + (3 \cdot 0) < 6$, or simply $0 < 6$. Since $0 < 6$ is true, we will shade in the same side of the line as $(0, 0)$. The solution appears in Figure 13.2. Since the line is *not* part of the solution, it is represented as a dotted line.

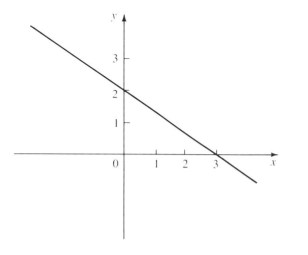

Figure 13.1

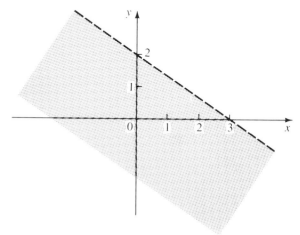

Figure 13.2

13.2 Graphing Linear Inequalities

b. $x - 7y \geq 0$

Solution After graphing $x - 7y = 0$, we must choose a point not on this line. Since $(0, 0)$ is on the line, we have to take another point, say $(0, 1)$. Then $0 - (7 \cdot 1) \geq 0$ becomes $-7 \geq 0$, which is not true. So we shade in the side opposite from $(0, 1)$. The solution appears in Figure 13.3. Since this inequality includes the possibility of equality $(\geq)$, the line is included in the solution and is drawn as a solid line.

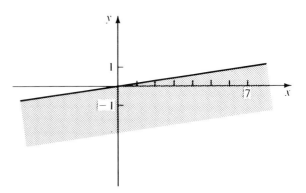

Figure 13.3

c. $x > -1$

Solution Many constraints in linear programming problems involve only one variable. The graph will then represent a vertical or horizontal line. The procedure is the same as before. We first graph $x = -1$ and then find a point, say $(0, 0)$, not on the line. Since $0 > -1$ is true, we shade in the right side of the graph. The solution is shown in Figure 13.4.

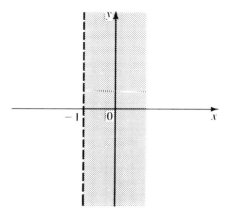

Figure 13.4

LINEAR PROGRAMMING

Most linear programming problems involve more than one constraint. When we are solving a system of two or more *equations*, we try to find the point where the lines meet. Similarly, with a system of inequalities we find the *intersection* of the solutions of each inequality.

EXAMPLE 13.6 Find the solution graphically to the system of inequalities.

a. $x - y \geq 0$
    $x < 3$

Solution Draw both inequalities on the same graph, and find their intersection. The solution is shown in Figure 13.5.

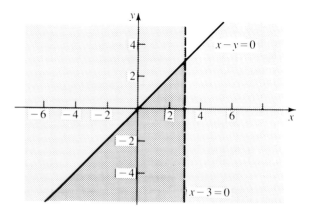

Figure 13.5

b. $2x + y \geq 4$
    $x - y < 2$
    $y \leq 3$

Solution In this case, we are looking for the intersection of all three inequalities. The solution is shown in Figure 13.6.

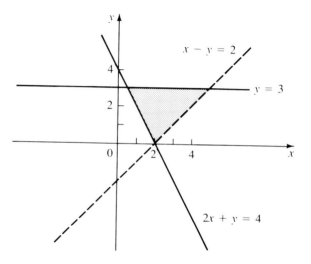

Figure 13.6

13.2 Graphing Linear Inequalities

Problem Set 13.2

In Problems 1–20, graph the linear inequality.

1. $x > 0$
2. $y < 2$
3. $y < 5$
4. $x \leq -3$
5. $x \leq -2$
6. $y > -1$
7. $x + y > 2$
8. $x - y \leq 7$
9. $2x - y \leq 4$
10. $3x + y > -2$
11. $3x + 2y > 0$
12. $-4x + 5y \leq 0$
13. $x \geq y$
14. $y \leq -x$
15. $3x < 2y + 1$
16. $-3x - 7 > 2y$
17. $2x + 6y - 3 \leq 0$
18. $4x - 7y + 2 < 5$
19. $2x + 3y < -5x - 2$
20. $3x - 7 > 4y - 3x$

In Problems 21–32, find the graphical solution to the system.

21. $x < 2$
    $y \geq 4$
22. $x \geq -3$
    $y < 2$
23. $x \geq -2$
    $y > 3$
24. $x < 10$
    $y \geq -2$
25. $x < 4$
    $x \geq -1$
26. $y \geq -5$
    $y < 7$
27. $x + y \geq -2$
    $y < 4$
28. $x < -2$
    $x - y \geq 4$
29. $x - 2y > 6$
    $3x + y \leq -4$
30. $2x + y \leq 7$
    $x - 5y > 10$
31. $x > 0$
    $x - 3y \leq 4$
    $y + 2x > 7$
32. $y \leq 0$
    $3x + 3y < 5$
    $-2x + 4y > 7$

In Problems 33–42, find the inequality or system of inequalities that yields the indicated graph.

33.

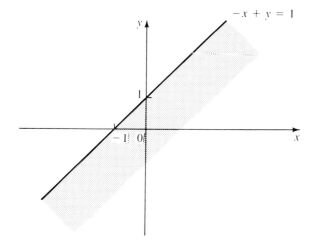

Figure 13.7

34.

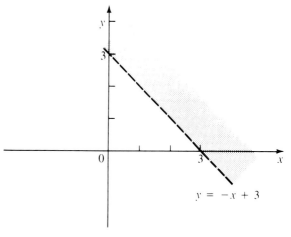

Figure 13.8

35.

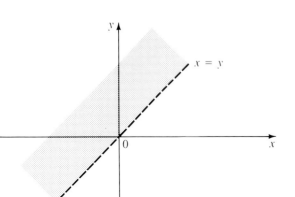

Figure 13.9

36.

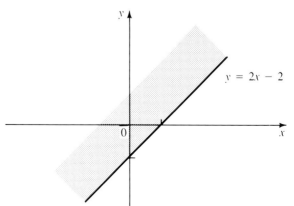

Figure 13.10

37.

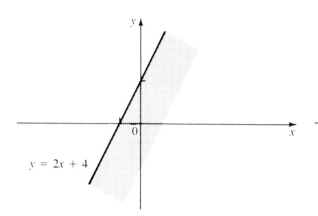

Figure 13.11

38.

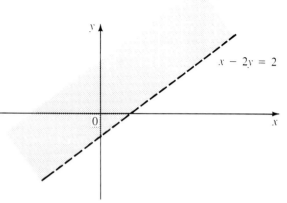

Figure 13.12

13.2 Graphing Linear Inequalities

39.

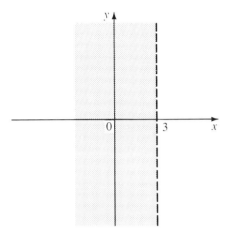

Figure 13.13

40.

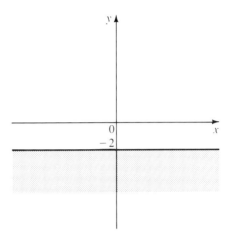

Figure 13.14

41.

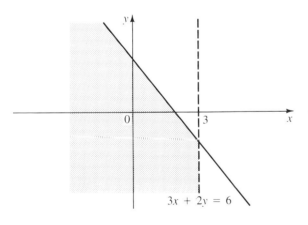

Figure 13.15

42.

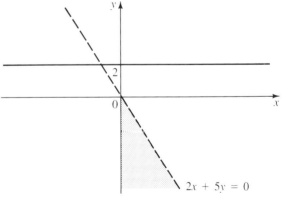

Figure 13.16

43. Find a system of two inequalities that has no solution.

44. Find a system of two inequalities that represents the second quadrant.

13.3 Graphical Linear Programming

We can solve linear programming problems that involve only two variables by graphing, taking x_1 and x_2 for x and y, respectively. As a first step, the intersection of all of the constraints is graphed, as illustrated in Section 13.2. The remaining steps are described in the next paragraph.

vertex

The optimal solution will always occur on the boundary, usually at a corner, or **vertex**, of the graphed region. There may be many optimal solutions. In other words, there may be many points which give the same value for the objective function. All we need to do is find *one* such optimal solution. One optimal solution will always occur at a vertex. There are usually not very many vertices in the graphed region. Each vertex is checked in the objective function, and the "best" result is the optimal solution.

Why does an optimal solution always occur at a vertex? This observation is easy to illustrate and will be shown after we have done Example 13.8.

EXAMPLE 13.7 Solve this linear programming problem by graphing:

maximize: $\quad z = 3x_1 + 2x_2$
constraints: $\quad x_i \geq 0$
$\quad\quad\quad\quad\quad x_1 + x_2 \leq 4$
$\quad\quad\quad\quad\quad 4x_1 - x_2 \geq -4$

Solution The graph of the region determined by the constraints is shown in Figure 13.17.

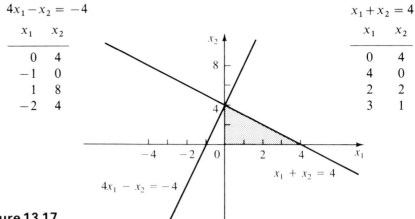

$4x_1 - x_2 = -4$

x_1	x_2
0	4
−1	0
1	8
−2	4

$x_1 + x_2 = 4$

x_1	x_2
0	4
4	0
2	2
3	1

Figure 13.17

There are three vertices for this region: $(0, 0)$, $(4, 0)$, and $(0, 4)$. We evaluate the objective function at each of these points:

13.3 Graphical Linear Programming

Vertex	$3x_1 + 2x_2$
$(0,0)$	$(3 \cdot 0) + (2 \cdot 0) = 0$
$(4,0)$	$(3 \cdot 4) + (2 \cdot 0) = 12$
$(0,4)$	$(3 \cdot 0) + (2 \cdot 4) = 8$

Since this problem is a maximization problem, the optimal solution is $\underline{12}$, which occurs at $\underline{x_1 = 4}$ and $\underline{x_2 = 0}$.

EXAMPLE 13.8 Minimize the objective function of Example 13.7 with the same constraints as Example 13.7.

Solution The region and vertices are the same, but we are now looking for a minimum. Thus,

$$\text{optimal solution:} \quad z = 0, \quad x_1 = 0, \quad x_2 = 0$$

In the previous two examples, the objective function was $z = 3x_1 + 2x_2$. The maximum value of z over the region was 12, and the minimum value of z over the region was 0. The lines $12 = 3x_1 + 2x_2$ and $0 = 3x_1 + 2x_2$ are parallel. In fact, all lines of the form $z = 3x_1 + 2x_2$ are parallel for all values of z. Figure 13.18 shows two of these lines.

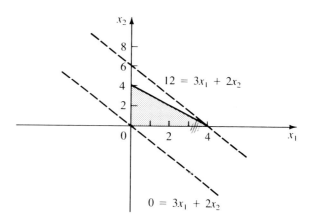

Figure 13.18

Imagine taking a line and "sliding" it along the graph until the maximum or minimum value of z in the region is attained. You should be able to see that these optimal values will always occur on the boundary of the region. If one of the lines forming the region is parallel to the objective function, then the optimal solution may occur not at one point but at all points on that side of the region. At least one of these points will be a vertex, so checking the vertices is always sufficient to find an optimal solution.

EXAMPLE 13.9 Solve this linear programming problem:

maximize: $z = 20x_1 + 40x_2$
constraints: $x_i \geq 0$
$2x_1 + x_2 \leq 6$
$x_1 + 3x_2 \leq 6$

Solution The region determined by the constraints is shown in Figure 13.19.

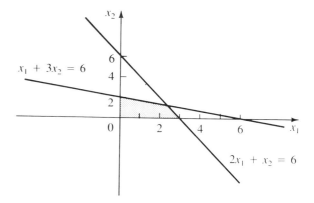

Figure 13.19

There are four vertices for this region: $(0, 0)$, $(3, 0)$, $(0, 2)$, and the intersection of the two lines, which can be found by the multiplication-addition method:

$$2x_1 + x_2 = 6$$
$$x_1 + 3x_2 = 6$$
$$\overline{2x_1 + x_2 = 6}$$
$$-2x_1 - 6x_2 = -12$$
$$\overline{-5x_2 = -6}$$
$$x_2 = \tfrac{6}{5}$$

$$x_1 + 3(\tfrac{6}{5}) = 6$$
$$x_1 + \tfrac{18}{5} = 6$$
$$x_1 = \tfrac{12}{5}$$

The other vertex is $(\tfrac{12}{5}, \tfrac{6}{5})$.

We evaluate the objective function at each vertex:

Vertex	$20x_1 + 40x_2$
$(0, 0)$	0
$(3, 0)$	60
$(0, 2)$	80
$(\tfrac{12}{5}, \tfrac{6}{5})$	96

13.3 Graphical Linear Programming

optimal solution: $z = 96$, $x_1 = \frac{12}{5}$, $x_2 = \frac{6}{5}$

EXAMPLE 13.10 Solve this linear programming problem:

minimize: $z = 4x_1 + 5x_2$
constraints: $2x_1 + x_2 \leq 8$
$2x_1 - 3x_2 \geq 0$
$x_1 + x_2 \geq 0$

Solution In this case, we are not restricting x_1 and x_2 to be positive, so the region is not restricted to the first quadrant. In Figure 13.20, we graph the intersection of the three constraints.

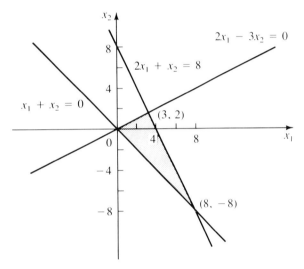

Figure 13.20

The vertices $(3, 2)$, $(8, -8)$, and $(0, 0)$ can be verified by using the multiplication-addition method to find the intersection of the appropriate lines (Problems 1 and 2). We evaluate the objective function next:

Vertex	$4x_1 + 5x_2$
(0, 0)	0
(3, 2)	22
(8, −8)	−8

optimal solution: $z = -8$, $x_1 = 8$, $x_2 = -8$

As illustrated in the previous examples, a vertex will always be the intersection of two lines. An *algebraic method* to solve these problems would consist of finding all such intersections and then finding the optimal vertex without drawing the graph.

Problem Set 13.3

1. Verify that $(3, 2)$ is a vertex in Example 13.10.
2. Verify that $(8, -8)$ is a vertex in Example 13.10.

In Problems 3–18, solve the linear programming problem by graphing.

3. maximize: $z = 10x_1 + 10x_2$
   constraints: $x_i \geq 0$, $x_1 \leq 3$, $x_2 \leq 3$
4. maximize: $z = 5x_1 + 20x_2$
   constraints: $x_i \geq 0$, $x_1 \leq 2$, $x_2 \leq 4$
5. maximize: $z = -3x_1 + 5x_2$
   constraints: $x_i \geq 0$, $x_1 \leq 5$, $x_2 \leq 4$
6. maximize: $z = 4x_1 - 7x_2$
   constraints: $x_i \leq 0$, $x_2 \geq 0$, $x_1 \geq -5$, $x_2 \leq 3$
7. minimize: $z = 20x_1 + 10x_2$
   constraints: $x_i \leq 0$, $x_1 + x_2 \leq -1$
8. minimize: $z = -2x_1 + 5x_2$
   constraints: $x_i \geq 0$, $x_1 + x_2 \leq 1$
9. maximize: $z = 6x_1 + 5x_2$
   constraints: $-3 \leq x_1 \leq 2$, $-4 \leq x_2 \leq 1$
10. maximize: $z = 10x_1 + 4x_2$
    constraints: $-6 \leq x_1 \leq -2$, $2 \leq x_2 \leq 4$
11. maximize: $z = 10x_1 - 10x_2$
    constraints: $x_i \geq 0$, $2x_1 + 3x_2 \leq 4$
12. maximize: $z = -5x_1 - 3x_2$
    constraints: $x_i \geq 0$, $x_1 + 4x_2 \leq 5$
13. maximize: $z = 4x_1 - 3x_2$
    constraints: $x_i \geq 0$, $-x_1 + x_2 \leq -1$
    $3x_1 + 4x_2 \leq 24$
14. minimize: $z = 2x_1 - x_2$
    constraints: $x_i > 0$, $3x_1 + 4x_2 \leq 18$
    $3x_1 + 5x_2 \leq 21$
15. minimize: $z = -30x_1 + 80x_2$
    constraints: $x_1 + x_2 \leq 4$, $2x_1 + x_2 \geq 4$
    $x_1 - x_2 \leq 0$
16. maximize: $z = 20x_1 + 40x_2$
    constraints: $x_i \geq 0$, $x_1 + x_2 \leq 4$
    $2x_1 + x_2 \geq 4$ $x_1 - x_2 \geq 0$
17. maximize: $z = 60x_1 + 35x_2$
    constraints: $x_i \geq 0$, $2x_1 + 5x_2 \leq 10$
    $x_1 + x_2 \leq 3$ $x_1 - x_2 \leq 0$
18. maximize: $z = 30x_1 + 100x_2$
    constraints: $x_i \geq 0$, $x_1 + x_2 \leq 5$
    $3x_1 + 8x_2 \leq 24$ $x_1 - x_2 \geq -2$
19. Minimize the region in Problem 13.
20. Maximize the region in Problem 14.
21. Maximize the region in Problem 15.
22. Minimize the region in Problem 16.

In Problems 23–28, solve the linear programming word problem from Section 13.1.

23. Two nonpositive numbers are restricted so that neither number is smaller than -10 and their sum is at least as large as -15. Find the values of these numbers that will make the difference of the first number and twice the second number as small as possible. (Problem 13)

24. Two nonnegative numbers are restricted so that their sum is never larger than 10. What values for these numbers will make the sum of five times the first number and three times the second number as large as possible? (Problem 12)

25. The Penguin Puppet Company makes two kinds of puppets: large and small. Working at peak efficiency, it can make no more than 100 large puppets a day and no more than 200 small puppets a day. The total number of puppets produced cannot exceed 150. If small puppets sell for $5 each and large puppets

sell for $7 each, find the maximum income possible. (Problem 15)

26. The Space Puzzle Company makes two kinds of puzzles: beginner and advanced. It can make no more than 80 beginner puzzles and 60 advanced puzzles per week, but it is limited to making no more than 100 total puzzles per week. If beginner puzzles sell for $6 each and advanced puzzles sell for $7 each, find the number of each type of puzzle that produces the maximum income. (Problem 16)

27. The Silver Bullet Company produces two kinds of guns: muzzle loader rifles and flintlock rifles. It has only $400 to spend; making a muzzle loader costs $40, and making a flintlock costs $50. The company only wants to make a maximum of nine guns per week, and it can only get materials for seven muzzle loaders at most and six flintlocks at most per week. The company wants to sell the muzzle loaders for $140 each and the flintlocks for $200 each. At these prices, the company will have no trouble selling all it makes. Find the maximum profit possible per week. (Problem 17)

28. Dinos Unlimited produces two kinds of stuffed dinosaurs: brontosaurus and tyrannosaurus. It can only spend $300 a day; making a brontosaurus costs $5, and making a tyrannosaurus costs $4. The company doesn't want to make more than 60 dinosaurs a day and can only get materials for 50 brontosaurus and 45 tyrannosaurus a day. A brontosaurus sells for $9, and a tyrannosaurus sells for $7. Find the maximum profit possible per day. (Problem 18)

29. Find a linear programming problem with the constraint $x_i \geq 0$ for which $(0,0)$ is *not* a minimum.

30. Find a linear programming problem with the constraint $x_i \geq 0$ for which $(0,0)$ is a maximum.

13.4 Simplex Method: Standard Case

The graphical method of solving linear programming problems that was covered in the previous section can only be used in problems involving no more than two variables. We need a general method that will solve *any* linear programming problem. The method that we will discuss in the next two sections is called the **simplex method** and is adaptable to computer programming.

simplex method

In the simplex method, the constraints must conform to a set of standard conditions. In this section, we will only consider problems that meet these conditions. This restriction is not a great handicap since a large number of linear programming applications normally fit these conditions. The standard conditions follow.

1. All variables must be nonnegative.
2. All numbers on the right of the inequality must be nonnegative.
3. All constraints other than 1 must be $\leq$.
4. It must be a *maximization* problem.

A linear programming problem rarely has variables that are negative, since the variables often represent the number of an item sold, ordered, produced, and so on. There is more likely to be an upper limit on a quantity ($\leq$) than a lower limit other than zero.

We will follow the steps of the simplex method with the following example:

maximize: $z = 2x_1 + 3x_2$
constraints: $x_i \geq 0$
$x_1 + 2x_2 \leq 6$
$x_1 + x_2 \leq 4$

Notice that this problem does follow the standard conditions. It is a maximization problem with all variables positive, all inequalities $\leq$, and nonnegative numbers on the right side of each inequality. Now, we will consider the steps.

First Step of the Simplex Method Change all inequalities involving $\leq$ into equations. This step can be done by adding a nonnegative variable to the left of each inequality to take up the "slack":

$$x_1 + 2x_2 + x_3 = 6$$
$$x_1 + x_2 + x_4 = 4$$

slack variables

Variables x_3 and x_4 are called **slack variables**. These variables were not part of the original problem; they contribute nothing to the objective function and will have coefficients of zero in the objective function. Our problem now looks like this:

maximize: $z = 2x_1 + 3x_2 + 0x_3 + 0x_4$
constraints: $x_i \geq 0$
$x_1 + 2x_2 + x_3 = 6$
$x_1 + x_2 + x_4 = 4$

initial simplex tableau

Second Step of the Simplex Method Construct the **initial simplex tableau**, consisting of the coefficients of the equations:

	x_1	x_2	x_3	x_4	B
x_3	1	2	1	0	6
x_4	1	1	0	1	4
	-2	-3	0	0	0

The first row consists of the variables in the objective function. The second and third rows consist of the coefficients of these variables in the two constraints. There will be one row in the simplex tableau for each constraint. The last column represents the number on the right side of each inequality. The entries in the last row are the negatives of the coefficients of z, and the last entry is always zero. The first column contains the slack variables used in the corresponding constraints. The initial solution is $z = 0$, $x_1 = 0$, and $x_2 = 0$.

13.4 Simplex Method: Standard Case

pivotal column

pivotal row

pivot

Third Step of the Simplex Method Find the pivot. The procedure is as follows:
1. Find the pivotal column. The **pivotal column** is the column that contains the most negative entry in the last row. If there is no negative entry in the last row, the problem is solved.
2. Find the pivotal row. To find the **pivotal row**, divide each number in the last column by the corresponding *positive* number in the pivotal column. The pivotal row is the row which produces the smallest quotient. If there is a tie for smallest quotient, take either row. If there are no positive numbers in the pivotal column, there is *no solution* to the problem.
3. Find the pivot. The **pivot** is the element in the pivotal row and pivotal column.

In our example, the second column is the pivotal column because -3 is the most negative number in the last row (see the following table). The ratios $\frac{6}{2} = 3$ and $\frac{4}{1} = 4$ tell us that the first row is the pivotal row since 3 is the smaller ratio. So the pivot is 2.

	x_1	x_2	x_3	x_4	B	
x_3	1	2	1	0	6	← Pivotal row
x_4	1	1	0	1	4	
	-2	-3	0	0	0	

↑
Pivotal column

Fourth Step of the Simplex Method Use matrix operations to change the pivot to 1 and every other entry in the pivotal column to 0, just as you did in finding the inverse of a matrix in Chapter 12. When this step is done, the slack variable in the first column of the pivot row is replaced by the variable in the pivot column. The result is shown next:

	x_1	x_2	x_3	x_4	B
x_2	$\frac{1}{2}$	1	$\frac{1}{2}$	0	3
x_4	$\frac{1}{2}$	0	$\frac{1}{2}$	1	1
	$-\frac{1}{2}$	0	$\frac{3}{2}$	0	9

Fifth (and Last) Step of the Simplex Method Repeat the previous two steps until the last row contains no negative entries. The pivotal column is now the first column. The ratios $3/(\frac{1}{2}) = 6$ and $1/(\frac{1}{2}) = 2$ tell us that the second row is the pivot row and $\frac{1}{2}$ is the pivot. After matrix operations, we are left with the following array:

	x_1	x_2	x_3	x_4	B
x_2	0	1	1	-1	2
x_1	1	0	-1	2	2
	0	0	1	1	10

Since there are no negative entries in the bottom row, the problem is solved. The optimal solution appears in the last column. The variables in the first column have the corresponding values in the last column ($x_2 = 2$, $x_1 = 2$) and the last entry is the value of the objective function ($z = 10$).

optimal solution: $z = 10, x_1 = 2, x_2 = 2$

What does the simplex method do? It starts with an initial solution ($x_1 = 0$, $x_2 = 0$ in our example) and then moves to an adjacent vertex. In our example, there are two vertices adjacent (next) to $(0, 0)$ (see Figure 13.21): $(0, 3)$ and $(4, 0)$.

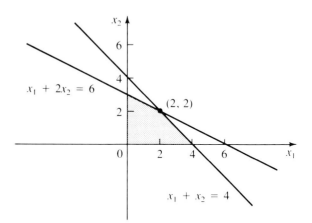

Figure 13.21

The simplex method goes to the adjacent vertex which most improves the objective function. Thus, $(0, 3)$ makes the objective function equal 9, and $(4, 0)$ makes z equal 8. So the simplex method goes to $(0, 3)$. For this reason, the x_2 column was used first. From $(0, 3)$, there are two adjacent vertices: $(0, 0)$ and $(2, 2)$. Point $(2, 2)$ is the optimal solution.

13.4 Simplex Method: Standard Case

EXAMPLE 13.11 Use the simplex method to solve:

$$\text{maximize:} \quad z = 2x_1 + x_2 + 2x_3$$
$$\text{constraints:} \quad x_i \geq 0$$
$$x_1 + x_3 \leq 2$$
$$3x_1 + 2x_2 \leq 6$$
$$x_1 + x_2 - x_3 \leq 4$$

Solution We need to introduce three slack variables to change each $\leq$ into an equation:

$$\text{maximize:} \quad z = 2x_1 + x_2 + 2x_3 + 0x_4 + 0x_5 + 0x_6$$
$$\text{constraints:} \quad x_i \geq 0$$
$$x_1 + x_3 + x_4 = 2$$
$$3x_1 + 2x_2 + x_5 = 6$$
$$x_1 + x_2 - x_3 + x_6 = 4$$

The initial simplex tableau looks like this:

	x_1	x_2	x_3	x_4	x_5	x_6	B
x_4	1	0	1	1	0	0	2
x_5	3	2	0	0	1	0	6
x_6	1	1	−1	0	0	1	4
	−2	−1	−2	0	0	0	0

The initial solution is $z = 0$, $x_1 = 0$, $x_2 = 0$, and $x_3 = 0$.

We can use either the first or the third column for the pivotal column since they both contain the most negative entry in the last row (-2). If we choose the third column, 1 becomes the pivot. The tableau is as follows:

	x_1	x_2	x_3	x_4	x_5	x_6	B
x_3	1	0	1	1	0	0	2
x_5	3	2	0	0	1	0	6
x_6	2	1	0	1	0	1	6
	0	−1	0	2	0	0	4

Notice that x_4 has been replaced by x_3 in the first column since the third column was the pivotal column.

The second column now becomes the pivotal column, and 2 is the pivot. The tableau simplifies as follows:

	x_1	x_2	x_3	x_4	x_5	x_6	B
x_3	1	0	1	1	0	0	2
x_2	$\frac{3}{2}$	1	0	0	$\frac{1}{2}$	0	3
x_6	$\frac{1}{2}$	0	0	1	$-\frac{1}{2}$	1	3
	$\frac{3}{2}$	0	0	2	$\frac{1}{2}$	0	7

Since there are no negative entries in the last row, we have reached the optimal solution: $z = 7$, $x_3 = 2$, $x_2 = 3$, and $x_6 = 3$. Since x_6 was not part of the original problem, it is not usually included in the optimal solution. What about x_1? The initial solution had $x_1 = 0$; and since x_1 has never been a pivotal column, its value is still 0. We are left with the following optimal solution:

optimal solution: $z = 7$, $x_1 = 0$, $x_2 = 3$, $x_3 = 2$

Problem Set 13.4

1. Why are slack variables nonnegative by definition?
2. Will the simplex method ever go to the same vertex twice? Why or why not?
3. Verify that the matrix operations in the demonstration example are correct.
4. Verify that the matrix operations in Example 13.11 are correct.

In Problems 5–18, introduce slack variables, and set up the initial simplex tableau.

5. maximize: $z = 6x_1 + 2x_2$
   constraints: $x_i \geq 0$, $x_1 + 3x_2 \leq 6$
6. maximize: $z = x_1 + 5x_2$
   constraints: $x_i \geq 0$, $4x_1 + 2x_2 \leq 8$
7. maximize: $z = 7x_1 + 3x_2$
   constraints: $x_i \geq 0$, $2x_1 + x_2 \leq 2$
8. maximize: $z = 10x_1 + 9x_2$
   constraints: $x_i \geq 0$, $4x_1 + 5x_2 \leq 10$
9. maximize: $z = 10x_1 + 13x_2$
   constraints: $x_i \geq 0$, $2x_1 + 5x_2 \leq 11$,
   $x_1 + 2x_2 \leq 5$
10. maximize: $z = 6x_1 + 5x_2$
    constraints: $x_i \geq 0$, $x_1 + x_2 \leq 7$,
    $2x_1 - x_2 \leq 5$
11. maximize: $z = 3x_1 + 4x_2$
    constraints: $x_i \geq 0$, $x_1 + 4x_2 \leq 10$,
    $3x_1 + 3x_2 \leq 7$
12. maximize: $z = 7x_1 - 2x_2$
    constraints: $x_i \geq 0$, $7x_1 + 3x_2 \leq 12$,
    $-x_1 + 4x_2 \leq 8$
13. maximize: $z = 5x_1 - 3x_2 + x_3$
    constraints: $x_i \geq 0$, $x_1 + 2x_2 \leq 4$,
    $x_2 + 4x_3 \leq 6$, $3x_1 - x_2 - x_3 \leq 3$
14. maximize: $z = 7x_1 + 6x_2 + 4x_3$
    constraints: $x_i \geq 0$, $2x_1 + 2x_2 - x_3 \leq 6$,
    $2x_1 + x_2 \leq 4$, $2x_2 + x_3 \leq 2$
15. Two nonnegative numbers are restricted so that twice the first number plus three times the second number is never larger than 12. Find the largest sum of these numbers under these conditions.
16. Two nonnegative numbers are restricted so that their sum is never larger than 10. What values for these numbers will make the sum of five times the first number and three times the second number as large as possible? (Section 13.1, Problem 12)
17. The Penguin Puppet Company makes two kinds of puppets: large and small. Working at

peak efficiency, it can make no more than 100 large puppets a day and no more than 200 small puppets a day. The total number of puppets produced cannot exceed 150. If small puppets sell for $5 each and large puppets sell for $7 each, find the maximum income possible. (Section 13.1, Problem 15)

18. The Magnetic Attraction produces large and small magnets at two factories, with the following weekly output:

Location	Large	Small
Port Huron	800	1000
Boca Raton	1250	900

Maximize the total number of magnets that can be produced in a ten-week period if neither factory can be open more than seven weeks, the number of large magnets cannot exceed 7500, and the number of small magnets cannot exceed 8500. (Section 13.1, Problem 20)

In Problems 19–32, use the simplex method to solve the indicated problem.

19. Problem 5   20. Problem 6
21. Problem 7   22. Problem 8
23. Problem 9   24. Problem 10
25. Problem 11   26. Problem 12
27. Problem 13   28. Problem 14
29. Problem 15   30. Problem 16
31. Problem 17   32. Problem 18

13.5 Simplex Method: General Case

Many linear programming applications follow the standard conditions stated in the previous section. And with a few modifications, the procedure of Section 13.4 will solve *any* linear programming problem. So let's consider each of the four standard conditions.

Condition 1: All variables must be nonnegative.

If variable x_1, for example, is not restricted to being nonnegative, we must introduce two new nonnegative variables, x_j and x_k, such that $x_1 = x_j - x_k$. By substituting $x_j - x_k$ for x_1 into z and all constraints, we can transform the problem into one with nonnegative variables. We must do this step for each variable which is not restricted to be nonnegative. The next example illustrates the technique.

■ **EXAMPLE 13.12** Change this problem into one with nonnegative variables:

$$\text{maximize:} \quad z = 2x_1 + 2x_2$$
$$\text{constraints:} \quad x_1 + x_2 \leq 3$$
$$2x_1 - x_2 \leq -4$$
$$x_1 - x_2 = -2$$

Solution There are no restrictions on x_1 or x_2, so we define $x_1 = x_3 - x_4$ and $x_2 = x_5 - x_6$, where $x_3, x_4, x_5,$ and x_6 are all nonnegative variables. If we substitute for x_1 and x_2 in z and all constraints, the problem then is as follows:

maximize: $z = 2x_3 - 2x_4 + 2x_5 - 2x_6$
constraints: $x_i \geq 0$
$$x_3 - x_4 + x_5 - x_6 \leq 3$$
$$2x_3 - 2x_4 - x_5 + x_6 \leq -4$$
$$x_3 - x_4 - x_5 + x_6 = -2$$

Condition 2: All numbers on the right of the inequality must be nonnegative.

This condition can easily be satisfied by multiplying any inequality or equation by -1 if necessary. The next example illustrates.

■ **EXAMPLE 13.13** Change each inequality or equation of Example 13.12 so that condition 2 is met.

Solution Multiplying by -1 changes the last two constraints to

$$-2x_3 + 2x_4 + x_5 - x_6 \geq 4$$
$$-x_3 + x_4 + x_5 - x_6 = 2$$

Remember that multiplying by a negative number reverses the direction of the inequality.

Condition 3: All constraints other than 1 must be $\leq$.

All constraints must be changed into equations. Recall that slack variables were used for $\leq$ inequalities. For $\geq$ expressions, we must *subtract* a positive number from the left side of the inequality to change the expression into an equation. This variable is called a **surplus variable**. By adding slack variables and subtracting surplus variables, we can change all constraints into equations.

surplus variable

In order for the simplex method to work, we must add a nonnegative variable called an **artificial variable** to the left side of each equation which does not contain a slack variable. Artificial variables must be zero in the optimal solution since they are added to *equations* without destroying equality. To force artificial variables to be zero, we want to give them very large coefficients (M) in objective functions that are to be minimized and very small coefficients ($-M$) in objective functions that are to be maximized. The next example illustrates these techniques.

artificial variable

■ **EXAMPLE 13.14** Add slack, surplus, and artificial variables to the problem in Example 13.13.

Solution
maximize: $z = 2x_3 - 2x_4 + 2x_5 - 2x_6 + 0x_7 + 0x_8 - Mx_9 - Mx_{10}$
constraints: $x_i \geq 0$
$$x_3 - x_4 + x_5 - x_6 + x_7 = 3$$
$$-2x_3 + 2x_4 + x_5 - x_6 - x_8 + x_9 = 4$$
$$-x_3 + x_4 + x_5 - x_6 + x_{10} = 2$$

13.5 Simplex Method: General Case

In the first equation, x_7 is a slack variable, and x_8 is a surplus variable in the second equation. Variables x_9 and x_{10} are artificial variables added to the last two equations because those equations don't contain slack variables.

With the introduction of artificial variables, the initial simplex tableau becomes a little more involved. Each constraint now contains either a slack or an artificial variable. The first column of the tableau contains the slack or artificial variables from the corresponding constraints. We add a new column to the right of this first column; the new column contains the coefficients of these slack or artificial variables from the objective function.

Now comes the hardest part of constructing the initial simplex tableau: finding the bottom row. To find the bottom row of the initial tableau for a maximization problem, we take the negatives of the coefficients from the objective function and add the products of the elements in this new (second) column by the corresponding elements in the column under discussion. An example should help to make this process clear.

■ **EXAMPLE 13.15**   Find the initial simplex tableau for the problem in Example 13.14.

Solution

		x_3	x_4	x_5	x_6	x_7	x_8	x_9	x_{10}	B
x_7	0	1	-1	1	-1	1	0	0	0	3
x_9	$-M$	-2	2	1	-1	0	-1	1	0	4
x_{10}	$-M$	-1	1	1	-1	0	0	0	1	2
		$-2+3M$	$2-3M$	$-2-2M$	$-2+2M$	0	M	0	0	$-6M$

The entries in the first column on the left represent the slack or artificial variables from the three constraints. Variable x_7 is a slack variable, and x_9 and x_{10} are artificial variables. The entries in the second column are the coefficients of x_7 (slack variable), x_9 (artificial variable), and x_{10} (artificial variable) from the objective function.

To find the first entry in the last row, we take -2 (since 2 is the coefficient of x_3 in the objective function) and add $[(0 \cdot 1)+(-M \cdot -2)+(-M \cdot -1)]$, or $0+2M+M = 3M$. So the first term becomes $-2+3M$. Similarly, to find the entry in the last row under x_9, we take M (since $-M$ is the coefficient of x_9 in the objective function) and add $[(0 \cdot 0)+(-M \cdot 1)+(-M \cdot 0)]$, or $0-M+0 = -M$. So this entry becomes $M+(-M)$, or 0. To find the last entry in the last row, we simply take the sum of the products of the second column and the last column: $[(0 \cdot 3)+(-M \cdot 4)+(-M \cdot 2)] = 0-4M-2M = -6M$.

So the initial solution has $z = -6M$. This result is not very good when we are trying to *maximize* z, but the simplex method will greatly improve our value for z after a few steps.

Condition 4: It must be a maximization problem.

The only change necessary now to alter the procedure for minimization problems is the procedure for calculating the bottom row. For minimization, we take the negative of the calculation for maximization—that is, we take the coefficients from the objective function and *subtract* the products of the elements in the new column by the corresponding elements in the column under discussion. The resulting bottom line will *not* be the negative of the bottom line for maximization, however, since the coefficients of the artificial variables in the objective function change from $-M$ to M. An example should help make this process clear.

■ **EXAMPLE 13.16** Find the initial simplex tableau to minimize the objective function of Example 13.12 with the same set of constraints.

Solution Since we are now minimizing z, the coefficients of the artificial variables are now M and the objective function becomes

minimize: $\quad z = 2x_3 - 2x_4 + 2x_5 - 2x_6 + 0x_7 + 0x_8 + Mx_9 + Mx_{10}$

The initial simplex tableau now is as follows:

		x_3	x_4	x_5	x_6	x_7	x_8	x_9	x_{10}	B
x_7	0	1	-1	1	-1	1	0	0	0	3
x_9	M	-2	2	1	-1	0	-1	1	0	4
x_{10}	M	-1	1	1	-1	0	0	0	1	2
		$2+3M$	$-2-3M$	$2-2M$	$-2+2M$	0	M	0	0	$-6M$

To find the first entry in the last row, we take 2 (the coefficient of x_3 in the objective function) and subtract $[(0 \cdot 1) + (M \cdot -2) + (M \cdot -1)]$, or $0 - 2M - M = -3M$. So the first term becomes $2 - (-3M)$, or $2 + 3M$. Similarly, to find the entry in the last row under x_9, we take M (the coefficient of x_9 in the objective function) and subtract $[(0 \cdot 0) + (M \cdot 1) + (M \cdot 0)]$, or $0 + M + 0 = M$. So this term becomes $M - M = 0$. To find the last entry in the last row, we simply take the negative of $[(0 \cdot 3) + (M \cdot 4) + (M \cdot 2)]$, or $-(0 + 4M + 2M) = -6M$.

The last entry in the last row does *not* represent the value for z, as it did in maximization problems. It represents the *negative* of the value for z. So our initial solution has $z = 6M$, not very good for the final answer in a minimization problem.

We can finally use the simplex method to solve the problem in Example 13.12. However, instead of having the original two variables (x_1 and x_2), we now have *eight* variables (x_3 through x_{10}). These additional variables make the process of finding a solution quite long and involved.

13.5 Simplex Method: General Case

Fortunately, most linear programming problems are not solved by hand but by computer. Computers can perform millions of calculations per second. The simplex method can be easily programmed by using a loop to repeat the matrix operations until a solution is attained. Computers can even be programmed to take a problem like the one of Example 13.12 and perform all of the changes which we have discussed in this section before performing the matrix operations.

Since there is no provision in the computer for an arbitrarily large number M, an actual number like 10,000 (or $-10,000$) is used for the coefficient of an artificial variable.

Problem Set 13.5

In Problems 1–12, adjust the constraints by introducing slack, surplus, and artificial variables when necessary. Identify each added variable as slack, surplus, or artificial.

1. $x_1 + x_2 \leq 4$
2. $3x_1 - 4x_2 \leq 7$
3. $2x_1 - 7x_2 = 2$
4. $-5x_1 + 6x_2 = 3$
5. $3x_1 + x_2 - 3x_3 \geq -6$
6. $2x_1 + 2x_2 - 4x_3 \geq 5$
7. $10x_1 - 3x_2 + 2x_3 \geq 4$
8. $x_1 - x_2 - 2x_3 = -5$
9. $8x_1 - 3x_2 + x_3 = -4$
10. $7x_1 + 3x_2 + 4x_3 \geq -3$
11. $2x_1 - 9x_2 \leq -11$
12. $3x_1 - 6x_2 + x_3 \leq -5$

For Problems 13–24, write the initial simplex tableau.

13. maximize: $z = 2x_1 + 6x_2$
    constraints: $x_i \geq 0$, $2x_1 - x_2 \leq 3$,
    $x_1 + 2x_2 \leq 2$

14. maximize: $z = 5x_1 - 3x_2$
    constraints: $x_i \geq 0$, $x_1 - 7x_2 \leq 4$,
    $3x_1 + x_2 \leq 1$

15. maximize: $z = 4x_1 - 2x_2$
    constraints: $x_i \geq 0$, $x_1 + x_2 \leq 4$,
    $2x_1 + x_2 = 8$

16. maximize: $z = 2x_1 + 7x_2$
    constraints: $x_i \geq 0$, $3x_1 + 7x_2 = 10$,
    $x_1 - x_2 \leq 1$

17. minimize: $z = x_1 + 3x_2$
    constraints: $x_i \geq 0$, $x_1 + x_2 = 5$,
    $2x_1 - x_2 \geq -3$

18. minimize: $z = 3x_1 + 4x_2$
    constraints: $x_i \geq 0$, $3x_1 - 6x_2 \geq -7$,
    $2x_1 + 3x_2 = 10$

19. maximize: $z = 7x_1 + 9x_2 + x_3$
    constraints: $x_i \geq 0$, $x_1 + 11x_2 + 6x_3 = 5$,
    $x_1 + 5x_2 - x_3 \leq 4$

20. maximize: $z = 18x_1 + 37x_2 + 11x_3$
    constraints: $x_i \geq 0$, $x_1 + x_2 - x_3 \leq 5$,
    $4x_1 - 5x_2 + 9x_3 = 3$

21. maximize: $z = 4x_1 + 12x_2 + x_3$
    constraints: $3x_1 + 7x_2 + 4x_3 = -3$,
    $x_1 + x_2 + 3x_3 \leq -5$

22. maximize: $z = 36x_1 + 30x_2 + 8x_3$
    constraints: $5x_1 - 6x_2 + 10x_3 \leq -5$,
    $2x_1 + 4x_2 + 2x_3 \geq -6$

23. minimize: $z = 14x_1 + 23x_2$
    constraints: $3x_1 - 7x_2 \leq -3$, $2x_1 + 5x_2 \geq 8$,
    $x_1 - 3x_2 = 11$

24. minimize: $z = 9x_1 + 2x_2$
    constraints: $2x_1 + 6x_2 = 16$,
    $-3x_1 - 5x_2 \leq -8$, $7x_1 + 4x_2 \geq 7$

In Problems 25–28, use the simplex method to find the optimal solution to the indicated problem.

25. Problem 13
26. Problem 14
27. Problem 15
28. Problem 16

29. Why must artificial variables be zero in the optimal solution?
30. Why are surplus variables nonnegative?
31. Can a constraint ever contain both a surplus and an artificial variable? If so, give an example.
32. Can a constraint ever contain both a surplus and a slack variable? If so, give an example.
33. Can a constraint ever contain both a slack and an artificial variable? If so, given an example.
34. Can a constraint ever contain neither a slack variable nor an artificial variable? If so, give an example.

CHAPTER SUMMARY

In this chapter, we studied linear programming, the process of minimizing or maximizing a linear quantity subject to some linear constraints. If there are no more than two variables involved, graphing is an effective method for solving these problems.

A more general procedure, the simplex method, is used to solve complex linear programming problems which satisfy the standard conditions. These standard conditions include maximization, nonnegative variables and constants, and only $\leq$ constraints. This procedure can be generalized to handle any linear programming problem, as we saw.

The basic steps of the simplex method are as follows:

1. Introduce slack variables, surplus variables, and artificial variables into the constraints when necessary.
2. Construct the initial simplex tableau.
3. Find the pivot.
4. Change the pivot to 1 and all other entries in the pivotal column to 0.
5. Repeat steps 3 and 4 until the bottom row contains no negative entries.

Many applications of linear programming were encountered in this chapter. But most real-world applications have too many variables to make hand calculations feasible. Fortunately, computer programs that will perform the simplex algorithm are easy to write and can then be used to solve very complex problems quickly.

REVIEW PROBLEMS

In Problems 1–4, graph the linear inequality.

1. $x + 3y \leq 3$
2. $2x - y \geq 10$
3. $3x - 2y > 10$
4. $-4x + y < 8$

Review Problems

In Problems 5–8, graph the region determined by the inequalities.

5. $x \geq 0$
   $y \geq 0$
   $x + 5y \leq 10$

6. $x \leq 0$
   $y \geq 0$
   $x - y \leq -5$

7. $3x - y \leq 12$
   $2x + y \geq 3$
   $x - y \leq 4$

8. $x + 2y \leq 2$
   $x - 2y \geq 2$
   $3x - 4y \leq 6$

In Problems 9–12, solve the linear programming problem by graphing.

9. maximize: $z = x_1 + 2x_2$
   constraints: $x_1 \geq 1$, $x_2 \geq 0$, $x_1 \leq 4$,
   $2x_1 + x_2 \leq 12$

10. minimize: $z = 3x_1 - x_2$
    constraints: $x_1 \leq 3$, $x_2 \geq -4$, $x_1 \geq 0$,
    $3x_1 + x_2 \leq 3$

11. minimize: $z = -2x_1 - 5x_2$
    constraints: $x_1 \geq 0$, $x_2 \geq 0$, $x_2 \leq -2x_1 + 1$,
    $10x_1 + x_2 \leq 10$

12. maximize: $z = 3x_1 + 4x_2$
    constraints: $x_1 \geq 0$, $x_2 \leq x_1 - 5$,
    $x_1 + x_2 \leq 15$

13. Two positive numbers are restricted such that ten times one number plus five times the other number cannot be larger than 20. Maximize the sum of the two numbers.

14. Two nonpositive numbers are restricted such that the sum of twice one number and three times the other number cannot be smaller than -12. Minimize the sum of the two numbers.

15. Maximize $6x_1 + 7x_2$ on the region in Problem 7.

16. Maximize $4x_1 + 5x_2$ on the region in Problem 10.

In Problems 17–20, use the simplex method to solve the problem.

17. maximize: $z = 10x_1 + 5x_2$
    constraints: $x_i \geq 0$, $x_1 + 2x_2 \leq 30$,
    $3x_1 - x_2 \leq 20$

18. maximize: $z = 6x_1 + 4x_2$
    constraints: $x_i \geq 0$, $4x_1 - x_2 \leq 13$,
    $2x_1 + 5x_2 \leq 23$

19. minimize: $z = 2x_1 + 2x_2 + x_3$
    constraints: $x_i \geq 0$, $x_1 + x_2 \leq 3$,
    $2x_1 - x_3 \leq 4$

20. minimize: $z = 5x_1 + 3x_2 + x_3$
    constraints: $x_i \geq 0$, $2x_1 + 2x_2 + 3x_3 \geq 0$,
    $x_1 - 2x_3 \leq 7$

In Problems 21–22, construct the initial simplex tableau.

21. The See Best Telescope Company produces 6-inch-, 8-inch-, and 10-inch-diameter telescopes, with the following monthly output at four locations:

Location	6 Inch	8 Inch	10 Inch
Houston	20	15	5
Cleveland	0	30	10
Pasadena	50	60	15
Greenbelt	30	0	0

The monthly cost of operating the four stores is $2000, $1500, $4000, and $1000, respectively. They must produce 200 6-inch telescopes, 700 8-inch telescopes, and 100 10-inch telescopes this year. Find the minimum cost this year needed to meet demand.

22. The Upper Limits produces toy rockets and airplanes. It can make no more than 1000 rockets and 800 airplanes per week. The total number of toys produced cannot exceed 1500. It costs $6 to make a rocket and $5 to make an airplane. Rockets sell for $10 apiece, and airplanes sell for $9 apiece. Maximize the weekly profit.

PROBABILITY AND STATISTICS

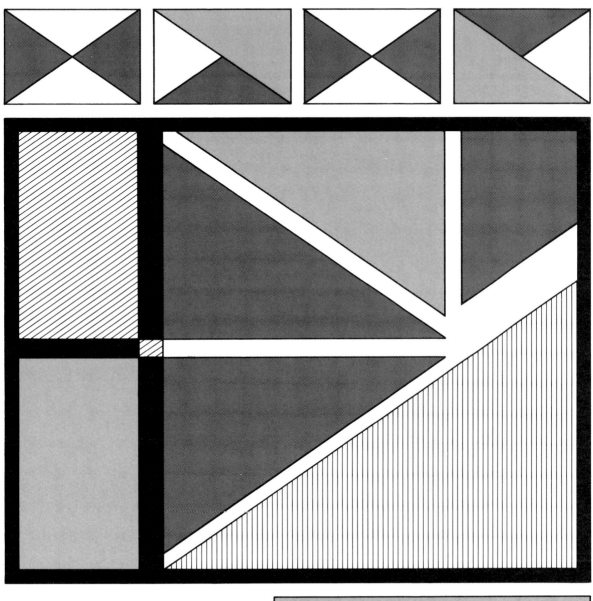

CHAPTER 14

probability

In this chapter, we will present an introduction to the concepts of probability and statistics and show how these topics are handled by computers. Whenever we take chances or "play the odds," we are dealing with the study of **probability**, the science of prediction where outcomes are uncertain. To use probability with computers will involve an introduction to random numbers.

statistics

We will next consider **statistics**, the gathering and analyzing of large amounts of numeric data. In statistics, we will study measures of the average and deviations from this average.

▬ 14.1 Permutations and Combinations

In this section, we will consider arrangements of data items from a given set. We will first consider arrangements where the *order* of the data items is important. For example, the telephone number 555–1234 consists of the same digits as 555–4231, but it is obviously not the same phone number. These types of arrangements are permutations of the data items.

permutation

Definition 14.1 A **permutation** of a set of data items is an arrangement of the data items in which the order of the items is important.

▬ **EXAMPLE 14.1** List all of the permutations of the letters C, A, and T, taking arrangements of three letters and repeating no letters.

Solution The arrangement CAT is different from ATC or TAC, for example, because order counts. The list of the six permutations follows:

CAT	ATC	TAC
CTA	ACT	TCA

Example 14.1 was not very difficult because we were only considering three letters. But we will need a general procedure to handle larger sets. Let's consider the problem from Example 14.1 again. We are looking for three-letter combinations repeating no letters. So we have three choices for the first letter: C, A, or T. We only have two choices for the second letter since we cannot repeat the first letter. There is only one choice left for the third letter since the other two letters have been used and cannot be repeated. The total number of permutations is then the product of $3 \cdot 2 \cdot 1$, or 6.

14.1 Permutations and Combinations

EXAMPLE 14.2 A group of nine players is to be assigned to the nine positions on a baseball team. How many permutations are there of these nine players?

Solution Obviously, no players can play two positions at the same time. There are nine choices for the first position, eight choices left for the second position, seven for the third, and so on. So the number of permutations is

$$9 \cdot 8 \cdot 7 \cdot 6 \cdot 5 \cdot 4 \cdot 3 \cdot 2 \cdot 1 = \underline{362{,}880}$$

Expressions like the one used in this example lead to the following definition.

n factorial

Definition 14.2 For any natural number n, ***n* factorial**, denoted $n!$, is the following product:

$$n(n-1)(n-2)\ldots 3 \cdot 2 \cdot 1$$

The $0!$ expression is defined to be 1.

EXAMPLE 14.3 Simplify the expression.

    a. $5!$ **b.** $\dfrac{6!}{3!}$

Solution

$$5! = 5 \cdot 4 \cdot 3 \cdot 2 \cdot 1 = \underline{120}$$

$$\frac{6!}{3!} = \frac{6 \cdot 5 \cdot 4 \cdot 3 \cdot 2 \cdot 1}{3 \cdot 2 \cdot 1} = 6 \cdot 5 \cdot 4 = \underline{120}$$

Permutations also occur where repetition is allowed. In this case, factorial calculations may not give the answer about how many permutations there are. The next example illustrates.

EXAMPLE 14.4 How many three-letter permutations are there of the letters C, A, and T if letters may be repeated?

Solution Order is still important, so ACC and CCA are different permutations. There are still three choices for the first letter, but since letters may be repeated, there are also three choices for each of the second and third letters. The total number of such permutations is then $3 \cdot 3 \cdot 3$, or $\underline{27}$.

EXAMPLE 14.5 The state of Michigan issues six-character license plates, where the first three characters are single-digit numbers and the last three characters are letters. How many different arrangements are possible for Michigan license plates?

Solution We are talking about permutations in this problem because order is important. Characters may be repeated, so the problem is similar to Example 14.4. There are

10 possible choices for single-digit numbers and 26 choices for letters. The total number of permutations then is

$$10 \cdot 10 \cdot 10 \cdot 26 \cdot 26 \cdot 26 = \underline{17{,}576{,}000}$$

EXAMPLE 14.6 How many permutations are there of seven items taken three at a time, if repetition is not allowed?

Solution In this case, we are not using all of the items. There are seven choices for the first item, six choices for the second item, and five choices for the third item. So we get

$$7 \cdot 6 \cdot 5 = \underline{210} \text{ permutations}$$

The answer in Example 14.6 is the same as $7!/(7-3)!$, which leads to the following result.

Theorem 14.1 The number of permutations of n objects taken r at a time without repetition is given by

$$\frac{n!}{(n-r)!}$$

Many times, order is *not* important in an arrangement. For example, if you were to pick three students out of a class to go on a field trip, the order in which the students were picked would not be important. Only the names of the students picked would be important. This example leads to the other type of arrangement that we will consider in this section: combinations.

combination

Definition 14.3 A **combination** of a set of data items is a collection of data items in which order is not important.

EXAMPLE 14.7 How many two-letter combinations of the letters C, A, and T are possible if no letters are repeated?

Solution Since we are looking for combinations, AT and TA are considered to be the same. So there are $\underline{3}$ combinations possible: AT, AC, and CT. Note that there are 6 permutations possible since we have to add TA, CA, and TC to the list.

In general, the number of combinations of n items taken r at a time equals the number of permutations of n items taken r at a time divided by the number of permutations of r items taken r at a time. In Example 14.7, there are six permutations of three items taken two at a time $[3!/(3-2)!]$. There are also two permutations of two items taken two at a time $(2!)$. So there are $[3!/(3-2)!]/2!$, or $6/2 = 3$, combinations of three items taken two at a time. Thus, we obtain the following formula.

14.1 Permutations and Combinations

Theorem 14.2 The number of combinations of n items taken r at a time without repetition is given by

$$\frac{n!}{(n-r)!r!}$$

which is denoted

$$\binom{n}{r}$$

EXAMPLE 14.8 Simplify the expression.

a. $\binom{4}{3}$ b. $\binom{7}{0}$

Solution

$$\binom{4}{3} = \frac{4!}{(4-3)!3!}$$
$$= \frac{24}{1 \cdot 6}$$
$$= \frac{24}{6}$$
$$= \underline{4}$$

$$\binom{7}{0} = \frac{7!}{(7-0)!0!}$$
$$= \frac{7!}{7! \cdot 1}$$
$$= \frac{7!}{7!}$$
$$= \underline{1}$$

EXAMPLE 14.9 Eight people apply to a company for jobs, and there are only five openings. In how many ways can the company fill these positions?

Solution We are trying to find the number of combinations of eight items taken five at a time. There are $8!/[(8-5)!5!]$ possible combinations. This expression is simplified as follows:

$$\frac{8 \cdot 7 \cdot 6 \cdot 5 \cdot 4 \cdot 3 \cdot 2 \cdot 1}{3 \cdot 2 \cdot 1 \cdot 5 \cdot 4 \cdot 3 \cdot 2 \cdot 1} = 8 \cdot 7 = \underline{56} \text{ combinations}$$

EXAMPLE 14.10 All 50 states submit applications to be considered for the locations of three computer installations, which must be placed in three different states. In how many ways can three states be selected?

Solution Assuming that the computer installations are identical, order of selection is not important. We are looking for the number of combinations of 50 items taken 3 at a time:

$$\frac{50!}{(50-3)!\,3!} = \frac{50!}{47!\,3!} = \frac{50 \cdot 49 \cdot 48 \cdot 47!}{47!\,3!}$$

$$= \frac{50 \cdot 49 \cdot 48}{6} = \underline{19{,}600} \text{ combinations}$$

We will use this knowledge of permutations and combinations in the next section when we discuss the probability of an event happening. Knowing the total number of possible outcomes is very important when we try to find the chances of one particular outcome occurring.

Problem Set 14.1

In Problems 1–12, evaluate the expression.

1. 6!
2. 7!
3. 10!
4. 0!
5. 6! + 5!
6. 8! − 6!
7. $\dfrac{8!}{4!}$
8. $\dfrac{7!}{4!}$
9. $\binom{8}{2}$
10. $\binom{9}{5}$
11. $\binom{7}{6}$
12. $\binom{4}{4}$

In Problems 13–18, rewrite the expression by using factorial notation.

13. $5 \cdot 4 \cdot 3 \cdot 2 \cdot 1$
14. $6 \cdot 5 \cdot 4 \cdot 3 \cdot 2 \cdot 1$
15. $7 \cdot 6 \cdot 5 \cdot 4$
16. $12 \cdot 11 \cdot 10 \cdot 9 \cdot 8$
17. $n \cdot (n-1) \cdot (n-2)$
18. $(n+1)(n)(n-1) \ldots 3 \cdot 2 \cdot 1$

19. How many permutations are there of seven items taken two at a time if no repetitions are allowed?
20. How many permutations are there of nine items taken seven at a time if no repetitions are allowed?
21. How many permutations are there of seven items taken five at a time if no repetitions are allowed?
22. How many permutations are there of nine items taken two at a time if no repetitions are allowed?
23. How many permutations are there of seven items taken two at a time if repetitions are allowed?
24. How many permutations are there of nine items taken seven at a time if repetitions are allowed?
25. How many combinations are there of seven items taken five at a time if repetitions are allowed?
26. How many combinations are there of nine items taken two at a time if no repetitions are allowed?
27. A password to gain access to the computer consists of four characters. The first three characters must be letters, and the fourth character must be a number. How many possible passwords are there?
28. A password to gain access to the computer consists of five characters. The first character must be a letter and the other four characters can be either letters or numbers. How many possible passwords are there?
29. Nine people are eligible for selection into the Hall of Fame, but only three people will actually be selected. In how many ways can these three people be selected?
30. Six people are running for election for four seats on the city council. In how many ways can these seats be filled?

31. First-, second-, and third-place prizes are to be awarded to three lucky contestants out of seven entrants. In how many ways can these prizes be awarded?

32. A winner and a runner-up are to be selected from a list of ten finalists. In how many ways can the awards be given?

33. The 50 states are to be ranked in a certain order. How many possible orders are there?

34. In how many ways can the top-ten basketball teams be ranked?

In Problems 35–38, determine how many arrangements of the numbers 1, 2, 7, 8, 9 are possible if exactly five numbers are used under the given conditions.

35. Order is unimportant, and no number may be repeated.

36. Order is important, and no number may be repeated.

37. The first digit must be a 9, order is important, and numbers may be repeated.

38. The third digit must be a 2 or an 8, order is important, and numbers may be repeated.

39. Write a flowchart for a procedure to calculate $n!$ for a given number n.

40. Write an algorithm for a procedure to calculate $n!$ for a given number n.

41. What is the largest value for n for which $n!$ can be stored by using no more than eight bits?

42. What is the largest value for n for which $n!$ can be stored by using no more than ten bits?

14.2 Introduction to Probability

Consider the act of tossing a coin. There are two possible outcomes: heads or tails. Trying to predict the chances of either outcome is a probability question. In this section, we will use probability theory to try and predict the chances of certain events happening.

If we have a fair coin, we know that the chances of getting either outcome are the same (50%). We will say that the probability of the coin coming up heads is 0.5, and the probability of the coin coming up tails is also 0.5. The probability of any event occurring is then a number ranging from 0 (no chance at all) through 1 (a certainty). If the chances of winning the lottery are 7%, then the probability of winning is 0.07, and the probability of losing is 0.93.

To determine the probability of an event occurring is not always easy. For example, what is the probability of rolling a pair of dice and getting a total of 7 on the two dice? In this case, we can roll a pair of dice many times and record the results. A good approximation to the probability of an event occurring can be calculated by dividing the number of times a certain outcome was achieved by the total number of times the experiment was performed. Suppose we rolled the pair of dice 100 times and noted that the dice totaled 7 on 18 of the rolls. We could then say that the probability of rolling a 7 is *approximately* 18%, or 0.18.

■ **EXAMPLE 14.11**   A coin was tossed 1000 times. Heads came up 523 times, and tails came up 477 times. From this observed data, calculate the probability of getting heads and tails.

Solution We will use the notation P(outcome) to designate the probability of the outcome. For this example, we get

$$P(\text{heads}) = \frac{523}{1000} = \underline{0.523}$$
$$P(\text{tails}) = \frac{477}{1000} = \underline{0.477}$$

Note that experiments like the one of Example 14.11 do not give exact probabilities, only approximations.

Now, consider tossing a fair six-sided die. There is an equal chance of rolling a 1, 2, 3, 4, 5, or 6. So the probability of rolling any particular number is $\frac{1}{6}$, or approximately 0.167. This observation leads to the concept of a sample space.

sample space

Definition 14.4 The **sample space** for an experiment is the set of all possible outcomes.

■ **EXAMPLE 14.12** Find the sample space for the experiment.
a. Tossing a coin

Solution sample space = $\underline{\{\text{heads, tails}\}}$

b. Rolling a die

Solution sample space = $\underline{\{1, 2, 3, 4, 5, 6\}}$

c. Choosing a day of the week

Solution sample space = $\underline{\{\text{Monday, Tuesday, Wednesday, Thursday, Friday, Saturday, Sunday}\}}$

d. Choosing a two letter combination of C, A, and T with no repetition

Solution sample space = $\underline{\{\text{CA, CT, AT}\}}$

e. Choosing a two-letter permutation of C, A, and T with no repetition

Solution sample space = $\underline{\{\text{AC, AT, CT, CA, TA, TC}\}}$

event

elementary event

An **event** is defined to be an experiment in which we are looking for a specific outcome. Any event for which we are looking for one specific outcome is called an **elementary event**. For example, rolling a die and getting a 5 is an elementary event. Rolling a die and getting some number smaller than 5 is *not* an elementary event since the outcomes 1, 2, 3, and 4 all satisfy this event.

If every possible outcome in the sample space has an equal chance of occurring, then to find the probability of each elementary event occurring, we simply divide 1 by the total number of outcomes in the sample space.

14.2 Introduction to Probability

EXAMPLE 14.13 Find the probability of rolling a die and getting a 5.

Solution $P(5) = \frac{1}{6} \approx \underline{0.167}$

EXAMPLE 14.14 Find the probability of choosing a letter combination of the letters C, A, and T with no repetition and getting CA.

Solution $P(CA) = \frac{1}{3} \approx \underline{0.333}$

Consider now the probability of rolling a die and getting a number smaller than 5. This event is not an elementary event. But to find its probability, we can use the following property of probability.

Addition Property If an event consists of n different elementary events, then the probability of that event occurring will be the *sum* of the probabilities of each elementary event.

So the probability of rolling a die and getting less than a 5 is:

$$P(D<5) = P(1) + P(2) + P(3) + P(4)$$
$$= \tfrac{1}{6} + \tfrac{1}{6} + \tfrac{1}{6} + \tfrac{1}{6}$$
$$= \tfrac{4}{6} \approx \underline{0.667}$$

If all outcomes are equally likely, the probability of an event occurring is simply the quotient of the number of outcomes that give the desired event and the total number of possible outcomes.

Now, let's consider some more examples.

EXAMPLE 14.15 Two people must be selected from a group of four for the planning committee. What is the probability that George, one of the four people, will be selected?

Solution We are looking for combinations of four items taken two at a time, so there are $\binom{4}{2}$, or 6, possible outcomes. There are 3 possible outcomes that include George: George and each of the other three people. So the probability of George being selected is

$$P(\text{George}) = \tfrac{3}{6} = \underline{0.5}$$

EXAMPLE 14.16 If two dice are tossed, what is the probability that the sum of the numbers on the two dice will add to 8?

Solution There are 6 possible outcomes for each die, so there are 36 possible outcomes when two dice are tossed (see Problem 7). If we represent the numbers from the two dice as an ordered pair, then (2, 6) represents a 2 on the first die and a 6 on the second die. The following rolls then add to 8:

$$(2, 6), (3, 5), (4, 4), (5, 3), (6, 2)$$

So the probability of getting a sum equal to 8 is

$$P(8) = \tfrac{5}{36} \approx \underline{0.139}$$

■ **EXAMPLE 14.17**   Five contestants enter a drawing for first, second, and third prizes in a contest. What are the chances that JoAnne, one of the five contestants, will finish either first or second?

Solution   In this problem, order is important, since there is a difference in the prizes. So we are looking for permutations of five items taken three at a time. There are $5 \cdot 4 \cdot 3$, or 60, possible outcomes. The number of outcomes for which JoAnne finishes first is $1 \cdot 4 \cdot 3$, or 12, since there is only one choice for first, four for second, and three for third. There are a similar number of possible outcomes for which JoAnne can finish second, by the same reasoning. So there are $12 + 12$, or 24, possible outcomes that can give JoAnne either first or second. Thus, the probability of JoAnne finishing either first or second is

$$P(\text{1st or 2nd}) = \tfrac{24}{60} = \underline{0.4}$$

In the previous examples, we were dealing with experiments in which every outcome was equally likely, which is not always the case. Consider the following examples.

■ **EXAMPLE 14.18**   Suppose a die is weighted so that 1 and 6 are twice as likely to be rolled as are the other numbers. What is the probability of rolling either a 4 or a 5?

Solution   There are still six possible outcomes, but 1 and 6 are twice as likely to occur as the other outcomes. One way of handling this situation is to say that there are eight possible outcomes, listing 1 and 6 twice. Then the probability of rolling a 1 or a 6 is each $\tfrac{2}{8}$, and the probability of rolling any other number is $\tfrac{1}{8}$. So the probability of rolling either a 4 or a 5 is

$$P(\text{4 or 5}) = \tfrac{1}{8} + \tfrac{1}{8} = \underline{0.25}$$

■ **EXAMPLE 14.19**   A basket contains four red balls, three yellow balls, and five purple balls. What are the chances of pulling out a red ball?

Solution   Although there are really only 3 possible outcomes—a red, a yellow, or a purple ball—we can consider that there are 12 possible outcomes since there are 12 balls in the basket. Four possible outcomes will result in a red ball, so the chances of pulling out a red ball are

$$P(\text{red}) = \tfrac{4}{12} \approx \underline{0.333}$$

random function

A computer can simulate probability problems like the ones we have considered in this section by using a **random function**, a function which will return a random number. Actually, a computer cannot generate truly random numbers. Some mathematical formula must be used. For example, we can use the following algorithm to generate four-digit "random" numbers:

1. Take any four-digit number.
2. Square this number, and take the middle four digits for the next random number.

14.2 Introduction to Probability

3. Repeat step 2 with the new random number.

For instance, if we start with the number 4321, the next number to be calculated is $(4321)^2$, or 18,671,041. The middle four numbers are 6710, so that is the next random number.

Many languages have a function—like RND in some versions of BASIC—which will return a random number. RND will return a real number between 0 and 1, not equal to 1. The expression RND*2 returns a real number between 0 and 2, not equal to 2. To simulate tossing a coin, we use the assignment statement C = INT(RND*2), which returns either a 0 or a 1 with equal probability. Here, 1 can represent heads, and 0 can represent tails. INT is a built-in function in many versions of BASIC which will take a real number and return the largest integer not greater than the real number. For example, INT(5.6) = 5, and INT(−3.2) = −4. Similarly, to simulate rolling a die, we use R = INT(RND*6) + 1 which returns either a 1, 2, 3, 4, 5, or 6 with equal probability.

The RND function allows the computer to be used to run experiments thousands of times when probability theory cannot be used to predict the chances of an event occurring.

Problem Set 14.2

In Problems 1–10, find the sample space for the experiment.

1. Picking a whole number larger than 4 and smaller than 8
2. Picking a natural number smaller than 10
3. Choosing a season
4. Choosing a month
5. Choosing a three-letter combination of A, E, I, O, and U, repeating no letters
6. Choosing a two-letter permutation of A, E, I, O, and U, repeating no letters
7. Rolling two dice and finding the total number of dots
8. Tossing two coins
9. Drawing a card from a normal deck of 52 cards
10. Choosing a color of the rainbow
11. Let the four people in Example 14.15 be George, Sam, Janet, and Carol. List all six possible combinations, and note the combinations that involve George.
12. Let the five people in Example 14.17 be JoAnne, Mary, Bill, Nancy, and Jason. List the 12 permutations that have JoAnne finishing first.
13. Find the first four "random" numbers generated by the algorithm from this section if the first number is 4545.
14. Find the first five "random" numbers generated by the algorithm from this section if the first number is 4000. Is this number a good number to start with?

A quarter and a dime are tossed. In Problems 15–20, find the probability of the given event occurring.

15. Two heads
16. Two tails
17. One head and one tail
18. Heads for the quarter
19. Heads for the quarter and tails for the dime
20. Heads for the quarter or tails for the dime

Two dice are tossed. In Problems 21–26, find the probability of the given event occurring. (*Hint:* Use Problem 7.)

21. Both dice show 1.
22. The sum of the dice equals 7.
23. The sum of the dice is less than 5.
24. Both dice show an even number.
25. Neither die show a 2.
26. Both dice show the same number.

A hat contains five red, six yellow, three orange, and seven green marbles. In Problems 27–32, find the probability of the given event occurring if one marble is pulled from the hat.

27. The marble is red.
28. The marble is green.
29. The marble is not red.
30. The marble is yellow or orange.
31. The marble is neither red nor orange.
32. The marble is not green.

Two dice are weighted so that 1 and 6 will never come up. In Problems 33–38, find the probability of the given event occurring when the two dice are rolled. (*Hint:* Find the sample space first.)

33. The sum of the dice is 3.
34. Both dice show an odd number.
35. Neither dice show a 5.
36. The sum of the dice is more than 5.
37. Both dice show an even number.
38. At least one die shows a 6.

Justin, Matt, Jean, Jenny, Ryan, and Lynnie are eligible for a drawing that will pick three of them to continue in the contest. In Problems 39–44, find the probability of the given event occurring.

39. Jenny is not drawn.
40. Ryan and Lynnie both are drawn.
41. Justin is drawn, but not Matt.
42. Justin, Matt, Jean, and Jenny are drawn.
43. Neither Jean nor Jenny is drawn.
44. Jean is drawn.

Jason entered three drawings in a contest; Jeffrey entered two drawings; and Shannon entered one. There were no other entries. These six drawings were put in a barrel, and two drawings were picked. The first drawing picked was awarded first prize, and the second drawing picked was awarded second prize. In Problems 45–50, find the probability of the given event occurring.

45. Jason receives both first and second prizes.
46. Shannon receives no prize.
47. Jeffrey receives second prize.
48. Jeffrey receives *only* second prize.
49. Shannon receives at least one prize.
50. Jason receives first prize and Jeffrey receives second prize.

When 400 people were asked which weekday was their favorite, 205 people said Friday, 112 people said Thursday, 23 people said Wednesday, 7 people said Tuesday, 3 people said Monday, and 50 people responded "None." From this data, for Problems 51–54, find the probability of the given event occurring.

51. A person prefers "Friday."
52. A person responds "None."
53. A person responds either "Friday" or "Monday."
54. A person does not respond "Thursday."
55. Write an algorithm for a procedure to simulate tossing a coin 500 times and counting the number of heads and the number of tails.
56. Write an algorithm for a procedure to simulate rolling a die 600 times and counting the number of 1s, 2s, 3s, 4s, 5s, and 6s.
57. Write an algorithm for a procedure to simulate rolling a pair of dice repeatedly until a total of 7 appears on the two dice. Print out the number of rolls necessary to achieve this event.

58. Write an algorithm for a procedure to toss a coin repeatedly until three heads in a row are attained. Print out the number of rolls necessary to achieve this event.

14.3 Summation Notation

Many important statistical measurements involve finding sums of large amounts of data. In this section, we will introduce special notation to designate these large sums.

Assume that we have this set of N data items: $\{X_1, X_2, \ldots, X_N\}$. The sum of these items is $X_1 + X_2 + \ldots + X_N$. We will use the Greek capital letter Σ (sigma) to represent the sum and X_i to represent an arbitrary item from the set. The value of i will range through all integral values from 1 (for the first item in the set) through N (for the last item in the set). This **summation**, or **sigma**, **notation** is illustrated next:

summation, sigma, notation

$$\sum_{i=1}^{N} X_i = X_1 + X_2 + \ldots + X_N$$

■ **EXAMPLE 14.20** Illustrate the summation notation for $X_1 = 2$, $X_2 = 4$, $X_3 = 7$, and $X_4 = 8$.

Solution
$$\sum_{i=1}^{4} = X_1 + X_2 + X_3 + X_4$$
$$= 2 + 4 + 7 + 8 = \underline{21}$$

Summation notation is particularly valuable when a formula is involved. For example, consider the formula $3i + 1$. The sum of the values of this expression as i takes on all integral values between 1 and 5 is as follows:

$$\sum_{i=1}^{5} (3i+1) = [(3 \cdot 1)+1] + [(3 \cdot 2)+1] + [(3 \cdot 3)+1] + [(3 \cdot 4)+1] + [(3 \cdot 5)+1]$$
$$= (3+1) + (6+1) + (9+1) + (12+1) + (15+1)$$
$$= 4 + 7 + 10 + 13 + 16 = \underline{50}$$

■ **EXAMPLE 14.21** Write the sum in sigma notation.

a. $4x_1 + 4x_2 + \ldots + 4x_N$

Solution $\sum_{i=1}^{N} 4x_i$

b. $(2x_3 + 7) + (2x_4 + 7) + \ldots + (2x_N + 7)$

Solution $\sum_{i=3}^{N} (2x_i + 7)$

c. $3+6+9+\ldots+30$

Solution In this sum, we are adding all multiples of 3 from $(3 \cdot 1)$ to $(3 \cdot 10)$. The sigma notation is then

$$\sum_{i=1}^{10} 3i$$

d. $2+6+10+\ldots+50$

Solution Each term in this sum is four more than the previous term. The sum $\sum_{i=1}^{13} 4i$ gives $4+8+12+\ldots+52$. Since each term in this sum is two more than the corresponding term in our sum, we must subtract 2 from the general formula $4i$. So we write

$$\sum_{i=1}^{13} (4i-2)$$

In general, if all of the terms of a sum differ from the previous term by the same amount, then i will be multiplied by that amount. Then we can decide what number must be subtracted or added to the expression to give the desired first term.

■ **EXAMPLE 14.22** Express this sum in sigma notation:

$$1^2 + 7^2 + 13^2 + \ldots + 67^2$$

Solution Each number differs from the previous number by 6 if we ignore the square, so the general term involves $6i$. Since $6 \cdot 1 = 6$ and the first number is 1, we must subtract 5 from $6i$ ($6-5=1$). Finally, each of the numbers is squared, so the general formula must be $(6i-5)^2$. Since $(6 \cdot 12) = 72$, the sigma notation is

$$\sum_{i=1}^{12} (6i-5)^2$$

■ **EXAMPLE 14.23** Find the sum.

a. $\sum_{i=1}^{4} (2i+4)$

Solution
$$\sum_{i=1}^{4} (2i+4) = [(2 \cdot 1)+4] + [(2 \cdot 2)+4] + [(2 \cdot 3)+4] + [(2 \cdot 4)+4]$$
$$= (2+4)+(4+4)+(6+4)+(8+4)$$
$$= 6+8+10+12 = \underline{36}$$

b. $\sum_{i=1}^{3} (2i^2 - 3i + 1)$

14.3 Summation Notation

Solution

$$\sum_{i=1}^{3}(2i^2-3i+1) = [(2\cdot 1^2)-(3\cdot 1)+1]+[(2\cdot 2^2)-(3\cdot 2)+1]$$
$$+[(2\cdot 3^2)-(3\cdot 3)+1]$$
$$= (2-3+1)+(8-6+1)+(18-9+1)$$
$$= 0+3+10 = \underline{13}$$

c. $\sum_{i=3}^{6}(i-1)^2$

Solution

$$\sum_{i=3}^{6}(i-1)^2 = (3-1)^2+(4-1)^2+(5-1)^2+(6-1)^2$$
$$= 2^2+3^2+4^2+5^2$$
$$= 4+9+16+25 = \underline{54}$$

d. $\left(\sum_{i=1}^{5}3i\right)^2$

Solution

$$\left(\sum_{i=1}^{5}3i\right)^2 = [(3\cdot 1)+(3\cdot 2)+(3\cdot 3)+(3\cdot 4)+(3\cdot 5)]^2$$
$$= (3+6+9+12+15)^2$$
$$= 45^2 = \underline{2025}$$

In Example 14.23c, each term is squared before being added to the sum, but in Example 14.23d, the sum is calculated first and then squared. There is a big difference between these two types of problems.

Sums similar to the ones discussed in this section are easily handled by computers using loops.

■ **EXAMPLE 14.24** Write a flowchart for a procedure to find the following sum:

$$\sum_{i=10}^{120}(6i-3)$$

Solution The flowchart appears in Figure 14.1.

■ **EXAMPLE 14.25** Write an algorithm for a procedure to input the number of congressional representatives in each of the 50 states and find the total number of representatives in the country.

Solution This problem is equivalent to finding

$$\sum_{i=1}^{50} X_i$$

where X_i represents the number of representatives in the ith state.

378 PROBABILITY AND STATISTICS

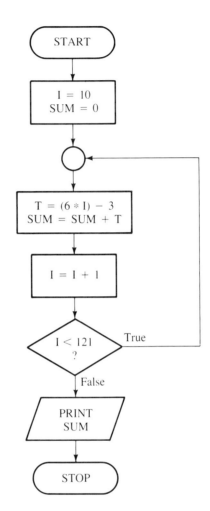

Figure 14.1

The algorithm is as follows:

1. SUM = 0 : Set sum to zero
2. I = 0 : Set counter to zero
3. INPUT N : Input number of reps
4. SUM = SUM + N : Add number of reps to sum
5. I = I + 1 : Add 1 to counter
6. IF I < 50 THEN Go to 3 : Repeat algorithm if all states have not been counted
7. PRINT SUM : Print final total

Problem Set 14.3

In Problems 1–10, write the sum in sigma notation.

1. $X_1 + X_2 + \ldots + X_{10}$
2. $1 + 2 + \ldots + 10$
3. $12 + 14 + 16 + \ldots + 40$
4. $3X_4 + 3X_5 + \ldots + 3X_{16}$
5. $(2X_1 - 3) + (2X_2 - 3) + \ldots + (2X_N - 3)$
6. $X_7^2 + X_8^2 + \ldots + X_N^2$
7. $(X_3 + X_4 + \ldots + X_N)^2$
8. $(X_1 + 4)^2 + (X_2 + 4)^2 + \ldots + (X_{17} + 4)^2$
9. $(3X_1^2 - 2X_1 + 6) + (3X_2^2 - 2X_2 + 6) + \ldots + (3X_N^2 - 2X_N + 6)$
10. $[(3X_1 - 1)^3 + (3X_2 - 1)^3 + \ldots + (3X_N - 1)^3]^2$

In Problems 11–20, find the sum.

11. $\sum_{i=1}^{5} i$
12. $\sum_{i=1}^{5} (i + 4)$
13. $\sum_{i=1}^{7} 3i$
14. $\sum_{i=1}^{8} (-7i)$
15. $\sum_{i=4}^{6} (2i - 2)$
16. $\sum_{i=1}^{7} (4 - 5i)$
17. $\sum_{i=1}^{3} [(i-2)^2 + 4]$
18. $\sum_{i=2}^{4} (3i^2 - 2i + 4)$
19. $\sum_{i=1}^{4} (3i^2 - 2i)^2$
20. $\left[\sum_{i=1}^{3} (2i^2 - 5i + 7)\right]^2$

In Problems 21–24, write an algorithm for a procedure to calculate the sum.

21. $\sum_{i=1}^{100} (4i - 2)$
22. $\sum_{i=1}^{200} (3i^2 - 4i + 7)$
23. $\sum_{i=50}^{250} (2i - 11)^2$
24. $\left[\sum_{i=100}^{1000} (4i + 6)\right]^2$

In Problems 25–28, write a flowchart for the algorithm in the indicated problem.

25. Problem 21
26. Problem 22
27. Problem 23
28. Problem 24

29. A class of 30 students took a test. Write an algorithm to find the square of the sum of the test scores ($[\sum_{i=1}^{30} T_i]^2$).
30. A class of 30 students took a test. Write an algorithm to find the sum of the squares of the test scores ($\sum_{i=1}^{30} T_i^2$).
31. Write an algorithm to find the sum of all even numbers from 2 to 100.
32. Write an algorithm to find the sum of $4 + 7 + \ldots + 34$.

In Problems 33–36, find the sum if $X_1 = 4$, $X_2 = -3$, $X_3 = 16$, and $X_4 = 8$.

33. $\sum_{i=1}^{4} X_i$
34. $\sum_{i=1}^{4} (2X_i + 3)$
35. $\sum_{i=1}^{4} X_i^2$
36. $\left(\sum_{i=1}^{4} X_i\right)^2$

37. Write an algorithm for a procedure to input a value for N and calculate the sum $1 + 2 + \ldots + N$.
38. The Valentine City Bank gives 5.5% (0.055) interest compounded annually to each of its 500 accounts. Write an algorithm for a procedure to find the sum of all accounts before interest is added, and the sum of all accounts after interest has been added (account = 1.055*account).

14.4 Measures of Central Tendency

Computers are very good at computing statistics on large amounts of information. In this section and the next, we will discuss the standard set of statistics that are frequently calculated by computers.

measures of central tendency

The first statistical measures that we will discuss are **measures of central tendency**, or measures of the average. The first measure we will consider is the mean.

mean

Definition 14.5 The **mean** of a set of numbers, denoted by the Greek letter μ, is the sum of all of the numbers in the set divided by the total number of numbers in the set.

In other words, the mean is simply the average as we have known it until now. We can use the sigma notation of the previous section to represent the mean of a set of data items $\{X_1, X_2, \ldots, X_N\}$. The mean is simply

$$\mu = \frac{\sum_{i=1}^{N} X_i}{N}$$

■ **EXAMPLE 14.26** Find the mean of the set of four test scores 70, 80, 63, and 89.

Solution
$$\mu = \frac{70 + 80 + 63 + 89}{4}$$
$$= \frac{302}{4} = \underline{75.5}$$

■ **EXAMPLE 14.27** Write a flowchart for a procedure to find the mean of a set of N numbers stored in an array.

Solution We will assume that the numbers have already been read in and stored in the array A. A loop is needed to go through the array and find the sum of all entries. After the loop is finished, the sum is divided by N. The flowchart for the procedure appears in Figure 14.2.

There are times when the mean is not the best measure of an average of a set of numbers. Consider these ages for people in a group: 98, 95, 97, 94, 90, 90, and 3. The mean age is 81, even though no age is near 81 and all but one age is well above 81. A better average for this group would be obtained from the median.

median

Definition 14.6 The **median** of a set of numbers is a number such that there are the same number of numbers larger than this number as there are numbers smaller than this number.

In other words, the median represents the middle score. The median is considered a better measure of the average than the mean in some circumstances because it is less affected by extremes (for instance 3 in the set of ages) than the mean. Thus, we sometimes talk about the median family income or the median wage.

14.4 Measures of Central Tendency

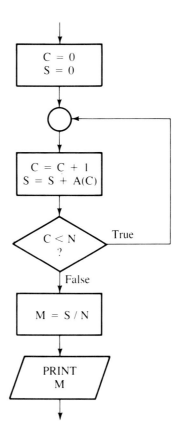

Figure 14.2

EXAMPLE 14.28 Find the median of the ages 98, 95, 97, 94, 90, 90, and 3.

Solution To find the median of a set of numbers, arrange the numbers in either decreasing or increasing order. The middle value is the median:

$$
\begin{array}{l}
98 \\
97 \\
95 \\
94 \quad \longleftarrow \text{ middle term} \\
90 \\
90 \\
3
\end{array}
$$

The median is 94.

If there are an odd number of numbers, then there will always be a middle value. If there are an even number of numbers, then there are two middle values. The mean of these two middle values is the median.

EXAMPLE 14.29
Find the median of the numbers 26, 13, 34, 43, 7, and 16.

Solution Arranging these numbers in decreasing order, we get

$$
\begin{array}{ll}
43 & \\
34 & \\
26 & \longleftarrow \quad \text{middle} \\
16 & \longleftarrow \quad \text{terms} \\
13 & \\
7 &
\end{array}
$$

The median is $(26 + 16)/2$, or $\underline{21}$.

In many computer languages, there is a built-in function, like FIX in BASIC, that will *truncate* a real number to an integer. For example, FIX(7.1) is 7, and FIX(10.8) is 10. If a set of N numbers has been ordered, then we find the following two numbers:

$M_1 = \text{FIX}((1+N)/2)$
$M_2 = \text{FIX}(1+(N/2))$

The mean of the numbers in positions M_1 and M_2 will be the median of the set of numbers.

For example, for a set of seven numbers (as in Example 14.28), we get

$M_1 = \text{FIX}((1+7)/2) = \text{FIX}(4) = 4$
$M_2 = \text{FIX}(1+(7/2)) = \text{FIX}(4.5) = 4$

The median age is then the fourth age in the ordered list. For Example 14.29 with $N = 6$, we get

$M_1 = \text{FIX}((1+6)/2) = \text{FIX}(3.5) = 3$
$M_2 = \text{FIX}(1+(6/2)) = \text{FIX}(4) = 4$

The median is the mean of the third and fourth numbers in the ordered set.

The third measure of central tendency that we will discuss is the mode.

mode

Definition 14.7 The **mode** of a set of numbers is the number which occurs most often in the set.

EXAMPLE 14.30
Find the mode of the numbers 53, 42, 66, 42, 42, 73, 82, 42, and 66.

Solution The number 42 occurs four times, and no other number occurs more than twice. So the mode is $\underline{42}$.

14.4 Measures of Central Tendency

The mode is used when you are trying to find the most "popular" number in a set. For example, consider the following statement: "When asked to choose which year in the 1970s was their favorite year, more people chose 1974 than any other year." This statement says that 1974 is the mode of the set of answers.

EXAMPLE 14.31 Find the mode of these numbers: 17, 63, 45, 26, 87, and 100.

Solution Since there are no repeated values, there is a six-way tie for the mode. In this case, where there are so many multiple values for the mode, the mode ceases to be a valuable statistic.

EXAMPLE 14.32 Find a set of seven numbers that satisfies these conditions:

$$\text{mean} = 70 \quad \text{median} = 50 \quad \text{mode} = 40$$

Solution Since the median is 50, the fourth (middle) number of our list must be 50, with three numbers smaller than 50 and three numbers larger than 50. Since the mode is 40, we should have more than one 40. Let's take all three numbers smaller than 50 to be 40. This assumption gives us the following list:

$$?? \quad ?? \quad ?? \quad 50 \quad 40 \quad 40 \quad 40$$

We now have to consider the mean, which is 70. The sum of these seven numbers must be 490 (7 · 70). The four numbers we already have add to 170, so the remaining three numbers must all be larger than 50 and add to 320 (490 − 170). Many sets of numbers will work. We will choose 120, 100, and 100. Although we are using 100 twice, 40 remains the mode since it appears three times in the list. The final list is now

$$120 \quad 100 \quad 100 \quad 50 \quad 40 \quad 40 \quad 40$$

Normally, the three measurements—mean, median, and mode—do not differ by very much, in which case any of them can be used as a measure of the average. None of these measures give any idea as to the variability or range of the numbers. They only give an idea of the average, or central tendency, of the numbers. The next section will discuss measures of the range and variability of the numbers.

Problem Set 14.4

In Problems 1–8, find the mean of the set of numbers.

1. 26, 25, 15, 14, 82, 73
2. 110, 51, 64, 53, 51, 27, 68
3. 76, 76, 76, 76, 76, 76

4. 10, 20, 35, 40, 50, 60, 70, 30, 30
5. 0, 10, 80, 0, 9, 93, 0, 93, 93
6. 40, 80, 80, 60, 40, 80, 40
7. 87, 93, 22, 76, 76, 65, 130, 59
8. 1, 2, 3, 4, 5, 6, 7, 8, 9, 10

In Problems 9–16, find the median of the set of numbers given in the indicated problem.

9. Problem 1
10. Problem 2
11. Problem 3
12. Problem 4
13. Problem 5
14. Problem 6
15. Problem 7
16. Problem 8

In Problems 17–24, find the mode of the set of numbers given in the indicated problem.

17. Problem 1
18. Problem 2
19. Problem 3
20. Problem 4
21. Problem 5
22. Problem 6
23. Problem 7
24. Problem 8

25. A student has test scores of 65, 73, 79, and 93. What must this student get on the fifth test to have an average (mean) of 80? (*Hint:* Total points divided by number of tests equals mean.)
26. A student had an average (mean) of 77 on four tests. If the last three test scores were 73, 80, and 84, what was the student's score on the first test?

In Problems 27–36, find a set of seven nonnegative numbers that satisfies the conditions (if possible). Answers may vary.

27. Mean = 70, median = 70, mode = 65
28. Mean = 75, median = 75, mode = 75
29. Mean = 70, median = 80, mode = 60
30. Mean = 100, median = 50, mode = 70
31. Mean = 100, median = 0, mode = 0
32. Mean = 20, median = 40, mode = 40
33. Mean = 87, median = 76, mode = 78
34. Mean = 40, median = 60, mode = 40
35. Mean = 60, median = 60, mode = 100
36. Mean = 100, median = 80, mode = 100

In Problems 37–42, which measure of central tendency is being discussed in each statement?

37. There were just as many employees with more than ten days vacation as there were employees with less than ten days vacation.
38. "Elusive Butterfly" is the most requested old song.
39. The average daily rainfall last month was 0.13 inch.
40. You are better off than half of the people in this country if you make $24,234 a year.
41. More people watched the All-American Bowl on New Year's Eve than any other football game that day.
42. If you divide the sum of the dots on all sides of a die by 6, the result is 3.5.

43. Give an example where the mode would be a very poor representative of the average.
44. In the list 20, 25, 25, 26, 10,000, is the mean or median more representative of the average?
45. Find a list of six numbers such that there are two modes, which are both different from the median.
46. Give an example where the mode is a better indication of the average than the median (if possible).

Another statistical measurement which is useful is the *mean square*, which is defined to be $[\sum_{i=1}^{N}(X_i)^2]/N$ for a set of data items $\{X_1, X_2, \ldots, X_N\}$. Use this definition in Problems 47–56.

47. Find the mean square of the numbers 71, 65, 54, 89, and 13.
48. Find the mean square of the numbers 90, 80, 70, 60, and 50.
49. Find the mean of the numbers in Problem 47.
50. Find the mean of the numbers in Problem 48.
51. Write an algorithm for a procedure to calculate the mean square of a set of N data items that have been stored in an array.
52. Write an algorithm for the procedure described in Problem 49.

53. Does the mean square give the same result as the square of the mean in Problems 47 and 49?

54. Does the mean square give the same result as the square of the mean in Problems 48 and 50?

55. Use sigma notation to write an expression for the square of the mean. Compare this expression with the expression for the mean square.

56. Can the mean square of a set of numbers ever be negative? If so, give an example. If not, state why not.

14.5 Measures of Dispersion

In the previous section, we discussed the mean, median, and mode: measures of the average. However, these measures tell us nothing about how varied the numbers are. The list 76, 76, 76, 76, 76 has a mean, median, and mode of 76. The list 87, 93, 22, 76, 76, 65, 130, 59 also has a mean, median, and mode of 76. There is obviously a big difference between these two lists of numbers. In this section, we will discuss ways of measuring differences between lists like these two. We will present three such **measures of dispersion**. The easiest one to find is the range.

measures of dispersion

range

Definition 14.8 The **range** of a set of numbers is the difference between the largest and smallest numbers in the set.

■ **EXAMPLE 14.33** Find the range of the set of numbers.
  a. 76, 76, 76, 76, 76

Solution range = 76 − 76 = $\underline{0}$

  b. 87, 93, 22, 76, 76, 65, 130, 59

Solution range = 130 − 22 = $\underline{108}$

The smaller the range, the closer together the numbers are. The range is a simple measure of how widely dispersed the set of numbers is. However, the range only takes into account the largest and smallest numbers in the set. We need a measure of dispersion which takes into account *every* number in the set.

A better measure of the dispersion is the variance, which takes into account each number in the set and the difference between that number and the mean. In Example 14.33a, no number differs from the mean (76). But in Example 14.33b, some numbers differ from the mean by as much as 54 (130 − 76 = 54). We would like a measure of the average (mean) deviation of numbers in the set from the mean.

We can't simply add all of the differences and divide by the number of values since the sum of the differences will always be zero. For example, the set of numbers 60, 80, and 100 has a mean of 80. The differences are 60 − 80, 80 − 80, and 100 − 80, or −20, 0, and 20. Their sum is 0.

variance

We could take the sum of the differences ignoring the minus sign (20+0 +20 = 40), and this procedure is sometimes used in statistics. However, a more widely used procedure is to square the differences before adding. This method also eliminates the minus signs, and it is how we will define variance.

Definition 14.9 The **variance**, denoted σ^2 (σ is the Greek lowercase letter sigma), of a set of N numbers is the sum of the squares of the differences of each number from the mean divided by N. In summation notation, we have

$$\sigma^2 = \frac{\sum_{i=1}^{N}(x_i - \mu)^2}{N}$$

where μ is the mean of the numbers and x_i is an arbitrary number from the set.

■ **EXAMPLE 14.34** Find the variance of this set of numbers: {87, 93, 22, 76, 76, 65, 130, 59}.

Solution The mean of these numbers is 76. We now calculate the variance.

i	x_i	$x_i - \mu$	$(x_i - \mu)^2$
1	87	$87 - 76 = 11$	$11^2 = 121$
2	93	$93 - 76 = 17$	$17^2 = 289$
3	22	$22 - 76 = -54$	$(-54)^2 = 2916$
4	76	$76 - 76 = 0$	$0^2 = 0$
5	76	$76 - 76 = 0$	$0^2 = 0$
6	65	$65 - 76 = -11$	$(-11)^2 = 121$
7	130	$130 - 76 = 54$	$54^2 = 2916$
8	59	$59 - 76 = -17$	$(-17)^2 = 289$

$$\sum_{i=1}^{8}(x_i - \mu)^2 = 6652$$

The variance is 6652/8, or $\underline{831.5}$

An alternative definition of the variance, which is relatively easy to program on a computer and works well with large sets of data, is

$$\sigma^2 = \frac{N\sum_{i=1}^{N}x_i^2 - \left(\sum_{i=1}^{N}x_i\right)^2}{N^2}$$

■ **EXAMPLE 14.35** Use the alternative definition of variance to find the variance of the set {87, 93, 22, 76, 76, 65, 130, 59}.

14.5 Measures of Dispersion

Solution We do not need to first calculate the mean, as was necessary in Example 14.34. We need only find the sum of the numbers and the sum of the squares of the numbers:

$$\sum_{i=1}^{8} x_i = 87 + 93 + 22 + 76 + 76 + 65 + 130 + 59$$
$$= 608$$

$$\sum_{i=1}^{8} x_i^2 = 87^2 + 93^2 + 22^2 + 76^2 + 76^2 + 65^2 + 130^2 + 59^2$$
$$= 7569 + 8649 + 484 + 5776 + 5776 + 4225 + 16,900 + 3481$$
$$= 52,860$$

Now, we use the alternative definition to find the variance:

$$\sigma^2 = \frac{N \sum_{i=1}^{8} x_i^2 - \left(\sum_{i=1}^{8} x_i\right)^2}{N^2} = \frac{(8 \cdot 52,860) - (608^2)}{8^2}$$

$$= \frac{422,880 - 369,664}{64} = \frac{53,216}{64}$$

$$= \underline{831.5}$$

It may seem cumbersome to calculate variance by using the alternative definition, but computers can perform these calculations quickly and easily. Using the first definition, we first have to find the mean; using the alternative definition, we can find the variance directly from the x_i's and N. Also, more round-off errors may occur when we use the first definition.

The variance by itself won't tell us a great deal about the variability of the set of data. But if variances of several sets of data are compared, the sets with the smaller variances contain data items that are less dispersed than the other sets.

Since all of the differences were squared in determining the variance in Definition 14.9, a truer measure of variability about the mean may be obtained by taking the square root of the variance.

standard deviation

Definition 14.10 The **standard deviation**, denoted σ, of a set of N numbers is the square root of the variance.

■ **EXAMPLE 14.36** Find the standard deviation of this set of numbers: $\{87, 93, 22, 76, 76, 65, 130, 59\}$.

Solution From Example 14.34, the variance is 831.5. So the standard deviation is $\sqrt{831.5}$, or approximately $\underline{28.8.}$

The standard deviation is a measure of how much an average number in the set differs from the mean. It is a very important statistic in discussing the variability of a set of numbers.

Another important concept in statistics is the difference between a number and the mean in terms of the standard deviation. For example, if the mean of a set of numbers is 70, the standard deviation is 10, and the particular number we are concerned with is 90, then we say that 90 is two standard deviations above the mean ($70 + 10 + 10 = 90$).

We can calculate how many standard deviations from the mean a number is by using the following formula:

$$\frac{\text{value} - \text{mean}}{\sigma}$$

■ **EXAMPLE 14.37** A set of test scores has a mean of 72 and a standard deviation of 14. Find how many standard deviations from the mean the given score is.

    a. 80

Solution $\quad \dfrac{80-72}{14} = \dfrac{8}{14} \approx \underline{0.57}$

We say that 80 is 0.57 standard deviation above the mean.

    b. 55

Solution $\quad \dfrac{55-72}{14} = \dfrac{-17}{14} \approx \underline{-1.22}$

Since we have a negative number, 55 is 1.22 standard deviations *below* the mean.

■ **EXAMPLE 14.38** A teacher bases her grades on the following scale for a student score of x, a class average of μ, and a standard deviation of σ:

    A $\quad x \geq \mu + \sigma$
    B $\quad \mu + 0.5\sigma \leq x < \mu + \sigma$
    C $\quad \mu - 0.5\sigma \leq x < \mu + 0.5\sigma$
    D $\quad \mu - \sigma \leq x < \mu - 0.5\sigma$
    F $\quad x < \mu - \sigma$

If the mean of the scores is 72 and the standard deviation is 14, then find the grades for the given score.

    a. 80

Solution From Example 14.37, 80 is 0.57 standard deviation above the mean. So $80 \geq \mu + 0.5\sigma$, but $x < \mu + \sigma$. So this score gets a grade of $\underline{B}$.

14.5 Measures of Dispersion

b. 55

Solution Similarly, 55 is 1.22 standard deviations below the mean. Thus, $55 < \mu - \sigma$. So 55 gets a grade of F.

Before concluding this section, we should point out that the formulas for variance and standard deviation presented in this section are used to gather statistics on the entire set of data items. Other formulas are used if we are only sampling data or using grouped data.

Problem Set 14.5

In Problems 1–8, find the range of the set of numbers.

1. 26, 25, 15, 14, 82, 73
2. 110, 51, 64, 53, 51, 27, 68
3. 76, 76, 76, 76, 76, 76
4. 10, 20, 30, 40, 50, 60, 70, 30, 30
5. 0, 10, 80, 0, 9, 93, 0, 93, 93
6. 40, 80, 80, 40, 40, 80, 40
7. 97, 53, 62, 43, 43, 77, 78, 79
8. 1, 2, 3, 4, 5, 6, 7, 8, 9, 10

In Problems 9–16, find the variance of the set of numbers given in the indicated problem.

9. Problem 1
10. Problem 2
11. Problem 3
12. Problem 4
13. Problem 5
14. Problem 6
15. Problem 7
16. Problem 8

In Problems 17–24, find the standard deviation of the set of numbers given in the indicated problem.

17. Problem 1
18. Problem 2
19. Problem 3
20. Problem 4
21. Problem 5
22. Problem 6
23. Problem 7
24. Problem 8

For Problems 25–30, use this set of test scores: {97, 86, 85, 84, 77, 76, 73, 71, 70, 66, 65, 60}.

25. What is the mean?
26. What is the range?
27. What is the variance?
28. What is the standard deviation?
29. How many students scored more than one standard deviation above the mean?
30. How many students scored more than one standard deviation below the mean?
31. On a college entrance exam, the mean was 75 and the standard deviation was 9. What scores can a student receive and still be within two standard deviations of the mean?
32. The mean age of people living at Leisure Time Retirement Home is 77 with a standard deviation of 13. What ages are within one standard deviation of the mean?
33. When will the standard deviation be zero?
34. When will the range be zero?

A set of numbers has a mean of 65 and a standard deviation of 7. In Problems 35–38, find how many standard deviations the given number is from the mean.

35. 55
36. 77
37. 170
38. 0

39. Use summation notation to denote the standard deviation, using the original definition of variance.
40. Use summation notation to denote the standard deviation, using the alternative definition of variance.

41. Use the alternative definition of variance to find the variance of {26, 25, 15, 14, 82, 73}, and compare your answer with the answer for Problem 9. Are they the same?
42. Use the alternative definition of variance to find the variance of {110, 51, 64, 53, 51, 27, 68}, and compare your answer with the answer for Problem 10. Are they the same?
43. Joe and Wilma each took a different chemistry test. Joe got a 72 on a test with a mean of 70 and a standard deviation of 4. Wilma got a 68 on a test with a mean of 66 and a standard deviation of 3. Which student scored farther above the mean in terms of standard deviation?
44. William and Catherine each took different standardized tests. William scored 50 on a test with a standard deviation of 7 and a mean of 56. Catherine scored 53 on a test with a standard deviation of 7 and a mean of 58. Which student scored closest to the mean in terms of standard deviation?

CHAPTER SUMMARY

In this chapter, we introduced the concepts of probability and statistics, and we showed how these topics are handled by computers.

We first considered how many ways r items can be arranged from a set of n items. This situation led to the idea of permutations, arrangements in which the order of the items within the arrangement is important, and combinations, arrangements in which the order of the items within the arrangement is not important. To simplify these calculations, we introduced the concept of n factorial, where $n!$ is defined to be the product of all natural numbers from 1 to n.

We then used probability to predict the chances of certain events occurring, where the probability of any event occurring ranges from 0 to 1.

The standard statistical measures of central tendency—mean, median, and mode—were discussed. Also presented were some statistical measures of dispersion: range, variance, and standard deviation. These definitions are summarized next for the set of data items $\{X_1, X_2, \ldots, X_N\}$:

Mean (μ)	$\dfrac{\sum_{i=1}^{N} X_i}{N}$
Median	The number M such that there are the same number of X_i's larger than M as there are X_i's smaller than M
Mode	The number which occurs most often in the set
Range	The difference between the largest and smallest numbers in the set
Variance (σ^2)	$\dfrac{\sum_{i=1}^{N} (X_i - \mu)^2}{N}$

Standard deviation (σ) $\sqrt{\dfrac{\sum_{i=1}^{N}(X_i-\mu)^2}{N}}$

REVIEW PROBLEMS

In Problems 1–4, simplify the expression.

1. $8!$
2. $\dfrac{7!}{4!}$
3. $\binom{9}{7}$
4. $\binom{10}{6}$

5. How many permutations are there of nine items taken five at a time if no repetitions are allowed?

6. How many permutations are there of nine items taken five at a time if repetitions are allowed?

7. How many combinations are there of nine items taken five at a time if repetitions are allowed?

8. How many combinations are there of nine items taken five at a time if no repetitions are allowed?

9. A three-character code is to be given to each computer user. If the middle character must be a number and the other two characters must be letters, how many possible codes are there?

10. A drawing is to be held to determine which 3 students in a class of 30 will get to use the computer today. In how many possible ways can the 3 students be selected?

In Problems 11–16, find the probability of the event occurring.

11. Tossing three coins and getting one head and two tails

12. Rolling three dice and getting three 4s

13. Pulling a blue rock from a bag that contains ten red, six blue, seven green, and two yellow rocks

14. Drawing a card from a shuffled deck of 52 cards and getting a 7

15. Receiving either a first or a second prize in a drawing involving three other people

16. Not choosing a short straw from a hat containing seven long, six medium, and two short straws.

17. Write an algorithm for a procedure to simulate rolling two dice 600 times and counting how many times the two dice show the same number.

18. Write a flowchart for a procedure to simulate tossing three coins 100 times and counting how many times one head and two tails were tossed.

In Problems 19–20, write the sum in summation notation.

19. $(3x_7-7)+(3x_8-7)+\ldots+(3x_{56}-7)$
20. $6+11+16+\ldots+51$

In Problems 21–22, find the sum.

21. $\sum_{i=3}^{7}(3i+8)$
22. $\sum_{i=1}^{5}(2i^2-4i+1)$

In Problems 23–24, find the mean, median, and mode of the set of numbers.

23. 34, 76, 82, 98, 63, 12
24. 55, 69, 102, 39, 69, 42, 17

25. Three employees have salaries of $14,567, $23,003, and $31,009. If the company wants the average (mean) salary for its four employees to be $25,000, then what must the salary of the fourth employee be?

26. Write a flowchart for a procedure to input three numbers and calculate the fourth number if the average of the four numbers must be 55.

In Problems 27–28, find a set of nine positive numbers that satisfies the given conditions (if possible).

27. Mean = 65, median = 70, mode = 75
28. Mean = 72, median = 68, mode = 34 and 70

In Problems 29–30, find the range, variance, and standard deviation of the set of numbers.

29. 34, 76, 82, 98, 63, 12
30. 55, 69, 102, 39, 69, 42, 17
31. Find the variance of the set of numbers in Problem 29 by using the alternative definition of variance.
32. Find the standard deviation of the set of numbers in Problem 30 by using the alternative definition of variance.

A set of numbers has a mean of 77 and a standard deviation of 10. In Problems 33–34, find how many standard deviations the given number is from the mean.

33. 27
34. 92
35. If the standard deviation of a set of numbers is zero, what does that tell you about the numbers?
36. In the simulations language SIMSCRIPT, the variance of a set of N data items is calculated in the following manner:

$$\sigma^2 = (\text{mean square}) - (\text{mean})^2$$

where the mean square is the mean (average) of the squares of the data items. Show that this formula is equivalent to the alternative definition of variance.

ANSWERS TO ODD-NUMBERED PROBLEMS

PROBLEM SET 1.1

1. N, I, Q, R 3. Q, R 5. R
7. I, Q, R 9. Q, R 11. R
13. None   15. Q, R 17. R
19. N, I, Q, R 21. I, Q, R 23. I, Q, R
25. N, I, Q, R
27. [number line with 3]
29. [number line with -1.5, 0]
31. [number line with 0, $\sqrt{3}$]
33. [number line with 0, $\frac{9}{4}$]
35. [number line with 0, $\frac{\pi}{3}$]
37. Not possible   39. $\frac{2}{3}$ (answers may vary)
41. Not possible   43. $-\frac{2}{3}$ (answers may vary)
45. Not possible
47. $\sqrt{2}$ (answers may vary)
49. 0.4   51. -0.43 53. 3.0
55. -1.41 57. -6.0 59. 0.0

PROBLEM SET 1.2

3. $-36 = -36$; associative property for $\cdot$
5. $1 = 1$; inverse property for $\cdot$
7. $-24 = -24$; distributive property
9. $5 = 5$; identity property for $\cdot$
11. $12 = 12$; commutative property for $\cdot$
13. Distributive property
15. Closure property for $+$
17. Inverse property for $+$
19. 10; identity property for $+$
21. $7 + 8$; associative property for $+$
23. 6; distributive property
25. 7; commutative property for $\cdot$
27. Real; closure property for $+$
29. -5; inverse property for $+$
31. No; $1 - (1 - 1) = 1$ and $(1 - 1) - 1 = -1$
33. Yes; $2 - 3 = -1$ (-1 is real)
35. No; $\frac{4}{5} = 0.8$ (0.8 is *not* in N)
37. Yes; $\frac{7}{1} = 7$; 1 is the "identity" for division
39. Yes; $a/a = 1, a \neq 0$; a is the "inverse" for division of a

41. $-20 \neq -16$ 43. $A * (B + C)$
45. $(A * B) * C$ 47. $A + (B * C)$

PROBLEM SET 1.3

1. 2^5 3. x^4 5. $a^4 b^2$
7. $(\frac{1}{4})^3 a^4$ 9. $3^2 x^4 y^2$ 11. 98
13. 8   15. $-\frac{1}{32}$ 17. 343
19. 64   21. 625   23. $\frac{16}{625}$
25. $8x^3$ 27. $-4z^2$ 29. x^6
31. $4x^2 y^2$ 33. $\frac{z^4}{xy^2}$ 35. $\frac{x^4 z^{12}}{y^4}$
37. $\frac{y^3}{x^{12} z^3}$ 39. $\frac{8y^6}{x^6}$ 41. $\frac{1}{x+y}$
43. $\frac{x^3}{y^{10} z^6}$
45. 8, 64, 512, 4096, and 32,768
47. 2, 4, 8, 16, 32, 64, 128, 256, 512, and 1024
49. $2^{30} = 1,073,741,824$
51. X^4
53. $2 * (X^(-3))$
55. $(X^3) - (3 * (X^2)) + (2 * X) - 1$
57. $(X^(-6)) * (Y^6) * (Z^2)$

PROBLEM SET 1.4

1. Yes   3. No   5. No   7. Yes
9. 6   11. -6 13. -4.5 15. 5
17. -4 19. 27
21. $9x^2 + x + 6$
23. $3x^2 y + 3xy^2 + 5xy - 7x + 5y$
25. $8x^2 y - xy^2 - 4xy + 12y^2$
27. $3y^3 + 8y^2 - 2y - 4$
29. $2x^4 - 6x^3 + 8x^2 + 10x$
31. $6x^2 - 7xy - 20y^2$
33. $10x^3 + 32x^2 + 9x - 18$
35. $2x^4 + x^3 - 18x^2 + 43x - 20$
37. $-4x^6 yz + 8x^5 y^4 - 12x^3 y^4 z^2 - 4x^3 y^2 z^5 + 14x^3 z^7 + 8x^2 y^5 z^4 - 28x^2 y^3 z^6 - 12y^5 z^6 + 42y^3 z^8$
39. Yes   41. 22   43. 4   45. 152
47. 66   49. $-12.\overline{857142} \left(-\frac{90}{7}\right)$
51. 3^2 53. $(2 \cdot 3)^2 + 5 - \frac{4}{2}$

Answers to Odd-Numbered Problems

PROBLEM SET 1.5

1. 3
3. 13
5. $\frac{9}{7}$
7. -6
9. $\frac{1}{3}$
11. $-\frac{5}{2}$
13. $\frac{2}{5}$
15. -2
17. $-\frac{5}{2}$
19. $\frac{-y+z+2}{2}$
21. $x<3$
23. $x>11$
25. $x\leq \frac{7}{3}$
27. $x\leq -28$
29. $x<\frac{16}{5}$
31. $x<-\frac{1}{3}$
33. $x\geq -8$
35. $x<-\frac{10}{3}$
37. $x>0$
39. $x\geq 11$
41. $w=\frac{P-2l}{2}$
43. $C=\frac{F-32}{1.8}$
45. $r=\sqrt[3]{\frac{3V}{4\pi}}$
47. Both sides are zero (equal)
49. Still equal
51. True
53. True
55. False
57. False
59. True

PROBLEM SET 1.6

1. $\sqrt[3]{5}$
3. $\frac{1}{\sqrt[7]{2.1}}$
5. $\sqrt[8]{5^3}$
7. $\sqrt[3]{x^4}$
9. $\sqrt{2x}$
11. $\frac{1}{\sqrt[7]{x^3}}$
13. $7^{1/2}$
15. $2^{4/3}$
17. $x^{1/5}$
19. $x^{-5/6}$
21. $b^{-1/2}$
23. $x^{-3/4}$
25. 5
27. 4
29. $2\sqrt{13}$
31. $7\sqrt{2}$
33. $4\sqrt{10}$
35. $3\sqrt[3]{5}$
37. $\frac{1}{2}$
39. 10
41. 27
43. 25
45. 16
47. $\frac{1}{125}$
49. $\frac{36}{25}$
51. $5\sqrt{2}$
53. $6\sqrt{5}$
55. 2
57. 4
59. $\sqrt{7}$
61. $\sqrt{10}$
63. 5
65. x
67. $\frac{\sqrt[n]{a}}{\sqrt[n]{b}}=\frac{a^{1/n}}{b^{1/n}}=\left(\frac{a}{b}\right)^{1/n}=\sqrt[n]{\frac{a}{b}}$
69. $n=3, a=-2: \sqrt[3]{(-2)^3}=\sqrt[3]{-8}=-2$
71. SQR (7)
73. SQR (X * (Y^2))
75. 3^(4/7)
77. (5 * X)^(−2/3)
79. (5^(2/3))^(1/4)

REVIEW PROBLEMS, CHAPTER 1

1. Q, R
3. N, I, Q, R
5. None
7. $-3, -5, -10$ (answers may vary)
9. Associative property for $+$:
$3+(5+7)=(3+5)+7$
$3+12=8+7$
$15=15$
11. Closure property for $\cdot$:
$3\cdot 5=15$ (15 is in R)
13. Distributive property:
$2(3+4)=(2\cdot 3)+(2\cdot 4)$
$2\cdot 7=6+8$
$14=14$
15. $\frac{27}{8}$
17. x^6
19. $\frac{9y^2}{25x^2}$
21. $2\sqrt[3]{5}$
23. $\frac{1}{9}$
25. 2
27. $10x^3-8xy^2-5x$
29. $x^6-4x^5+9x^4-13x^3+38x^2-51x+28$
31. $\frac{13}{2}$
33. -3
35. $x>\frac{1}{2}$
37. $x<\frac{3}{5}$
39. $w=\frac{V}{lh}$
41. (A * (X^2)) + (B * X) + C
43. SQR (A − ((B^3) * C))

PROBLEM SET 2.1

1. Solution
3. Solution
5. Solution
7. Not a solution
9. Not a solution
11. Solution
13. $D=\{-2, 0, 2, 6\}; R=\{-3, 1, 5\}$
15. $D=\{2, 3, 4, 5\}; R=\{2\}$
17. $D=\{2, 3, 8\}; R=\{-6, -4, 0, 5\}$
19. $D=\{1, 2, 3\}; R=\{1, 2, 3\}$
21. -29

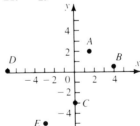

31. First quadrant
33. y axis
35. x axis
37. Fourth quadrant
39. Second quadrant
41. Fourth quadrant
43. First or fourth quadrant
45. x axis
47. Third quadrant
49. First or second quadrant

51.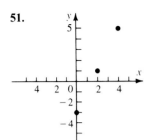

53. $(0, \frac{3}{2}), (3, 0)$

PROBLEM SET 2.2

1. Function
3. Function
5. Not a function
7. Not a function
9. A *function* is a relation in which each element in the domain is paired with exactly one element in the range.
11. 1
13. 10
15. $-3x+1$
17. 2
19. 38
21. $16a^2 - 8a + 3$
23. -1
25. $-\frac{1}{7}$
27. $\dfrac{1}{z-2}$
29. 2
31. 1
33. $\sqrt{t+4}$
35. 5
37. 16
39. -27
41. 212
43. 9π
45. 3.27
47. Function; $f(x) = 2x$
49. Not a function
51. Function; $p(x) = $ age
53. DEF FNF(C) = (1.8 * C) + 32
55. DEF FNA(R) = 3.14 * (R^2)
57. DEF FNM(Y) = 1.09 * Y
59. DEF FNL(X) = (4 * X) + 8
61. DEF FNY(A, B, C) = (A * B) + (A * C)

PROBLEM SET 2.3

1. Linear
3. Linear
5. Not linear
7. Not linear
9. Linear
11. Not linear
13. Not linear

15.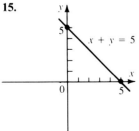

17.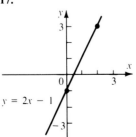

19.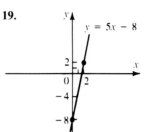

21.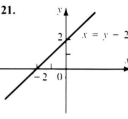

23.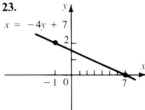

25.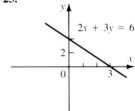

27. **29.**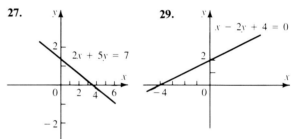

31.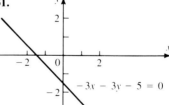

33.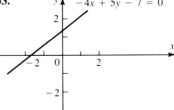

Answers to Odd-Numbered Problems

35. x intercept: $-\frac{3}{4}$; y intercept: 3
37. x intercept: -2; y intercept: $\frac{2}{3}$
39. x intercept: $\frac{3}{2}$; y intercept: -3
41. x intercept: $-\frac{5}{3}$; y intercept: $\frac{5}{2}$
43. No, except $y = 0$
45. Yes, a horizontal line
47.
49.

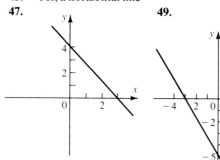

51.

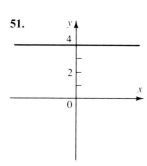

53. An infinite number of lines
55. DEF FNY(X) = (2 * X) − 7
57. DEF FNY(X) = 4 − (10 * X)
59. $y = \dfrac{c - ax}{b}$

PROBLEM SET 2.4

1.

3. 1
5. $\frac{2}{3}$
7. -1.9
9. $\frac{7}{3}$
11. -12
13. -0.66
15. Undefined
17. $\frac{10}{49}$
19. 0
21. $\frac{43}{19}$
23. $x = 2$
25. $y = 5$
27. $y = 4$
29. $x = 5$
31. $y = 0$
33. Undefined
35. $-\frac{3}{2}$
37. $-\frac{3}{2}$
39. $AB: -\frac{5}{2}; AC: -\frac{1}{4}; BC: \frac{1}{2}$
41. All three slopes are $\frac{2}{3}$
43. $(0, 3), (1, 1),$ and $(2, -1)$; slope $= -2$
45. $m_1 = \frac{5}{4}; m_2 = 1; m_3 = \frac{3}{2}$
47. $m_1 = 0; m_2 =$ undefined$; m_3 = -\frac{7}{9}$
49. DEF FNM(X1, Y1, X2, Y2) = (Y2 − Y1)/(X2 − X1)
51. They are parallel to each other
53. An infinite number of lines

PROBLEM SET 2.5

1. $2x + 3y = -7$
3. $2x - y = -5$
5. $x - 7y = 2$
7. $2x - y = 1$
9. $2x - 5y = 9$
11. $5x - y = 9$
13. $x + y = -1$
15. $3x - y = -5$
17. $y = -3$
19. $x = 4$
21. $x + y = 4$
23. $4x - y = 14$
25. $x = 4$
27. $y = 1$
29. $x + y = -12$
31. $2x - y = -5$
33. $3x + y = 7$
35. $4x + y = -7$
37. $y = -3$
39. $x = 0$
41. Slope $= 3$; y intercept $= 4$
43. Slope $= -5$; y intercept $= -6$
45. Slope $= \frac{1}{2}$; y intercept $= -\frac{7}{2}$
47. Slope $= -2$; y intercept $= 7$
49. Slope $=$ undefined; y intercept $=$ none
51. $2x - y = 1$
53. $2x - 5y = 43$
55. $y = 2$
57. $x + 3y = -7$
59. $7x - 3y = 35$
61. $2x - y = 5$
63. $10
65. $P(x) = 100x + 200$

REVIEW PROBLEMS, CHAPTER 2

1. and 3.

5. (0, 4) and (6, 0)
7. 25
9. $4h^2 + 3h + 3$
11. 2
13.

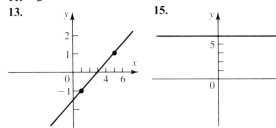

17.

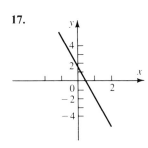

19. x intercept: 5; y intercept: $\frac{5}{3}$
21. $-\frac{2}{5}$
23. $\frac{4}{3}$
25. $4x - y = 9$
27. $10x + y = 24$
29. $y = 1$
31. $3x + y = 6$
33. $4x - 3y = -12$
35. $7x - y = 31$
37. $4x + y = -9$
39. Slope: $\frac{2}{3}$; y intercept: -3
41. $10.30
43. DEF FNP(X) = 2*(X + 4)

PROBLEM SET 3.1

1. Not a solution
3. Solution
5. Solution
7. Solution
9. Not a solution
11. $(-7, 2)$
13. $(-3, 4)$
15. $(2, 8)$
17. $(2, 1)$
19. Actual answer: $(0.7, 0.1)$
21. $(0, 3)$
23. $(-5, 2)$
25. Actual answer: $(-\frac{20}{19}, \frac{42}{19})$
27. $(-1, 5)$
29. No solution
31. Same line
33. They are the same line
35. The two lines don't meet. They are parallel.

37. Actual answer: $(\frac{8}{9}, \frac{35}{9})$
39. $(-\frac{1}{3}, 1)$
41. $x + y = -1; 2x + y = 1$

PROBLEM SET 3.2

1. $(4, 3)$
3. $(1, -1)$
5. $(3, 5)$
7. $(4, 2)$
9. $(8, -1)$
11. $(4, -4)$
13. $(2, 1)$
15. $(-\frac{38}{13}, -\frac{9}{13})$
17. $(-6, -2)$
19. $(19, 15)$
21. $(-\frac{7}{2}, 0)$
23. $(\frac{75}{64}, \frac{23}{32})$
25. $(8, -1)$
27. No; solve each equation for the variable: $(4, -3)$
29. $40°$
31. $(-\frac{3}{4}, \frac{11}{2})$
33. $(9, -3)$
35. $(9, -3)$

PROBLEM SET 3.3

1. $(2, 0)$
3. $(-1, 2)$
5. $(0, -6)$
7. $(5, -1)$
9. $(-\frac{14}{9}, -\frac{13}{3})$
11. $(7, -2)$
13. $(-6, -3)$
15. $(-5, -3)$
17. $(2, 1)$
19. $(\frac{23}{49}, \frac{79}{98})$
21. $(1, 7)$
23. $(9, \frac{25}{2})$
25. $(\frac{23}{49}, \frac{79}{98})$
27. You wouldn't
29. You are left with $0 = 0$ as both variables drop out
31. You are left with $0 = -8$ as both variables drop out
33. $m = \frac{5}{3}; b = -\frac{22}{3}$
35. $17x + 2y - 36 = 0$
37. $(2, 2)$
39. $y = \dfrac{cd - af}{bd - ae}$
41. When $bd - ae = 0$
43. $(3, -1)$

PROBLEM SET 3.4

1. Independent, inconsistent
3. Independent, consistent
5. Dependent, consistent
7. Dependent, consistent
9. Dependent, consistent
11. Independent, inconsistent
13. Dependent, consistent
15. Independent, inconsistent
17. Independent, inconsistent
19. No solution
21. Infinite number of solutions
23. $(\frac{31}{13}, \frac{2}{13})$
25. Infinite number of solutions
27. No solution
29. No solution

Answers to Odd-Numbered Problems

31. No; if the system is dependent, then it has an infinite number of solutions. If the system is inconsistent, then it has no solution.
33. No; all that matters is that the statement is *not true*
35. -6
37. -6
39. Infinite number of solutions

PROBLEM SET 3.5

1. 21, 22
3. $x - 5 = 12$
5. $3x = 24$
7. $x + y = 45$
9. $x = y + 3$
11. 16 and 26
13. 1 and 3
15. 300 in.^{2}
17. 4, 8, and 8
19. 29¢/lb for corn; 31¢/lb for beans
21. 7.5 ml of X, 2.5 ml of Y
23. 3 gal of orange juice; 2 gal of grapefruit juice
25. 4 touchdowns; 2 field goals
27. 11 nickels; 8 dimes
29. $3500 for tuition; $1500 for room and board
31. 5 h driving; 4 h walking
33. $a = \frac{32}{5}; b = -\frac{8}{15}$
35. Infinite number of solutions
37. High = 94; low = 10
39. 400 miles
41. (2, 800)

REVIEW PROBLEMS, CHAPTER 3

1. (4, 2)
3. $(-1, 2)$
5. Inconsistent and independent
7. Inconsistent and independent
9. (3, 9)
11. $(\frac{24}{5}, -\frac{4}{5})$
13. (2, 0)
15. $(-2, -4)$
17. (1, 2)
19. $(1, -1)$
21. (14, 10)
23. $(\frac{19}{7}, \frac{24}{7})$
25. $\frac{6}{5}$ and $\frac{26}{5}$
27. 8 oz of first drink; 28 oz of second drink
29. $y = \frac{4}{7}x - 4$

PROBLEM SET 4.1

1. -4 and 4
3. $-5\sqrt{2}$ and $5\sqrt{2}$
5. -3 and 9
7. $-5 \pm 2\sqrt{15}$
9. $-5 \pm \sqrt{17}$
11. 2 and 6
13. $\dfrac{2 \pm 2\sqrt{3}}{7}$
15. $\dfrac{5}{2}$
17. No real solution
19. $\dfrac{1 \pm 2\sqrt{3}}{4}$
21. 0
23. 0
25. 9
27. 36
29. $\dfrac{49}{4}$
31. $\dfrac{b^2}{4a^2}$
33. 0 and -4
35. $-2 \pm \sqrt{10}$
37. $1 \pm \sqrt{7}$
39. $-5 \pm \sqrt{30}$
41. 1
43. $-2 \pm 2\sqrt{3}$
45. $2 \pm \sqrt{7}$
47. $-3 \pm \sqrt{13}$
49. No real solution
51. No real solution
53. $x = \dfrac{-b \pm \sqrt{b^2 - 4ac}}{2a}$
55. Division by zero (if $a = 0$, the equation is linear)

PROBLEM SET 4.2

1. 0
3. 0
5. 2
7. 1
9. 0
11. No real solution
13. No real solution
15. $\dfrac{9 \pm \sqrt{33}}{2}$
17. 5
19. No real solution
21. $-2 \pm 2\sqrt{5}$
23. 3 and -3
25. No real solution
27. No real solution
29. 3 and 4
31. $\dfrac{-3 \pm \sqrt{21}}{2}$
33. -1 and -4
35. $x = \dfrac{-b}{2a}$
37. $\dfrac{c}{a}$
39. $5x^2 - 5x - 4 = 0$
41. DEF FNR(A, B, C) = $(-B + SQR((B^2) - (4*A*C))/(2*A)$
    DEF FNS(A, B, C) = $(-B - SQR((B^2) - (4*A*C))/(2*A)$
43. 60 desks

PROBLEM SET 4.3

1. $V(1, 0), x = 1$
3. $V(-2, -3), x = -2$
5. $V(6, 4), x = 6$
7. $V(4, 3), x = 4$
9. $V(-3, 17), x = -3$
11. $V(-2, -2), x = -2$
13. $V(-\frac{5}{2}, -\frac{61}{4}), x = -\frac{5}{2}$
15. $V(3, -14), x = 3$
17. $V(\frac{3}{4}, -\frac{19}{4}), x = \frac{3}{4}$

19.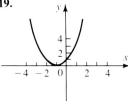

21.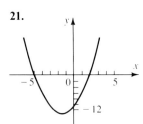

23.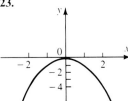

25.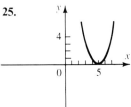

27.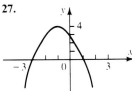

29.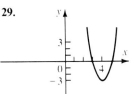

31.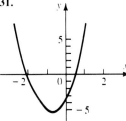

33.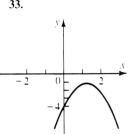

35. It becomes a linear function
37. The vertex is on the y axis and the equation is of the form $f(x) = ax^2 + c$
39. $y = (x-3)^2 + 4$; $y = 2(x-3)^2 + 4$
41. 3
43. 12.25 h
45. 7200 ft

47. $h = \dfrac{-b}{2a}$; $k = \dfrac{4ac - b^2}{4a}$

49. DEF FNY(X) = (2*(X^2)) − (3*X) + 4

PROBLEM SET 4.4

1.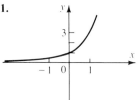

3.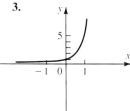

5.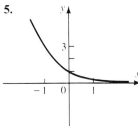

7.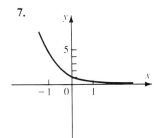

9.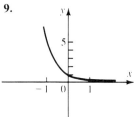

11.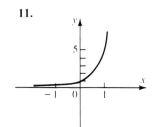

13.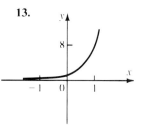

15.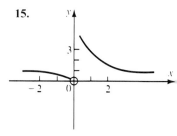

Answers to Odd-Numbered Problems

17. $y = 1$ **19.** $y = 0$ (undefined at $x = 0$)   **49.**   **51.**

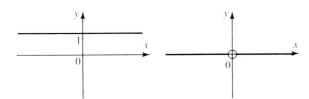
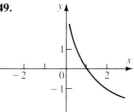
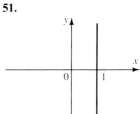

21. $(-4)^{-2} = \frac{1}{16}$; $(-4)^{-1} = -\frac{1}{4}$; $(-4)^0 = 1$; $(-4)^{1/2}$ is undefined; $(-4)^1 = -4$; $(-4)^2 = 16$

23. Never **25.** None **27.** 6^x
29. $2(7^x)$ **31.** 5^{x^2} **33.** 2^{1-x}
35. 1 **37.** e^{-3x-1} **39.** e
41. e^2 **43.** 20,000 **45.** 180,000
47. 50
49. DEF FNA(P, T, I) = P * EXP(T * I)

53. $x = 0$ (undefined at $y = 0$)

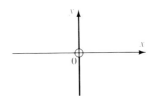

PROBLEM SET 4.5

1. $\log_3 y = x$ **3.** $\log_a y = x$
5. $\log_3 2 = m$ **7.** $\log_2 16 = 4$
9. $\log_{8/27} \frac{4}{9} = \frac{2}{3}$ **11.** $x = 4^y$
13. $52 = 11^y$ **15.** $t = 4^{-7}$
17. $7^2 = 49$
19. $81 = 3^4$ **21.** 4 **23.** -1
25. 3 **27.** $\frac{3}{2}$ **29.** 9
31. $\frac{1}{64}$ **33.** $\frac{1}{8}$ **35.** 6
37. 4 **39.** 8

55. $x = 1$
57. $x \leq 0$
59.

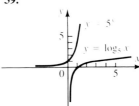

41. **43.**

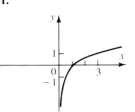
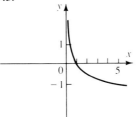

61. DEF FNL(X) = LOG(X)/LOG(10)
63. DEF FNE(X) = EXP(X * LOG(2))
65. 7
67. 4.3984375
69. 100
71. DEF FNP(T) = 100 * LOG((3 * T) + 10)/LOG(10)
73. 10 decibels

45. **47.**

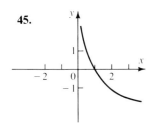
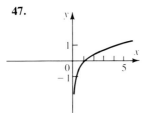

PROBLEM SET 4.6

1. Arithmetic **3.** Geometric
5. Neither **7.** Neither
9. Arithmetic **11.** $a_n = 3n - 2$
13. $a_n = -6n + 33$ **15.** $a_n = 0.1n + 0.9$
17. $a_n = (-1)^{n-1}$ **19.** $a_n = 2^{n-1}$
21. $a_n = 9(-\frac{1}{3})^{n-1}$ **23.** $a_n = n^3$
25. $a_n = a_{n-1} + 3$ **27.** $a_n = a_{n-1} - 6$

29. $a_n = a_{n-1} + 0.1$
31. $a_n = -1 \cdot a_{n-1}$
33. $a_n = 2 \cdot a_{n-1}$
35. $a_n = a_{n-1} \cdot (-\frac{1}{3})$
37. $a_n = (\sqrt[3]{a_{n-1}} + 1)^3$
39. 87
41. $3 \cdot (\frac{1}{9})^{42}$
43. 10^{-42}
45. -23.2
47. 0, 2
49. $-4, 2$
51. 1, 1, 1, 1, 1, ...
53. 4, 11, 18, 25, 32
55. $-3, 6, -12, 24, -48$
57. 1, 1, 2, 3, 5, 8, 13, 21, 34, 55, 89

31.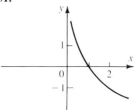

REVIEW PROBLEMS, CHAPTER 4

1. 1 and -5
3. $-4 \pm 3\sqrt{2}$
5. No real solution
7. $\frac{7}{3}$ and $\frac{1}{3}$
9. $\dfrac{-1 \pm \sqrt{13}}{2}$
11. $V(4, 5), x = 4$
13. $V(-4, -20), x = -4$
15. $V(-2, -16), x = -2$
17. $V(-\frac{2}{3}, -\frac{49}{3}), x = -\frac{2}{3}$

19.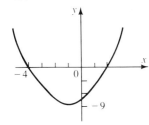
21.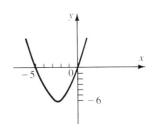

23.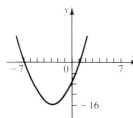
25.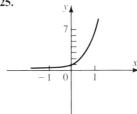

27.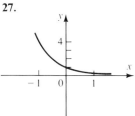
29.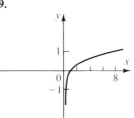

33. 2
35. -2
37. $\frac{1}{25}$
39. $a_n = 5n - 3$
41. $a_n = 4(-5)^{n-1}$
43. $a_n = 4a_{n-1}$
45. $a_n = a_{n-1} - 6$
47. $(\frac{1}{2})^{109}$
49. DEF FNF(X) = 3 * EXP((2 * X) + 1)

PROBLEM SET 5.1

1.
Face Value	Place Value
2	10
5	1

3.
Face Value	Place Value
1	1000
7	100
8	10
9	1

5.
Face Value	Place Value
1	10000
3	1000
0	100
0	10
7	1

7.
Face Value	Place Value
2	10
6	1
7	.1
1	.01

9.
Face Value	Place Value
4	100
5	10
3	1
1	.1
2	.01
0	.001
8	.0001

Answers to Odd-Numbered Problems

11. $(4 \cdot 10) + (3 \cdot 1)$
13. $(1 \cdot 1000) + (2 \cdot 100) + (9 \cdot 10) + (8 \cdot 1)$
15. $(1 \cdot 100000) + (2 \cdot 10000) + (0 \cdot 1000) + (0 \cdot 100) + (0 \cdot 10) + (7 \cdot 1)$
17. $(3 \cdot 10) + (1 \cdot 1) + (1 \cdot 0.1) + (4 \cdot 0.01)$
19. $(5 \cdot 100) + (4 \cdot 10) + (3 \cdot 1) + (6 \cdot 0.1) + (7 \cdot 0.01) + (0 \cdot 0.001) + (8 \cdot 0.0001)$
21. 99999
23. 10000
25.
$10^{-10} = .0000000001$ $10^1 = 10$
$10^{-9} = .000000001$ $10^2 = 100$
$10^{-8} = .00000001$ $10^3 = 1000$
$10^{-7} = .0000001$ $10^4 = 10,000$
$10^{-6} = .000001$ $10^5 = 100,000$
$10^{-5} = .00001$ $10^6 = 1,000,000$
$10^{-4} = .0001$ $10^7 = 10,000,000$
$10^{-3} = .001$ $10^8 = 100,000,000$
$10^{-2} = .01$ $10^9 = 1,000,000,000$
$10^{-1} = .1$ $10^{10} = 10,000,000,000$
$10^0 = 1$

PROBLEM SET 5.2

1.
Face Value	Place Value
1	8
0	4
0	2
1	1

3.
Face Value	Place Value
1	32
1	16
0	8
1	4
1	2
1	1

5.
Face Value	Place Value
1	4
1	2
0	1
1	.5

7.
Face Value	Place Value
1	16
1	8
0	4
1	2
1	1
1	.5
1	.25
0	.125
1	.0625

9.
Face Value	Place Value
1	100
0	10
1	1
1	.1
1	.01

11. $(1 \cdot 8) + (1 \cdot 4) + (1 \cdot 2) + (0 \cdot 1)$
13. $(1 \cdot 32) + (0 \cdot 16) + (1 \cdot 8) + (1 \cdot 4) + (1 \cdot 2) + (0 \cdot 1)$
15. $(1 \cdot 2) + (1 \cdot 1) + (0 \cdot 0.5) + (1 \cdot 0.25) + (1 \cdot 0.125)$
17. $(1 \cdot 10) + (1 \cdot 1) + (0 \cdot 0.1) + (1 \cdot 0.01) + (0 \cdot 0.001) + (1 \cdot 0.0001)$
19. $(1 \cdot 8) + (0 \cdot 4) + (1 \cdot 2) + (0 \cdot 1) + (0 \cdot 0.5) + (1 \cdot 0.25) + (0 \cdot 0.125) + (1 \cdot 0.0625)$

21.
Decimal	Binary	Decimal	Binary
13_{10}	1101_2	19_{10}	10011_2
14_{10}	1110_2	20_{10}	10100_2
15_{10}	1111_2	21_{10}	10101_2
16_{10}	10000_2	22_{10}	10110_2
17_{10}	10001_2	23_{10}	10111_2
18_{10}	10010_2	24_{10}	11000_2

23.
$2^{10} = 10000000000_2$ $2^{-1} = .1_2$
$2^9 = 1000000000_2$ $2^{-2} = .01_2$
$2^8 = 100000000_2$ $2^{-3} = .001_2$
$2^7 = 10000000_2$ $2^{-4} = .0001_2$
$2^6 = 1000000_2$ $2^{-5} = .00001_2$
$2^5 = 100000_2$ $2^{-6} = .000001_2$
$2^4 = 10000_2$ $2^{-7} = .0000001_2$
$2^3 = 1000_2$ $2^{-8} = .00000001_2$
$2^2 = 100_2$ $2^{-9} = .000000001_2$
$2^1 = 10_2$ $2^{-10} = .0000000001_2$
$2^0 = 1_2$

25. 1111111_2 **27.** 1000000_2 **29.** 256_{10}
31. 10100_2 **33.** 4_{10} **35.** 14_{10}

PROBLEM SET 5.3

1.

Face Value	Place Value
4	64
3	8
7	1

3.

Face Value	Place Value
1	512
2	64
0	8
7	1

5.

Face Value	Place Value
3	8
5	1
7	.125
5	.015625

7.

Face Value	Place Value
1	8
1	4
0	2
1	1
1	.5
0	.25
1	.125

9.

Face Value	Place Value
7	8
7	1
7	.125
7	.015625
7	.001953125

11. $237_8 = (2 \cdot 64) + (3 \cdot 8) + (7 \cdot 1)$
13. $1475_8 = (1 \cdot 512) + (4 \cdot 64) + (7 \cdot 8) + (5 \cdot 1)$
15. $123.4_8 = (1 \cdot 64) + (2 \cdot 8) + (3 \cdot 1) + (4 \cdot 0.125)$
17. $101101.101_2 = (1 \cdot 32) + (0 \cdot 16) + (1 \cdot 8) + (1 \cdot 4)$
$+ (0 \cdot 2) + (1 \cdot 1) + (1 \cdot 0.5) + (0 \cdot 0.25) + (1 \cdot 0.125)$
19. $1234.567_8 = (1 \cdot 512) + (2 \cdot 64) + (3 \cdot 8) + (4 \cdot 1)$
$+ (5 \cdot 0.125) + (6 \cdot 0.015625) + (7 \cdot 0.001953125)$

21.

Decimal	Octal
13	15
14	16
15	17
16	20
17	21
18	22
19	23
20	24
21	25
22	26
23	27
24	30

23. $(8^5)_{10} = 100000_8$
$(8^4)_{10} = 10000_8$
$(8^3)_{10} = 1000_8$
$(8^2)_{10} = 100_8$
$(8^1)_{10} = 10_8$
$(8^0)_{10} = 1_8$
$(8^{-1})_{10} = 0.1_8$
$(8^{-2})_{10} = 0.01_8$
$(8^{-3})_{10} = 0.001_8$
$(8^{-4})_{10} = 0.0001_8$
$(8^{-5})_{10} = 0.00001_8$

25. 7777777_8
27. 1000000_8
29. 32,768
31. 100000_8
33. 777_8
35. 159_8
37. If an octal number ends in 0, 2, 4, or 6, then it is even. Otherwise, it is odd.
39. 128

PROBLEM SET 5.4

1.

Face Value	Place Value
5	16
6	1

3.

Face Value	Place Value
1	256
4	16
E	1

Answers to Odd-Numbered Problems

5. | Face Value | Place Value |
    |---|---|
    | 3 | 16 |
    | A | 1 |
    | E | .0625 |
    | 4 | .00390625 |

7. | Face Value | Place Value |
    |---|---|
    | 1 | 8 |
    | 0 | 4 |
    | 1 | 2 |
    | 1 | 1 |
    | 1 | .5 |
    | 0 | .25 |
    | 1 | .125 |

9. | Face Value | Place Value |
    |---|---|
    | A(10) | 256 |
    | 7 | 16 |
    | 9 | 1 |
    | C(12) | .0625 |
    | D(13) | .00390625 |
    | F(15) | .000244140625 |

11. $134_{16} = (1 \cdot 256) + (3 \cdot 16) + (4 \cdot 1)$
13. $23ED_{16} = (2 \cdot 4096) + (3 \cdot 256) + (14 \cdot 16) + (13 \cdot 1)$
15. $74.9_{16} = (7 \cdot 16) + (4 \cdot 1) + (9 \cdot 0.0625)$
17. $345.77_8 = (3 \cdot 64) + (4 \cdot 8) + (5 \cdot 1) + (7 \cdot 0.125) + (7 \cdot 0.015625)$
19. $AD.ECB_{16} = (10 \cdot 16) + (13 \cdot 1) + (14 \cdot 0.0625) + (12 \cdot 0.00390625) + (11 \cdot 0.000244140625)$

21. | Decimal | Hexadecimal |
    |---|---|
    | 21 | 15 |
    | 22 | 16 |
    | 23 | 17 |
    | 24 | 18 |
    | 25 | 19 |
    | 26 | 1A |
    | 27 | 1B |
    | 28 | 1C |
    | 29 | 1D |
    | 30 | 1E |
    | 31 | 1F |
    | 32 | 20 |
    | 33 | 21 |
    | 34 | 22 |

23. $16^3 = 1000_{16}$
    $16^2 = 100_{16}$
    $16^1 = 10_{16}$
    $16^0 = 1_{16}$
    $16^{-1} = 0.1_{16}$
    $16^{-2} = 0.01_{16}$
    $16^{-3} = 0.001_{16}$

25. $FFFFFFF_{16}$
27. 1000000_{16}
29. 4096
31. 1000000_{16}
33. $2F_{16}$
35. 308_{16}
37. It is even if it ends in 0, 2, 4, 6, 8, A, C, or E. Otherwise, it is odd.
39. 256

PROBLEM SET 5.5

1. 14_{10}
3. 1.375_{10}
5. 27.1875_{10}
7. 299_{10}
9. 398.875_{10}
11. 2122.447265625_{10}
13. 3385_{10}
15. 1591.3125_{10}
17. 17234.8984375_{10}
19. 110_{10}
21. 9_{10}
23. 109_{10}
25. 205_{10}
27. 193_{10}
29. 1127_{10}
31. 1253_{10}
33. 122_{10}
35. 573_{10}
37. 5103_{10}
39. 125_{10}
41. 63_{10}
43. 63_{10}
45. 4095_{10}
47. 11111111_2
49. Every three binary digits corresponds to one octal digit

PROBLEM SET 5.6

1. 111001_2
3. 100001010_2
5. 1101101000_2
7. 45_8
9. 325_8
11. 1025_8
13. $2E_{16}$
15. $B3_{16}$
17. $3D7_{16}$
19. 1113_4
21. 110001_2
23. 11001110_2
25. 1110110011_2
27. 67_8
29. 224_8
31. 1353_8
33. $4B_{16}$
35. $D6_{16}$
37. $3A9_{16}$
39. 1001_4
41. $0.0\overline{1001}_2$
43. $0.01\overline{1100}_2$
45. 0.001_2
47. $0.0\overline{6314}_2$
49. $0.\overline{7}_8$
51. $0.4\overline{3146}_8$
53. $0.\overline{3}_{16}$
55. $0.5\overline{9}_{16}$

57. 0.4_{16}
59. $0.\overline{12}_4$
61. $100101.\overline{00011}_2$
63. $111001010.1\overline{0110}_2$
65. $76.2\overline{3146}_8$
67. $55.\overline{34631}_8$
69. $2C.E\overline{6}_{16}$
71. $100000000_2, 400_8, 100_{16}$; 256 is a power of 2 and 16 but is not a power of 8

PROBLEM SET 5.7

1. 15_8
3. 157_8
5. 6111_8
7. 6.55_8
9. 3.6424_8
11. E_{16}
13. $136B_{16}$
15. $4E.A_{16}$
17. $37.ED8_{16}$
19. $C55.35_{16}$
21. $2^{-1} = 0.5$
    $2^{-2} = 0.25$
    $2^{-3} = 0.125 = 8^{-1}$
    $2^{-4} = 0.0625$
    $2^{-5} = 0.03125$
    $2^{-6} = 0.015625 = 8^{-2}$
23. Group the binary digits by twos from the point and convert each group to the appropriate base 4 digit (0, 1, 2, or 3) ($2^2 = 4$)
25. 12111.22_4
27. 55.66_8 or 45.84375_{10} (base 8 definitely easier)
29. 8

PROBLEM SET 5.8

1. 110011100_2
3. 11010001100101_2
5. 11011.111101_2
7. 100100.101110000000111_2
9. $1100101110111.100000010001011_2$
11. 1110110_2
13. 10010110111_2
15. 10011110.11_2
17. 101100011.00101001_2
19. $10110011111110.101111001101_2$
21. $3B_{16}$
23. $72A_{16}$
25. $14.AC_{16}$
27. 47.598_{16}
29. $53.82EE_{16}$
31. 244_8
33. 13554_8
35. 217.7_8
37. 570.15_8
39. 355.2551034_8
41. Change each base 4 digit to two binary digits
43. 110110.11_2
45. Yes; change each pair of base 4 digits to one hexadecimal digit ($4^2 = 16$)
47. $10010101101_2, 1197_{10}, 2255_8$
49. $1111111111111111_2, 65535_{10}, 177777_8$

REVIEW PROBLEMS, CHAPTER 5

1. 23_{10}
3. 11.6875_{10}
5. 285_{10}
7. 35.306640625_{10}
9. 1004_{10}
11. 377.6875_{10}
13. 101011001_2
15. $0.\overline{0110}_2$
17. $1101.01\overline{0110}_2$
19. 1067_8
21. $0.\overline{6314}_8$
23. $53.2\overline{3146}_8$
25. 542_{16}
27. $32.A\overline{6}_{16}$
29. 555_8
31. 33.5_8
33. $1F7_{16}$
35. $36.C_{16}$
37. $E7_{16}$
39. 256.562_8
41. $317_8, 11001111_2, 207_{10}$
43. $11111111111111111111111111111111_2$, 37777777777_8, $FFFFFFFF_{16}$; $4,294,967,295_{10}$ ($2^{32} - 1$)

PROBLEM SET 6.1

1. 1111_2
3. 100000_2
5. 1000010_2
7. 1001.00_2
9. 10000.0001_2
11. 110100_2
13. 1010101_2
15. 10100.00_2
17. 10000.0000_2
19. 10011.10101_2
21. 1001000_2
23. 111.100_2
25. 100.10001_2
27. 1011001_2
29. 100011.0001_2
31. 100111_2
33. No
37. 01001_2 (1's complement of 10110)
41. 11010_2 ($a_2 + a_2 = a0_2$)
43. 01001110_2
45. 00101110_2

PROBLEM SET 6.2

1. 257_8
3. 1060_8
5. 3765_8
7. 30.11_8
9. 16605_8
11. CEB_{16}
13. $152B_{16}$
15. 11545_{16}
17. $AD8.C28_{16}$
19. $1CF35_{16}$
21. 5551_8
23. 456.7776_8
25. 520.773_8
27. $8C.77_{16}$
29. $433.78C_{16}$
31. 1454_8
33. $147FF_{16}$
35. 1220_8
37. 5551_8
39. 65322_8
41. 4

PROBLEM SET 6.3

1. 64_{10}
3. 876_{10}

Answers to Odd-Numbered Problems

5. 76593_{10}
7. 54.12_{10}
9. 876.90543_{10}
11. 22_{10}
13. 805_{10}
15. 76540_{10}
17. 4.0946_{10}
19. 667.2104_{10}
21. 585_{10}
23. 6.943_{10}
25. 0_{10}
27. -768_{10}
29. -184_{10}
31. 111_{10}
33. 37725_{10}
35. 0_{10}
37. -267_{10}
39. -22.493_{10}
41. 67.893_{10}
43. -345.8_{10}
45. No
47. 2766_{10}
49. 1132_{10}

PROBLEM SET 6.4

1. 0_2
3. 010_2
5. 00100_2
7. 0010.00010_2
9. 000.00_2
11. 0010_2
13. 0101011_2
15. 0010.001001_2
17. 100.000_2
19. 00111010.11011_2
21. 10_2
23. 0.001_2
25. -1010_2
27. 11_2
29. 0.10_2
31. 11_2
33. -0.011_2
35. -10000_2
37. 0_2
39. -110_2
41. Because the top number is not changed
43. 011101_2 (the 1's complement of 100010_2)
45. 110_2
47. 5_{10}
49. 00101_2
51. 11111_2
53. -111010111_2

PROBLEM SET 6.5

1. 46_8
3. 4564_8
5. 657.722_8
7. 000.01_8
9. 5416.71_8
11. $2C_{16}$
13. $E76C_{16}$
15. $ED.CC_{16}$
17. $F000.00_{16}$
19. 576.82_{16}
21. 112_8
23. 14.15_8
25. 0_8
27. 765.4_8
29. -25.3775_8
31. $A51_{16}$
33. $1F3.1_{16}$
35. $C79.7_{16}$
37. $-DF6.FC_{16}$
39. $88E1.292_{16}$
41. 2217_8
43. FCD_{16}
45. 11_{10}
47. $1C.77_{16}$
49. $A51_{16}$

PROBLEM SET 6.6

1.

·	0	1	2	3	4	5	6	7
0	0	0	0	0	0	0	0	0
1	0	1	2	3	4	5	6	7
2	0	2	4	6	10	12	14	16
3	0	3	6	11	14	17	22	25
4	0	4	10	14	20	24	30	34
5	0	5	12	17	24	31	36	43
6	0	6	14	22	30	36	44	52
7	0	7	16	25	34	43	52	61

3. 100_2
5. 100111110_2
7. 111.0111_2
9. 11.110011_2
11. 110010.1001011_2
13. 644_8
15. 6073_8
17. 24.4750_8
19. 120021.5764_8
21. 11227.4614063_8
23. $B52_{16}$
25. $36F08_{16}$
27. $29.5B8_{16}$
29. $A9.21F3_{16}$
31. $203.DAC1D6_{16}$
33. 13061_8
35. $2BB6_{16}$
37. A zero is added to the right of the number
39. Three zeros are added to the right of the number
41. $.001_2$
43. 1118_{10}

PROBLEM SET 6.7

1. 10_2
3. $101_2, R1100_2$
5. $101_2, R1001_2$
7. $100_2, R1001_2$
9. 10101_2
11. $100_2, R10001_2$
13. $11_2, R100100_2$
15. $5_8, R1_8$
17. $35_8, R11_8$
19. $14_8, R12_8$
21. $60_8, R166_8$
23. $156_8, R26_8$
25. $6_{16}, R9_{16}$
27. $12_{16}, R9F_{16}$
29. $6_{16}, RD_{16}$
31. $D26_{16}, R574_{16}$
33. $100_2, R1001_2$
35. $7 (111_2)$
37. Same as multiplying by 10_8: adds a zero to the right
39. $.001_2$
41. $6_{16}, R9_{16}$
43. $101_2, R1100_2$
45. $28_{10}, R7_{10}$

REVIEW PROBLEMS, CHAPTER 6

1. 100100_2
3. 110.00111_2
5. 100011.0000_2
7. 37.03_8
9. $176.A5_{16}$
11. 2435_{10}
13. 100.010_2
15. -10.0110_2
17. 5.05_8
19. $-1AB.8B_{16}$
21. 1110.011_2
23. 10011100.1011_2
25. 17154.762_8
27. $1398.72D_{16}$

29. 11010_2, R11_2 31. 1010_2, R11_2
33. 175_8 35. 5_{16}, R$8F_{16}$
37. 22.25_{10} 39. 2222_{10}
41. 156_8

PROBLEM SET 7.1

1. accuracy: 1; precision: 1
3. accuracy: 2; precision: 1
5. accuracy: 2; precision: 0.1
7. accuracy: 1; precision: 0.1
9. accuracy: 4; precision: 0.1
11. accuracy: 1; precision: 0.0001
13. accuracy: 3; precision: 0.1
15. accuracy: 6; precision: 0.0001
17. accuracy: 8; precision: 0.001
19. accuracy: 2; precision: 1000
21. accuracy: 4; precision: 1
23. Zero is a significant digit whenever it is between two nonzero digits or is a trailing digit in a number containing a decimal point
25. Never 27. 23.617 29. 131349.24
31. The two numbers have the same precision (0.01)
33. 10000110_2; 7 significant digits

PROBLEM SET 7.2

1. 10^2 3. 3.12×10^1
5. 4.57×10^2 7. 1.2×10^{-2}
9. 3.098 11. -3.5×10^1
13. 3.2617×10^4 15. 4×10^{-6}
17. -3.14159×10^3 19. 0
21. 1.4×10^5 23. 1000000000
25. 2.1 27. 3120
29. 0.38 31. 0.0000001002
33. 1870000000000 35. -0.00654
37. 3456000 39. 1.234
41. 3240 43. 10^{100}
45. 10^{100} 47. 5.775×10^1

PROBLEM SET 7.3

1. 00010111 3. 10011111 5. 00000001
7. 01111000 9. 10101101 11. 00011011
13. 00000000 15. 11001110 17. 01110100
19. 11000110 21. 11101001 23. 01000010
25. 10010100 27. 3.17E+00 29. -1.4E+01
31. -1.25E-02 33. 1.56123E+02
35. 1.0000001E+06 37. $-.1351$E+01
39. .51E+02 41. .23E-01
43. $-.1536$E+03 45. .3300497E+05
47. 0 1000100 11000000 00000000 00000000
49. 1 1000011 11100000 00000000 00000000
51. 0 1000110 10001111 00110011 00110011
53. 0 0111011 10100011 11010111 00001010
55. 1 0111110 10000000 00000000 00000000
57. 127_{10}

PROBLEM SET 7.4

1. 14.38 3. $-.25$ 5. 2614.3
7. 3.1 9. -26.0 11. .00023
13. -35.00 15. 2768.4 17. .9999999
19. 2.79 21. -33.6 23. 47.27
25. 279.3 27. $-.111111111$
29. 3.25253 31. .2793939
33. .5000000 35. .0000200
37. 100011_2 (no conversion error)
39. 1001001_2 (no conversion error)
41. 1111111_2 (no conversion error)
43. $.01_2$ (no conversion error)
45. $.0100110_2$ (conversion error = $.296875_{10}$)
47. 11010.00_2 (conversion error = 26_{10})
49. $.09765625_{10}$ (approximately)
51. 001101111001 53. 000110000000
55. 00110101001110000000
57. BCD: 010100110101 12 bits
    Binary: 1000010111_2 10 bits
59. $29,600,000

PROBLEM SET 7.5

1. 16 3. 4 5. 4 7. -2
9. 6 11. 512 13. 197 15. -25
17. .35 19. 144 21. 3 23. 4
25. 16 27. -4 29. -1
31. $3+9+12$
33. $(-4+((4\hat{\,}2)-(4*1*2))\hat{\,}0.5)/(2*1)$
    $(-4-((4\hat{\,}2)-(4*1*2))\hat{\,}0.5)/(2*1)$
35. $(3\hat{\,}2)+(5\hat{\,}2)$
37. $2*(3+5*(2-7))$
39. $(10000*.06)/12$
41. 64 and 256
    Order makes a difference in exponentiation
43. Yes
    $4/2*2 = 2*2 = 4$ (left to right)
    $4/2*2 = 4/4 = 1$ (right to left)
45. Neither

Answers to Odd-Numbered Problems

47. 4/2 * 3 (answers may vary)
49. 3 + 2 − 1 (answers may vary)

PROBLEM SET 7.6

1. Yes
3. Yes
5. No; left side must be one variable
7. Yes
9. No; left side must be one variable
11. No; left side must be a variable
13. No; left side must be one variable
15. No; left side must be one variable
17. No; left side must be one variable
19. No; left side must be one variable
21. $E = A + B + C + D$
23. $A = (3 * X\char`^2) + (4 * X) - 5$
25. $T = (5/9) * (F - 32)$
27. $Y = (X + 1) * (X - 3)\char`^2$
29. $E = 2\char`^X$
31. $S = ((X\char`^3) * Y * (Z\char`^5))\char`^0.5$
33. $T = (6 * X) + (2 * Y)$
35. $I = (B * Y)/12$
37. −3
39. −5
41. 10
43. 38
45. 10
47. 4
49. Yes

PROBLEM SET 7.7

1. 1100100011001001
3. 1111001101011100111110010
5. 11110101
7. 11000111110101100100000011100011110001011100000111010100
9. 110110011111001011000100111110010
11. 101001111010111110101111010100100
13. 1010001101010011101100000101000
15. 1010111110100011101101001010111101000101010010110110010
17. 010100100101000001010000010100001
19. 101000011011001110100011101010010110101001
21. "ABC"
25. 127
27. Binary: 0 1000011 1010010000000000000000000
    EBCDIC: 11110101010010111111000111110010110101
29. DIM

REVIEW PROBLEMS, CHAPTER 7

1. Accuracy: 3; precision: .01
3. Accuracy: 5; precision: .001
5. Accuracy: 3; precision: 1
7. 6.789×10^2
9. 4.03×10^{-4}
11. 1.55×10^2
13. 57000000
15. .00001023
17. 00110001
19. 11001001
21. $2.54E-01$
23. $2.3456E+01$
25. $.1234E+02$
27. $.2034E+01$
29. 45.08
31. .234
33. 5.0
35. −.0216
37. .3984375 (conversion error)
39. 001101100111100001000
41. $C = X * Y$
43. 23
45. 4
47. Yes
49. 11000011110010000100000011110001
51. 1011000101001011010111010100111101101011010100110101110

PROBLEM SET 8.1

1. {1, 2, 3, 4, 5, 6, 7, 8, 9}
3. {−1, 0, 1, 2, 3}
5. $\{1, \frac{1}{2}, \frac{1}{3}, \frac{1}{4}, \ldots\}$
7. $\{0_2, 1_2, 10_2, 11_2\}$
9. {−1, 0, 1, 2}
11. {Mercury, Venus, Earth, Mars, Jupiter, Saturn, Uranus, Neptune, Pluto}
13. {a, c, d, e, h, i, l, m, x}
15. $\{13_8, 14_8, 15_8, 16_8, 17_8, 20_8, 21_8, 22_8\}$
17. $\{x \mid x \in N \text{ and } x < 5\}$
19. $\{x \mid x \text{ is a number system used by computers}\}$
21. $\{x \mid x \in N \text{ and } x \text{ is even}\}$
23. $\{x \mid x \in N \text{ and } x \text{ is a nonnegative integral power of 2}\}$
25. $\{x \mid x \in Q \text{ and } x > 0\}$
27. $\{x \mid x \text{ is an overdrawn customer at Michigan State Bank}\}$
29. $\{x \mid x \text{ is a binary number and } x < 1000_2\}$
31. Equivalent
33. Equal and equivalent
35. Equal and equivalent
37. Neither
39. Equivalent
41. Equal and equivalent
43. Neither
45. Yes
47. { }

PROBLEM SET 8.2

1. $B \subseteq A$
3. $A \subseteq B, B \subseteq A$
5. $A \subseteq B$
7. $A \subseteq B, B \subseteq A$
9. $A \subseteq B$
11. $B \subseteq A$
13. $A \subseteq B, B \subseteq A$
15. $B \subseteq A$

410 Answers to Odd-Numbered Problems

17. $A \subseteq B$ 19. $\varnothing, \{4\}$
21. $\varnothing, \{\text{savings}\}, \{\text{checking}\}, \{\text{savings, checking}\}$
23. $\varnothing, \{1\}, \{2\}, \{3\}, \{4\}, \{1,2\}, \{1,3\}, \{1,4\}, \{2,3\}, \{2,4\},$
    $\{3,4\}, \{1,2,3\}, \{1,2,4\}, \{1,3,4\}, \{2,3,4\}, \{1,2,3,4\}$
25. 64   27. $\varnothing$ 29. None
31. $\varnothing$, {east}, {west}, {north}, {east, west},
    {east, north}, {west, north}
33. $\varnothing, \{1\}, \{2\}$ 35. 63
37. $1 \in A$ 39. $B \nsubseteq A$
41. $4 \notin A$ 43. $B \subseteq U$
45. $U \nsubseteq B$ 47. Yes
49. Yes

9.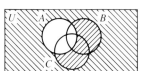

11.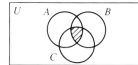

13.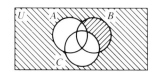

15.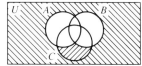

PROBLEM SET 8.3

1. $\{1, 2, 3, 4, 5, 7, 8, 9\}$ 3. $\{8\}$
5. $\{1, 3, 7\}$ 7. $\{1, 2, 4, 5, 8, 10\}$
9. $\{1, 3, 5, 7, 9\} = A$ 11. $\{1, 2, 3, 4, 5, 6, 7, 8, 9\}$
13. $\{5\}$ 15. $\{1\}$
17. $\{2, 4, 6, 8, 10\}$ 19. $\{6, 10\}$
21. $\{1, 2, 3, 4, 5, 6, 7, 8, 9, 10\} = U$
23. $\{2, 4, 5, 9\}$ 25. $\{2, 4, 6, 8, 9, 10\}$
27. $\{c, t, l\}$ 29. $\{o, a, l\}$
31. $A \cap B = A$ 33. U
35. When $A = B$ 37. Always true
39. Only when $A = U = \{\ \}$
41. Always true
43. Only when A and B are disjoint sets
45. A
47. No; $U = \{1, 2, 3, 4\}, A = \{1, 2, 3\}, B = \{2, 4\}$;
    $A - B = \{1, 3\}, B - A = \{4\}$

17.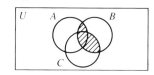

19.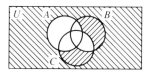

21.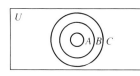

23.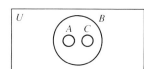

PROBLEM SET 8.4

1.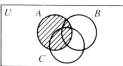

3.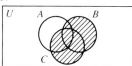

25.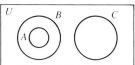

27.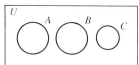

5.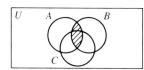

7.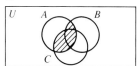

29.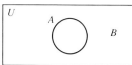

31.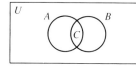

Answers to Odd-Numbered Problems

33. $\{1, 5\}$
35. $\{5\}$
37. $\{3, 6, 7, 9, 10, 11, 12, 13, 14, 15, 17\}$
39. $\{1, 2, 3, 4, 5, 7, 8, 10, 11, 16\}$
41. $\{13, 14, 15, 17\}$
43. $\{6, 9, 12, 13, 14, 15, 17\}$
45. 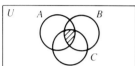   47.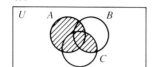

PROBLEM SET 8.5

1. Commutative property for union
3. Closure property for intersection
5. DeMorgan property
7. Identity property for intersection
9. Associative property for union
11. Idempotent property for intersection
13. Complement property for union
15. Distributive property for union over intersection
17. $\{1, 2, 3, 4, 5\} \cap (\{1, 2, 4, 6\} \cap \{3, 4, 5, 7\})$
    $= \{1, 2, 3, 4, 5\} \cap \{4\} = \{4\}$
    $(\{1, 2, 3, 4, 5\} \cap \{1, 2, 4, 6\}) \cap \{3, 4, 5, 7\}$
    $= \{1, 2, 4\} \cap \{3, 4, 5, 7\} = \{4\}$
19. $\{1, 2, 3, 4, 5\} \cup \{1, 2, 3, 4, 5\} = \{1, 2, 3, 4, 5\}$
21. $\overline{\{1, 2, 3, 4, 5\}} \cup \overline{\{1, 2, 4, 6\}} = \{6, 7\} \cup \{3, 5, 7\}$
    $= \{3, 5, 6, 7\}$
    $\overline{\{1, 2, 3, 4, 5\} \cap \{1, 2, 4, 6\}} = \overline{\{1, 2, 4\}}$
    $= \{3, 5, 6, 7\}$
23. $\{1, 2, 3, 4, 5\} \cup (\{1, 2, 4, 6\} \cap \{3, 4, 5, 7\})$
    $= \{1, 2, 3, 4, 5\} \cup \{4\} = \{1, 2, 3, 4, 5\}$
    $(\{1, 2, 3, 4, 5\} \cup \{1, 2, 4, 6\}) \cap (\{1, 2, 3, 4, 5\} \cup \{3, 4, 5, 7\}) = \{1, 2, 3, 4, 5, 6\} \cap \{1, 2, 3, 4, 5, 7\}$
    $= \{1, 2, 3, 4, 5\}$
25. The union (intersection) of two sets will be a subset of the universal set
27. Yes $(0 + 0 = 0)$
29. $\{p, q\} \cup (\{p, q\} \cup \{\ \ \}) = \{p, q\} \cup \{p, q\} = \{p, q\}$
    $(\{p, q\} \cup \{p, q\}) \cup \{\ \ \} = \{p, q\} \cup \{\ \ \} = \{p, q\}$
31. $\{p, q\} \cup \{\ \ \} = \{\ \ \} \cup \{p, q\} = \{p, q\}$
    $\{p, q\} \cap \{\ \ \} = \{\ \ \} = \{p, q\}$
33. $\{p, q\} \cap \{p, q\} = \{p, q\}$

35.
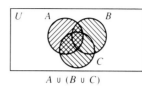 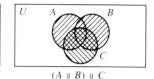
$A \cup (B \cup C)$ $(A \cup B) \cup C$

37.
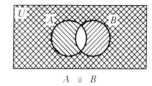 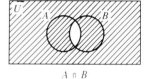
$A \cup B$ $A \cap B$

39.
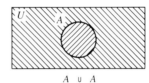 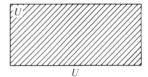
$A \cup A$ U

REVIEW PROBLEMS, CHAPTER 8

1. $\{-6, -5, -4, -3, -2, -1, 0, 1, 2\}$
3. $\{0_8, 1_8, 2_8, 3_8, 4_8, 5_8, 6_8, 7_8\}$
5. $\{x \mid x \text{ is a nonnegative integral power of } 8\}$
7. $\{x \mid x \text{ is a bank that offers IRAs}\}$
9. Equal and equivalent
11. Neither
13. $\varnothing, \{-4\}, \{1\}, \{9\}, \{-4, 1\}, \{-4, 9\}, \{1, 9\}, \{-4, 1, 9\}$
15. $\varnothing, \{1\}, \{2\}, \{3\}, \{4\}, \{1, 2\}, \{1, 3\}, \{1, 4\}, \{2, 3\}, \{2, 4\}, \{3, 4\}, \{1, 2, 3\}, \{1, 2, 4\}, \{1, 3, 4\}, \{2, 3, 4\}$
17. $\varnothing$ 19. $\{5\}$ 21. $\{2, 4, 6, 7, 8\}$
23. $\{2, 4, 6, 7\}$ 25. $\{1, 5\}$ 27. $\{8\}$
29. $\{4, 6, 7\}$ 31. $\{2, 4, 6, 7, 8\}$

33. 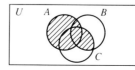   35.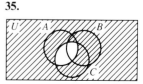

37. 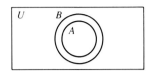 **39.**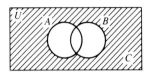

41. Closure property for intersection
43. Distributive property for union over intersection
45. Complement property for union
47.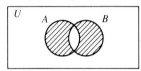

49. $(A \cup B) \cap (\overline{A \cap B})$

PROBLEM SET 9.1

1. Yes
3. No; imperative sentences are not statements
5. No; exclamatory sentences are not statements
7. Yes
9. No; interrogative sentences are not statements
11. Yes
13. No; this is not even a sentence
15. Compound
17. Simple
19. Compound
21. p: I invested my money in stocks.
    q: I went bankrupt.
    $p \wedge q$
23. p: The baseball team is having a good season.
    $\sim p$
25. p: The algebra class is boring.
    q: The computer class is boring.
    $p \wedge \sim q$
27. p: The tire is flat.
    q: The road is rough.
    $p \vee q$
29. The moon is not blue.
31. There is enough time to finish the job.
33. No one believes me.
35. Landing on the moon was dramatic, or flying the space shuttle is a challenge.
37. Landing on the moon was not dramatic.
39. Flying the space shuttle is not a challenge, and landing on the moon was dramatic.
41. It is not the case that either landing on the moon was dramatic or flying the space shuttle is a challenge.

PROBLEM SET 9.2

1. | p | $\sim p$ | $\sim(\sim p)$ |
    |---|---|---|
    | T | F | T |
    | F | T | F |

3. | p | q | $\sim q$ | $p \wedge \sim q$ |
    |---|---|---|---|
    | T | T | F | F |
    | T | F | T | T |
    | F | T | F | F |
    | F | F | T | F |

5. | p | p | $p \vee p$ |
    |---|---|---|
    | T | T | T |
    | F | F | F |

7. | p | p | $p \wedge \sim p$ |
    |---|---|---|
    | T | F | F |
    | F | T | F |

9. | p | q | $q \vee p$ | $p \wedge (q \vee p)$ |
    |---|---|---|---|
    | T | T | T | T |
    | T | F | T | T |
    | F | T | T | F |
    | F | F | F | F |

11. | p | q | $p \wedge q$ | $\sim(p \wedge q)$ |
    |---|---|---|---|
    | T | T | T | F |
    | T | F | F | T |
    | F | T | F | T |
    | F | F | F | T |

13. | p | q | $p \vee q$ | $\sim(p \vee q)$ |
    |---|---|---|---|
    | T | T | T | F |
    | T | F | T | F |
    | F | T | T | F |
    | F | F | F | T |

Answers to Odd-Numbered Problems

15.

p	q	r	$q \wedge r$	$p \wedge (q \wedge r)$
T	T	T	T	T
T	T	F	F	F
T	F	T	F	F
T	F	F	F	F
F	T	T	T	F
F	T	F	F	F
F	F	T	F	F
F	F	F	F	F

17.

p	q	r	$p \wedge q$	$(p \wedge q) \wedge r$
T	T	T	T	T
T	T	F	T	F
T	F	T	F	F
T	F	F	F	F
F	T	T	F	F
F	T	F	F	F
F	F	T	F	F
F	F	F	F	F

19.

p	q	$\sim q$	$p \vee \sim q$	$q \wedge (p \vee \sim q)$	$p \wedge [q \wedge (p \vee \sim q)]$
T	T	F	T	T	T
T	F	T	T	F	F
F	T	F	F	F	F
F	F	T	T	F	F

21.

p	q	$\sim p$	$\sim q$	$p \wedge \sim q$	$\sim p \vee q$	$(p \wedge \sim q) \wedge (\sim p \vee q)$
T	T	F	F	F	T	F
T	F	F	T	T	F	F
F	T	T	F	F	T	F
F	F	T	T	F	T	F

23.

p	q	$\sim p$	$\sim q$	$\sim p \wedge \sim q$	$\sim(\sim p \wedge \sim q)$
T	T	F	F	F	T
T	F	F	T	F	T
F	T	T	F	F	T
F	F	T	T	T	F

25.

p	q	r	$q \wedge r$	$p \vee (q \wedge r)$
T	T	T	T	T
T	T	F	F	T
T	F	T	F	T
T	F	F	F	T
F	T	T	T	T
F	T	F	F	F
F	F	T	F	F
F	F	F	F	F

27.

p	q	r	$\sim p$	$\sim r$	$q \vee \sim r$	$\sim p \wedge (q \vee \sim r)$
T	T	T	F	F	T	F
T	T	F	F	T	T	F
T	F	T	F	F	F	F
T	F	F	F	T	T	F
F	T	T	T	F	T	T
F	T	F	T	T	T	T
F	F	T	T	F	F	F
F	F	F	T	T	T	T

29.

p	q	r	$p \vee r$	$\sim(p \vee r)$	$q \wedge \sim(p \vee r)$
T	T	T	T	F	F
T	T	F	T	F	F
T	F	T	T	F	F
T	F	F	T	F	F
F	T	T	T	F	F
F	T	F	F	T	T
F	F	T	T	F	F
F	F	F	F	T	F

31.

p	q	r	$p \vee q$	$p \vee r$	$(p \vee q) \wedge (p \vee r)$
T	T	T	T	T	T
T	T	F	T	T	T
T	F	T	T	T	T
T	F	F	T	T	T
F	T	T	T	T	T
F	T	F	T	F	F
F	F	T	F	T	F
F	F	F	F	F	F

33. 16
35. The conjunction is associative
37. The disjunction distributes over the conjunction
39. The DeMorgan propety holds for logic
41. Yes
43. This statement is true when at least one of the statements $p, q, r, s,$ or t is true
45. This statement is false if either the bank is open or today is not a holiday
47. This statement is false if overtime workers do get a bonus at our company

PROBLEM SET 9.3

1.
p	q	$p \land q$	$\sim(p \land q)$
T	T	T	F
T	F	F	T
F	T	F	T
F	F	F	T

3.
p	q	$\sim p$	$\sim q$	$p \land \sim q$	$\sim p \land q$	$(p \land \sim q) \lor (\sim p \land q)$
T	T	F	F	F	F	F
T	F	F	T	T	F	T
F	T	T	F	F	T	T
F	F	T	T	F	F	F

5.
p	q	$\sim p$	$\sim p \bar\land q$
T	T	F	T
T	F	F	T
F	T	T	F
F	F	T	T

7.
p	q	$p \lor q$	$p \veebar q$	$(p \lor q) \lor (p \veebar q)$
T	T	T	F	T
T	F	T	T	T
F	T	T	T	T
F	F	F	F	F

9.
p	q	$\sim p$	$\sim q$	$\sim p \bar\land \sim q$	$(\sim p \bar\land \sim q) \lor p$
T	T	F	F	T	T
T	F	F	T	T	T
F	T	T	F	T	T
F	F	T	T	F	F

11.
p	p	$p \bar\land p$
T	T	F
F	F	T

13.
p	q	$p \veebar q$	$p \bar\land q$	$\sim(p \bar\land q)$	$(p \veebar q) \land \sim(p \bar\land q)$
T	T	F	F	T	F
T	F	T	F	F	F
F	T	T	F	F	F
F	F	F	T	F	F

15.
p	q	r	$q \veebar r$	$p \bar\land (q \veebar r)$
T	T	T	F	T
T	T	F	T	F
T	F	T	T	F
T	F	F	F	T
F	T	T	F	T
F	T	F	T	T
F	F	T	T	T
F	F	F	F	T

17.
p	q	r	$p \veebar q$	$p \bar\land r$	$(p \veebar q) \land (p \bar\land r)$
T	T	T	F	F	F
T	T	F	F	T	F
T	F	T	F	F	F
T	F	F	F	T	F
F	T	T	F	T	F
F	T	F	F	T	F
F	F	T	T	T	T
F	F	F	T	T	T

19.
p	q	$\sim q$	$p \bar\land \sim q$	$\sim(p \bar\land \sim q)$	$\sim(p \bar\land \sim q) \bar\land q$
T	T	F	T	F	T
T	F	T	F	T	T
F	T	F	T	F	T
F	F	T	T	F	T

21. $(p \land q) \lor \sim(p \land q)$
23. c 25. t 27. t
29. c 31. t 33. c
35. The NAND operation is the negation of the conjunction

Answers to Odd-Numbered Problems

37. OR is true when *at least one* of the two statements is true; Exclusive OR is true when *exactly one* of the two statements is true

39. No

PROBLEM SET 9.4

1.

p	q	$\sim q$	$p \to \sim q$
T	T	F	F
T	F	T	T
F	T	F	T
F	F	T	T

3.

p	q	$p \to q$	$\sim(p \to q)$
T	T	T	F
T	F	F	T
F	T	T	F
F	F	T	F

5.

p	q	$p \wedge q$	$(p \wedge q) \to q$
T	T	T	T
T	F	F	T
F	T	F	T
F	F	F	T

7.

p	q	$p \underline{\vee} q$	$p \vee q$	$\sim(p \vee q)$	$(p \underline{\vee} q) \to \sim(p \vee q)$
T	T	F	T	F	T
T	F	T	T	F	F
F	T	T	T	F	F
F	F	F	F	T	T

9.

p	q	r	$\sim q$	$r \underline{\vee} p$	$\sim q \bar{\wedge} (r \underline{\vee} p)$	$p \to [\sim q \bar{\wedge} (r \underline{\vee} p)]$
T	T	T	F	F	T	T
T	T	F	F	F	T	T
T	F	T	T	F	T	T
T	F	F	T	F	T	T
F	T	T	F	F	T	T
F	T	F	F	F	T	T
F	F	T	T	F	T	T
F	F	F	T	T	F	T

11.

p	q	$\sim p$	$\sim p \leftrightarrow q$
T	T	F	F
T	F	F	T
F	T	T	T
F	F	T	F

13.

p	q	$\sim p$	$\sim q$	$\sim p \leftrightarrow \sim q$
T	T	F	F	T
T	F	F	T	F
F	T	T	F	F
F	F	T	T	T

15.

p	q	$p \wedge q$	$p \to q$	$(p \wedge q) \leftrightarrow (p \to q)$
T	T	T	T	T
T	F	F	F	T
F	T	F	T	F
F	F	F	T	F

17.

p	q	$p \leftrightarrow q$	$p \underline{\vee} q$	$(p \leftrightarrow q) \wedge (p \underline{\vee} q)$
T	T	T	F	F
T	F	F	T	F
F	T	F	T	F
F	F	T	F	F

19.

p	q	r	$\sim q$	$\sim r$	$p \wedge \sim r$	$r \leftrightarrow \sim q$	$(p \wedge \sim r) \to (r \leftrightarrow \sim q)$
T	T	T	F	F	F	F	T
T	T	F	F	T	T	T	T
T	F	T	T	F	F	T	T
T	F	F	T	T	T	F	F
F	T	T	F	F	F	F	T
F	T	F	F	T	F	T	T
F	F	T	T	F	F	T	T
F	F	F	T	T	F	F	T

21.

p	q	$\sim p$	$\sim q$	$q \to p$	$\sim p \to \sim q$	$(q \to p) \leftrightarrow (\sim p \to \sim q)$
T	T	F	F	T	T	T
T	F	F	T	T	T	T
F	T	T	F	F	F	T
F	F	T	T	T	T	T

23. Inverse: If the interest rates are not low, then we cannot buy a house. Converse: If we can buy a house, then the interest rates are low. Contrapositive: If we cannot buy a house, then the interest rates are not low.
25. Inverse: If computers use binary numbers, then computers are not dumb. Converse: If computers are dumb, then computers don't use binary numbers. Contrapositive: If computers are not dumb, then computers use binary numbers.
27. Inverse: $p \to \sim q$; Converse: $q \to \sim p$; Contrapositive: $\sim q \to p$
29. $\sim(p \wedge \sim q)$
31. Order does not matter
33. Equivalent to $p \leftrightarrow q$
35. Not equivalent
37. Not equivalent
39. Equivalent

PROBLEM SET 9.5

1. Commutative property for $\wedge$
3. NAND property
5. Exclusive OR property
7. Biconditional property
11. Tautology property
13. Idempotent property for $\wedge$
15. Double negation property
17. DeMorgan property
19. Contradiction property
23. Yes 25. No 27. No 29. No
31. No 33. $\sim p \wedge q$ 35. q
37. $(p \wedge q) \vee \sim(p \wedge q)$ 39. $p \wedge \sim q$

PROBLEM SET 9.6

1. Valid 3. Valid 5. Valid
7. Invalid 9. Valid 11. Valid
13. Valid 15. Valid 17. Valid
19. Invalid
21. No C's are D's. No D's are C's.
23. No B's are D's. No D's are B's.
25. No D's are A's. No A's are D's.
27. Invalid 29. Invalid 31. Valid
33. Invalid 35. Invalid 37. Valid
39. H: All cats are blue.
    All blue animals smoke.
    C: All cats smoke.

REVIEW PROBLEMS, CHAPTER 9

1. Yes
3. No; imperative sentences are not statements
5. Simple
7. Compound

9.
p	q	$\sim q$	$p \wedge \sim q$	$\sim(p \wedge \sim q)$
T	T	F	F	T
T	F	T	T	F
F	T	F	F	T
F	F	T	F	T

11.
p	q	$q \wedge p$	$p \vee (q \wedge p)$
T	T	T	T
T	F	F	T
F	T	F	F
F	F	F	F

13.
p	q	$q \to p$	$p \vee (q \to p)$
T	T	T	T
T	F	T	T
F	T	F	F
F	F	T	T

15.
p	q	$p \leftrightarrow q$	$p \wedge q$	$(p \leftrightarrow q) \to (p \wedge q)$
T	T	T	T	T
T	F	F	F	T
F	T	F	F	T
F	F	T	F	F

17.
p	q	$\sim p$	$p \to q$	$q \to \sim p$	$(p \to q) \wedge (q \to \sim p)$
T	T	F	T	F	F
T	F	F	F	T	F
F	T	T	T	T	T
F	F	T	T	T	T

19.
p	q	$\sim p$	$\sim q$	$\sim q \vee p$	$\sim p \vee (\sim q \vee p)$
T	T	F	F	T	T
T	F	F	T	T	T
F	T	T	F	F	T
F	F	T	T	T	F

Answers to Odd-Numbered Problems

21.

p	q	$p \to q$	$q \to p$	$(p \to q) \leftrightarrow (q \to p)$
T	T	T	T	T
T	F	F	T	F
F	T	T	F	F
F	F	T	T	T

23.

p	q	r	$p \vee q$	$(p \vee q) \bar{\wedge} r$
T	T	T	T	F
T	T	F	T	T
T	F	T	T	F
T	F	F	T	T
F	T	T	T	F
F	T	F	T	T
F	F	T	F	T
F	F	F	F	T

25.

p	q	r	$\sim p$	$\sim r$	$\sim p \wedge \sim r$	$(\sim p \wedge \sim r) \bar{\vee} q$
T	T	T	F	F	F	F
T	T	F	F	T	F	F
T	F	T	F	F	F	T
T	F	F	F	T	F	T
F	T	T	T	F	F	F
F	T	F	T	T	T	F
F	F	T	T	F	F	T
F	F	F	T	T	T	F

27. Converse: $\sim q \to \sim p$; inverse: $p \to q$; contrapositive: $q \to p$
29. Valid   **31.** Invalid
33. Invalid   **35.** Valid
37. A tautology is a statement that is always true

PROBLEM SET 10.1

1. 1   **3.** 1   **5.** 1   **7.** 1   **9.** 1

11.

A	B	$A \cdot B$	$A + (A \cdot B)$
1	1	1	1
1	0	0	1
0	1	0	0
0	0	0	0

13.

A	B	$\bar{A}$	$\bar{A} + B$
1	1	0	1
1	0	0	0
0	1	1	1
0	0	1	1

15.

A	B	$A + B$	$A \cdot B$	$(A + B) \cdot (A \cdot B)$
1	1	1	1	1
1	0	1	0	0
0	1	1	0	0
0	0	0	0	0

17.

A	B	C	$A + B$	$(A + B) + C$
1	1	1	1	1
1	1	0	1	1
1	0	1	1	1
1	0	0	1	1
0	1	1	1	1
0	1	0	1	1
0	0	1	0	1
0	0	0	0	0

19.

A	B	$A + B$	$A \cdot (A + B)$
1	1	1	1
1	0	1	1
0	1	1	0
0	0	0	0

21.

A	B	$A + B$
1	1	1
1	0	1
0	1	1
0	0	0

Equivalent to 22

23.

A	B	$B \cdot A$
1	1	1
1	0	0
0	1	0
0	0	0

Equivalent to 24

25.

A	B	C	B·C	A+(B·C)
1	1	1	1	1
1	1	0	0	1
1	0	1	0	1
1	0	0	0	1
0	1	1	1	1
0	1	0	0	0
0	0	1	0	0
0	0	0	0	0

Equivalent to 26

27.

A	B	C	A·B	A·C	(A·B)+(A·C)
1	1	1	1	1	1
1	1	0	1	0	1
1	0	1	0	1	1
1	0	0	0	0	0
0	1	1	0	0	0
0	1	0	0	0	0
0	0	1	0	0	0
0	0	0	0	0	0

Equivalent to 28

29.

A	0	A+0
1	0	1
0	0	0

$(A+0 = A)$

31. $(A \cdot B)+(\overline{A} \cdot \overline{B})$

33. The biconditional statement

PROBLEM SET 10.2

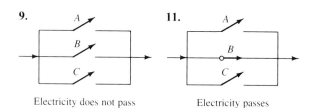

9. Electricity does not pass

11. Electricity passes

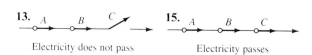

13. Electricity does not pass

15. Electricity passes

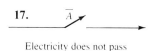

17. Electricity does not pass

19. No **21.** Yes **23.** No
25. Yes **27.** No
29. Yes; when A and B are closed

31. → A → $\overline{A}$ →

33. Electricity passes whenever at least one of the switches A, B, or C is closed

PROBLEM SET 10.3

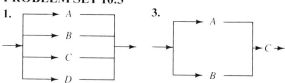

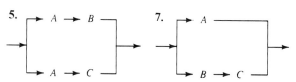

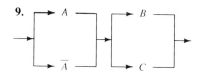

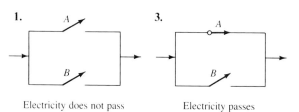

Electricity does not pass   Electricity passes

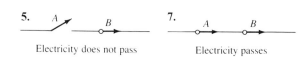

Electricity does not pass   Electricity passes

11.
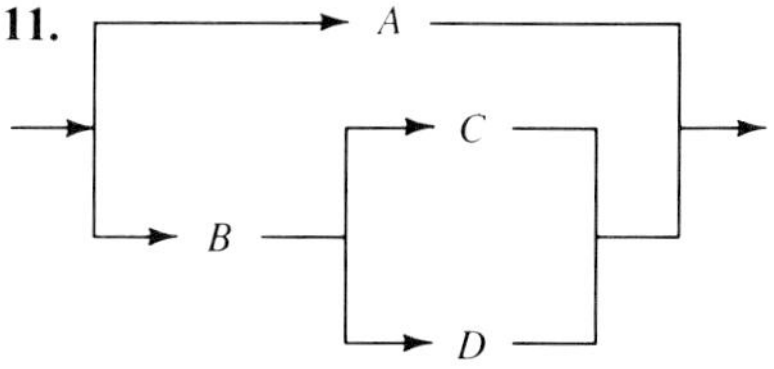

13.
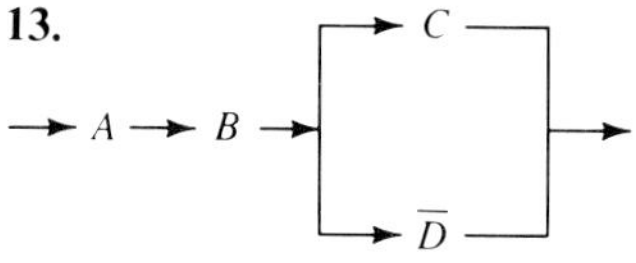

15.
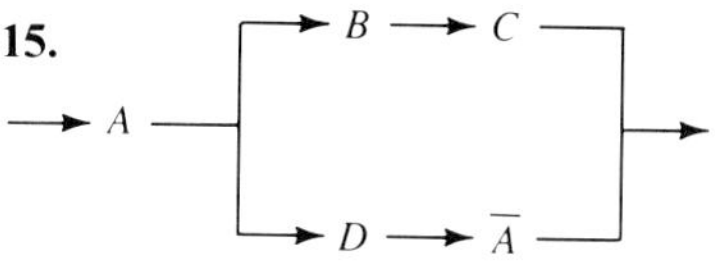

17.
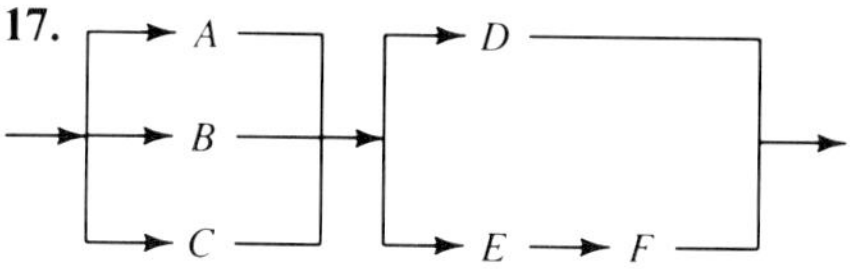

19.
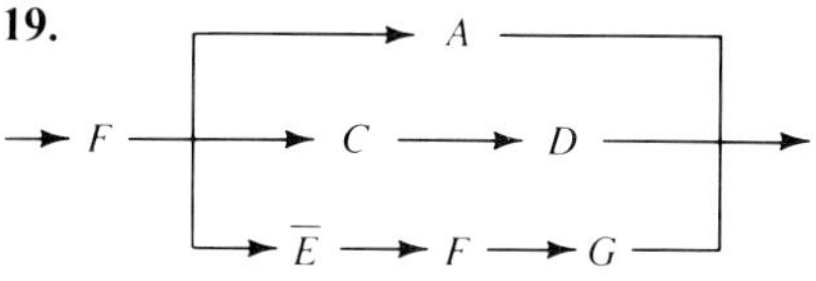

21. $ABCD$
23. $A(B+CD)$
25. $A(BC+D\overline{A})$
27. $A(B+\overline{A}+D)(B+C)$
29. $AB+(C+D)+\overline{A}\overline{C}+(E+D)$
31.
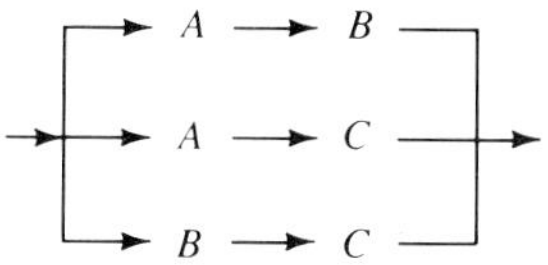

33.
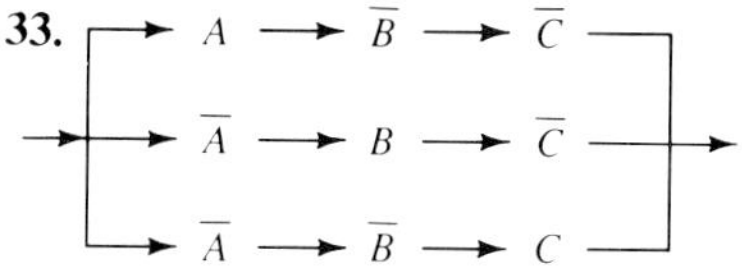

35.
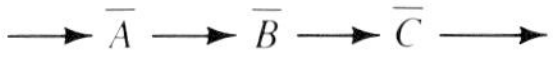

37.
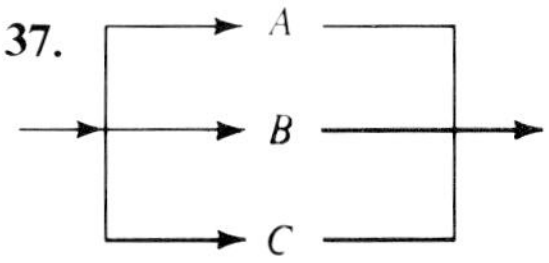

39.
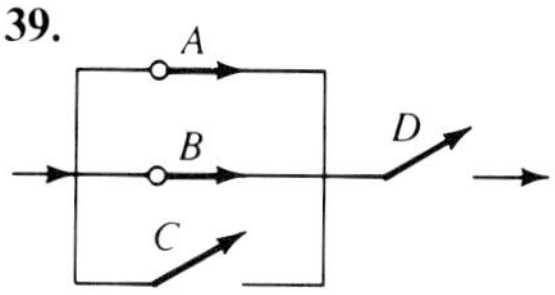

Electricity does not pass

41.
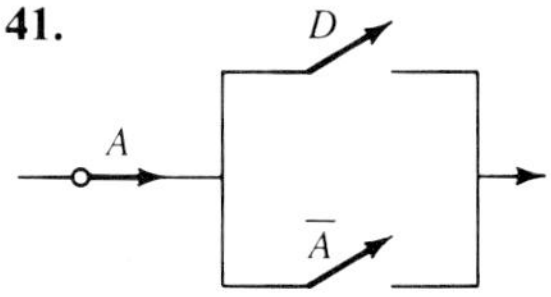

Electricity does not pass

43.
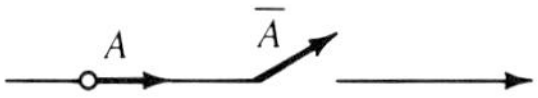

Electricity does not pass

45.
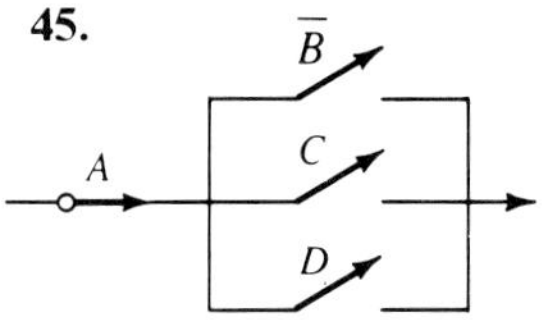

Electricity does not pass

47.
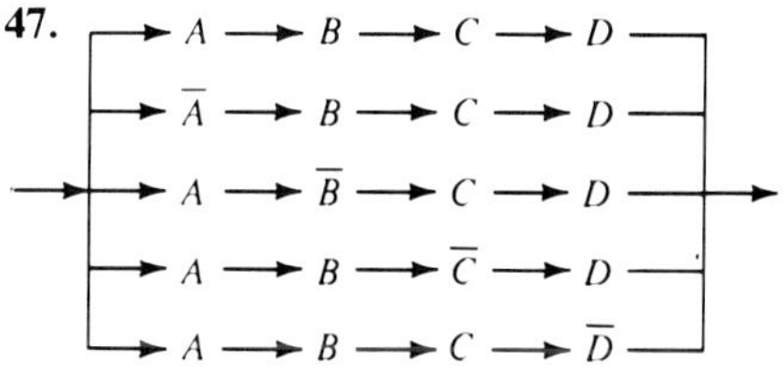

49.
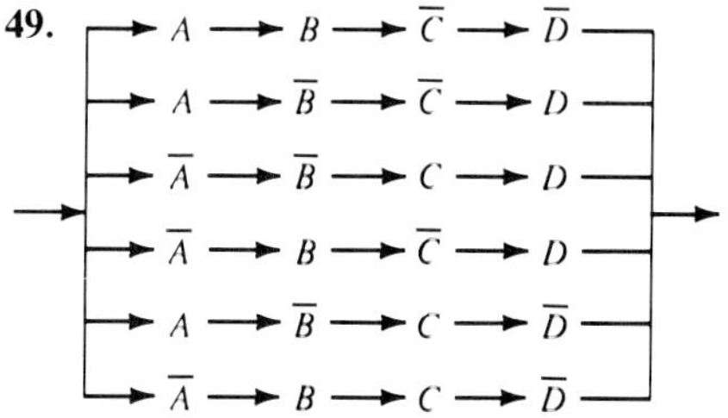

PROBLEM SET 10.4

1. Associative property for multiplication
3. De Morgan property
5. Distributive property for multiplication over addition
7. Identity property for multiplication
9. Idempotent property for addition
11. Commutative property for addition
13. Complement property for multiplication
15. One property

17.

A	B	C	AB	BC	$A(BC)$	$(AB)C$
1	1	1	1	1	1	1
1	1	0	1	0	0	0
1	0	1	0	0	0	0
1	0	0	0	0	0	0
0	1	1	0	1	0	0
0	1	0	0	0	0	0
0	0	1	0	0	0	0
0	0	0	0	0	0	0

19.

A	A	AA
1	1	1
0	0	0

21.

A	0	$A0$
1	0	0
0	0	0

23.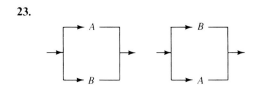

25. → A → B → C →   → A → B → C →

27. → A —∘ 1 →

29.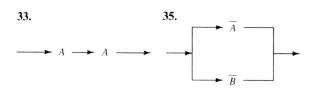

31.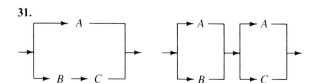

33. → A → A →

35. (circuit with $\bar{A}$ and $\bar{B}$ in parallel)

37. $A + BC$ [contains fewer switches (3)]

39. → $\bar{A}$ → $\bar{B}$ →

41.

A	$\bar{A}$	$\bar{\bar{A}}$
1	0	1
0	1	0

43. A 45. 1

PROBLEM SET 10.5

1. Distributive property for multiplication over addition
3. Commutative property for multiplication
5. Commutative property for addition
7. Complement property for addition
9. Commutative property for addition
11. Complement property for addition
13. A 15. 1   17. 1   19. 0
21. 1   23. BA 25. $A+C$
27. $A+BC$ 29. A 31. AD
33. 1   35. $B(\bar{A}+C)$
37. $A(B+C)+BC$ 39. $(A \cdot \bar{B}) \to (\bar{A} \cdot B)$
41. No   43. No   45. $0 \cdot 1$

PROBLEM SET 10.6

1. 1001   3. 0000
5. 10000001   7. 1101
9. 1111   11. 11001111
13. 0100   15. 011100
17. 00100010

Answers to Odd-Numbered Problems

19.

21.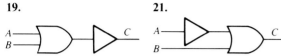

23.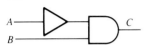

25.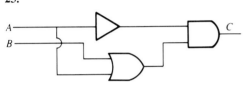

27.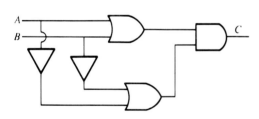

29.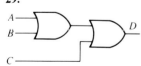

31. $(A \cdot B) + \bar{B}$ **33.** $A \cdot \bar{B}$
35. Sum: 1; carry: 0 **37.** 0 and 1
39. 1 and 1 **41.** 1 and 0 **43.** 0 and 1
45. $\{[(A+B) \cdot (\overline{A \cdot B})] + C\} \cdot \{\overline{[(A+B) \cdot (\overline{A \cdot B})] \cdot C}\}$
$\{[(A+B) \cdot (\overline{A \cdot B})] \cdot C\} + (A \cdot B)$

47.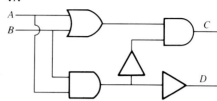

49. $(A+B) \cdot (\overline{A \cdot B})$

REVIEW PROBLEMS, CHAPTER 10

1.

A	B	$\bar{A}$	$B+\bar{A}$	$A(B+\bar{A})$
1	1	0	1	1
1	0	0	0	0
0	1	1	1	0
0	0	1	1	0

3.

A	B	C	$\bar{C}$	AB	$B\bar{C}$	$AB+BC$
1	1	1	0	1	0	1
1	1	0	1	1	1	1
1	0	1	0	0	0	0
1	0	0	1	0	0	0
0	1	1	0	0	0	0
0	1	0	1	0	1	1
0	0	1	0	0	0	0
0	0	0	1	0	0	0

5. Yes **7.** Yes

9.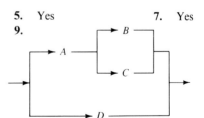

11.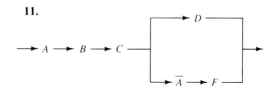

Answers to Odd-Numbered Problems

13. $AB(C+D)$
15. $(ED + \overline{A} + BC)F$
17.
19. DeMorgan property
21. Associative property for multiplication
23. $A(B+C)$
25. $AC + B\overline{C}$
27. $A(B+C)$
29.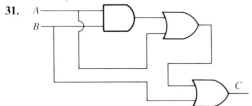
31.
33.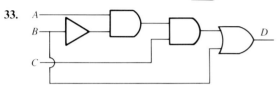
35. $(A+B) \cdot \overline{A}$
37. $(A \cdot B) + (\overline{A} \cdot B)$
39.

PROBLEM SET 11.1

1. (a) Drive to gas station.
   (b) Turn off engine.
   (c) Take off gas cap.
   (d) Put gas nozzle into gas tank.
   (e) Pump gas.
   (f) Put gas nozzle back.
   (g) Put gas cap back on.
   (h) Pay for gas.
   (i) Start engine.
   (j) Drive off!
3. (a) Input the three coefficients (A, B and C).
   (b) Find the first root.
   $$R = \frac{-B + \sqrt{(B^2 - 4AC)}}{2A}$$
   (c) Find the second root.
   $$S = \frac{-B - \sqrt{(B^2 - 4AC)}}{2A}$$
   (d) Print out the two roots (R and S).
5. (a) Input the number of feet and inches (F and I).
   (b) Convert to total inches (T = 12F + I).
   (c) Print out the total inches.
7. (a) Input field goals and free throws (F and G).
   (b) Calculate the total points (T = 2F + G).
   (c) Print the total points.
9. (a) Input the base and height (B and H).
   (b) Calculate the area (A = 0.5BH).
   (c) Print out the area.
11. (a) Input a binary number (B).
    (b) Convert to a decimal number (D).
    (c) Print out the decimal number.
13. (a) Input an octal number (O).
    (b) Convert to a binary number.
    (c) Print out the binary number.
15. 10 INPUT A, B, C
    20 R = (−B + (B^2 − 4 * A * C)^0.5) / (2 * A)
    30 S = (−B − (B^2 − 4 * A * C)^0.5) / (2 * A)
    40 PRINT R, S
17. 10 INPUT F, I
    20 T = (12 * F) + I
    30 PRINT T
19. 10 INPUT F, G
    20 T = (2 * F) + G
    30 PRINT T
21. 10 INPUT B, H
    20 A = 0.5 * B * H
    30 PRINT A
23. The sum cannot be found before the variables A and B are inputted
25. Centimeters must be changed to inches before inches can be printed out
27. 10 INPUT A, B
    20 S = A + B
    30 PRINT S
29. 10 INPUT C
    20 I = C / 2.54
    30 PRINT I

PROBLEM SET 11.2

1. (a) Input price (P).
   (b) Calculate new price [NP = (P − .10P) * 1.05].
   (c) Print out new price (NP).
3. 10 INPUT P
   20 NP = (P − (.10 * P)) * 1.05
   30 PRINT NP
5. Terminal symbol
7. Input/output symbol
9. Calculation symbol

Answers to Odd-Numbered Problems

11.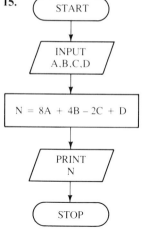

13.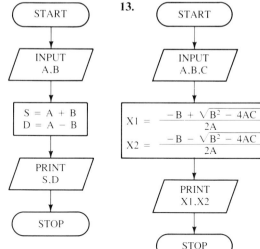

15.

(flowchart: START → INPUT A,B,C,D → N = 8A + 4B − 2C + D → PRINT N → STOP)

17. (a) Input the two numbers (A and B).
    (b) Calculate the sum (S = A + B).
    (c) Calculate the difference (D = A − B).
    (d) Print out the sum and difference (S and D).

19. (a) Input the coefficients (A, B, and C).
    (b) Calculate the first value.   (c) Calculate the second value.

$$R = \frac{-B + \sqrt{B^2 - 4AC}}{2A} \qquad S = \frac{-B - \sqrt{B^2 - 4AC}}{2A}$$

    (d) Print out the two values.

21. (a) Input the four digits (A, B, C, and D).
    (b) Convert to a decimal number.
        N = 8A + 4B + 2C + D
    (c) Print out the decimal number.

23. 10 INPUT A, B
    20 S = A + B
    30 D = A − B
    40 PRINT S, D

25. 10 INPUT A, B, C
    20 R = (−B + (B^2 − 4*A*C)^0.5)/(2*A)
    30 S = (−B − (B^2 − 4*A*C)^0.5)/(2*A)
    40 PRINT R, S

27. 10 INPUT A, B, C, D
    20 N = (8*A) + (4*B) + (2*C) + D
    30 PRINT N

29. The values of the variables must be inputted before calculations involving these variables can take place. Otherwise, the computer will not know the values of the variables when the calculation is performed.

PROBLEM SET 11.3

1.

(flowchart: INPUT N → N>0? T: A=1 / F: → N=0? T: A=0 / F: → N<0? T: A=−1 / F: →)

3. (a) INPUT X, Y
   (b) IF X > Y THEN Y = X
               ELSE X = Y
   (c) Y = X − 5
   (d) X = Y − 3
   (e) PRINT X, Y
5. (a) INPUT A, S
   (b) IF A = 65 THEN PRINT A
7. (a) INPUT N
   (b) IF N < 0 THEN PRINT N
9. (a) INPUT A, B, C
   (b) S = A + B + C
   (c) IF S > 0 THEN PRINT "Positive"
   (d) IF S < 0 THEN PRINT "Negative"
   (e) IF S = 0 THEN PRINT "Zero"
11. (a) INPUT A, B, C
    (b) IF (A > B) AND (A > C) THEN PRINT A
    (c) IF (B > A) AND (B > C) THEN PRINT B
    (d) IF (C > A) AND (C > B) THEN PRINT C
13. (a) INPUT A, S
    (b) IF A > 32 THEN S = S + 1000
                ELSE S = S − 500
    (c) PRINT A, S
15. (a) INPUT L, W, P
    (b) A = L * W
    (c) IF A > 100 THEN C = (100 * P) + (A − 100) * (2 * P)
                ELSE C = A * P
    (d) PRINT A, C

17.

19.

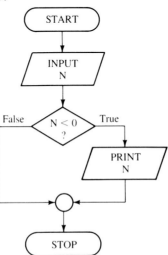

21.

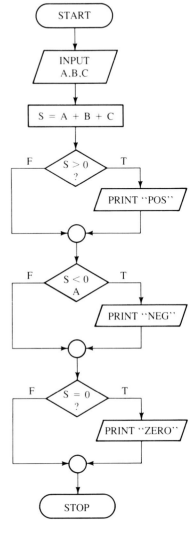

Answers to Odd-Numbered Problems

23.

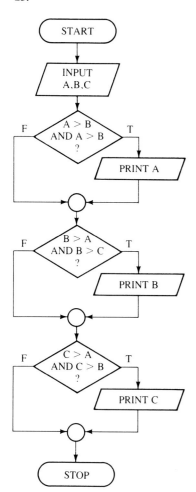

25.

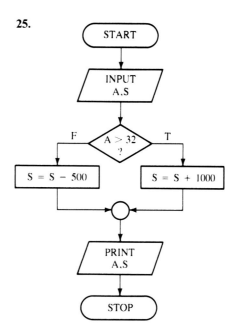

27.

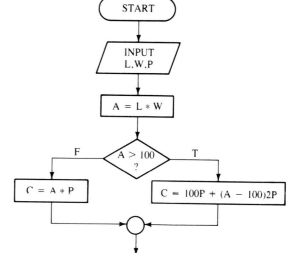

29. P OR Q
31. (P OR Q) AND (P OR R)
33. 9 and 7
35. 6 and 3
37. -3 and -3
39. (a) INPUT X, Y
    (b) IF X ≤ Y THEN IF X < Y
                     THEN X = Y + 2
                     ELSE Y = X − 3
              ELSE X = Y
    (c) PRINT X, Y

PROBLEM SET 11.4

1.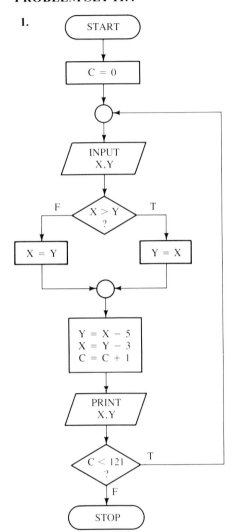

3. H = 98, L = 34
5. (a) C = 0
   (b) REPEAT
   　　(i) T = 0
   　　(ii) SUM = 0
   　　(iii) REPEAT
   　　　　(i) INPUT S
   　　　　(ii) T = T + 1
   　　　　(iii) SUM = SUM + S
   　　　　(iv) UNTIL (T = 10)
   　　(v) C = C + 1
   　　(vi) A = SUM / 10
   　　(vii) PRINT A
   (c) UNTIL (C = 76)
7. (a) C = 0
   (b) DO-WHILE (C < 587)
   　　(i) INPUT H, C
   　　(ii) G = H / C
   　　(iii) PRINT G
   　　(iv) C = C + 1
   (c) END DO-WHILE
9. (a) F = 32
   (b) DO-WHILE (F ≤ 212)
   　　(i) C = (5/9) * (F − 32)
   　　(ii) PRINT F, C
   　　(iii) F = F + 1
   (c) END DO-WHILE
11. (a) INPUT H, W
    (b) DO-WHILE (NOT EOF)
    　　(i) IF H > 40
    　　　　THEN P = (40 * W) + (H − 40) * (2 * W)
    　　　　ELSE P = H * W
    　　(ii) PRINT P
    　　(iii) INPUT H, W
    (c) END DO-WHILE
13. (a) N = 1
    (b) REPEAT
    　　(i) INPUT X, Y, Z
    　　(ii) A = (X + Y + Z) / 3
    　　(iii) PRINT A
    　　(iv) N = N + 1
    (c) UNTIL (N > 77)
15. (a) REPEAT
    　　(i) W = 1
    　　　　T = 0
    　　(ii) INPUT S
    　　(iii) REPEAT
    　　　　(i) T = T + S
    　　　　(ii) INPUT S
    　　　　(iii) W = W + 1
    　　　　(iv) UNTIL (W > 52)
    　　(v) PRINT T
    (b) UNTIL (EOF)
17. (a) INPUT A, B, C
    (b) REPEAT
    　　(i) P = A + B + C

Answers to Odd-Numbered Problems

         (ii) PRINT P
         (iii) INPUT A, B, C
    (c) UNTIL (EOF)
19.   10   FOR C = 1 TO 121
       20   INPUT X, Y
       30   If X > Y THEN Y = X
                     ELSE X = Y
       40   Y = X − 5
       50   X = Y − 3
       60   PRINT X, Y
       70   NEXT C

REVIEW PROBLEMS, CHAPTER 11

1. (a) INPUT A, B, C
    (b) S = A + B + C
    (c) P = A * B * C
    (d) PRINT S, P
3. (a) INPUT T, S
    (b) M = (13 * T) + (17 * S)
    (c) PRINT M
5. (a) INPUT S, A
    (b) IF A > 35 THEN S = S + 4000
                ELSE S = S − 2300
    (c) PRINT S
7. (a) INPUT A, B, C
    (b) IF A > B AND A < C THEN PRINT A
    (c) IF A < B AND A > C THEN PRINT A
    (d) IF B > A AND B < C THEN PRINT B
    (e) IF B < A AND B > C THEN PRINT B
    (f) IF C > A AND C < B THEN PRINT C
    (g) IF C < A AND C > B THEN PRINT C

9. **11.**

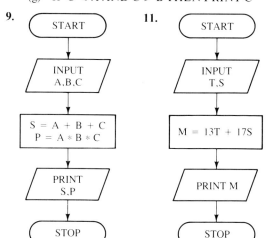

13.

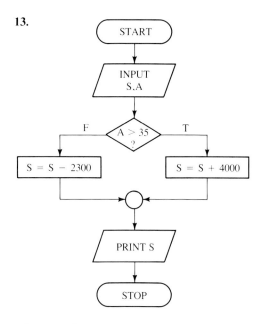

15. (a) C = 0
     (b) T = 0
     (c) DO-WHILE (C < 561)
          (i) INPUT N, M
          (ii) T = T + M
          (iii) PRINT N, M
          (iv) C = C + 1
     (d) END DO-WHILE
     (e) PRINT T

17. (a) INPUT S, A
     (b) DO-WHILE (NOT EOF)
         (i) IF A > 35
             THEN S = S + 4000
             ELSE S = S − 2300
         (ii) PRINT S
         (iii) INPUT S, A
     (c) END DO-WHILE

19. (a) C = 0
     (b) T = 0
     (c) REPEAT
          (i) INPUT N, M
          (ii) T = T + M
          (iii) PRINT N, M
          (iv) C = C + 1
     (d) UNTIL (C = 561)
     (e) PRINT T

21. (a) INPUT S, A
(b) REPEAT
  (i) IF A > 35
    THEN S = S + 4000
    ELSE S = S − 2300
  (ii) PRINT S
  (iii) INPUT S, A
(c) UNTIL (EOF)

PROBLEM SET 12.1

1. 1×5 **3.** 1 **5.** 4 **7.** 5
9. 2 **11.** -2 **13.** 3 **15.** 2, 3
17. Not possible **19.** 1 **21.** 4
23. No **25.** No
27. [17,075 18,091 15,043 20,261 24,000]
29. $\begin{bmatrix} 3 & 4 & 5 & 4 \\ 2 & 5 & 7 & 4 \\ 3 & 3 & 2 & 2 \\ 5 & 4 & 1 & 6 \\ 3 & 3 & 3 & 3 \end{bmatrix}$

PROBLEM SET 12.2

1. $\begin{bmatrix} -4 & 1 & 1 \\ 5 & 5 & 7 \\ 4 & 7 & 2 \end{bmatrix}$ **3.** Not possible

5. $\begin{bmatrix} 1 & 1 & -1 \\ 2 & 5 & 2 \\ 3 & 5 & 2 \end{bmatrix}$ **7.** $\begin{bmatrix} -3 & 1 & 1 \\ 5 & 6 & 7 \\ 4 & 7 & 3 \end{bmatrix}$

9. $\begin{bmatrix} -3 & 1 & 1 \\ 5 & 6 & 7 \\ 4 & 7 & 3 \end{bmatrix}$ **11.** $\begin{bmatrix} 1 & 1 & -1 \\ 2 & 5 & 2 \\ 3 & 5 & 2 \end{bmatrix}$

13. $\begin{bmatrix} -4 & -1 & 3 \\ 1 & -3 & 3 \\ -2 & -3 & 0 \end{bmatrix}$ **15.** Not possible

17. $\begin{bmatrix} -28 & 0 & 14 \\ 21 & 7 & 35 \\ 7 & 14 & 7 \end{bmatrix}$ **19.** $\begin{bmatrix} 0 & -2 & 2 \\ -4 & -8 & -4 \\ -6 & -10 & -2 \end{bmatrix}$

21. $\begin{bmatrix} -4 & 3 & 1 \\ 0 & 1 & 2 \\ 2 & 5 & 1 \end{bmatrix}$ **23.** $\begin{bmatrix} 3 & 2 & 1 \\ 0 & -1 & -2 \end{bmatrix}$

25. $\begin{bmatrix} -10 & 0 & 6 \\ 9 & 5 & 15 \\ 3 & 6 & 5 \end{bmatrix}$ **27.** Not possible

29. $\begin{bmatrix} 12 & -7 & -5 \\ 4 & 5 & -2 \\ 0 & -5 & -1 \end{bmatrix}$ **31.** Yes

33. No
35. Yes; the matrix of all zeros
37. Net yards:

	Quarter			
Team	1	2	3	4
1	70	73	25	−4
2	11	38	82	99

39. $A = \begin{bmatrix} 1 & 2 \\ 2 & 1 \end{bmatrix}$ (Answers may vary) **41.** Yes

PROBLEM SET 12.3

1. $\begin{bmatrix} 29 & 11 & 8 \\ 1 & 3 & 6 \end{bmatrix}$ **3.** $\begin{bmatrix} 3 & 1 & 4 \\ 2 & 0 & -1 \end{bmatrix}$

5. Not possible **7.** Not possible

9. $\begin{bmatrix} 17 & 7 & 30 \\ 14 & 4 & 15 \\ 12 & 2 & 5 \end{bmatrix}$ **11.** $\begin{bmatrix} 35 & 5 \\ 2 & -2 \\ 39 & -9 \end{bmatrix}$

13. $\begin{bmatrix} 7 & 17 & 27 \\ 2 & 4 & 6 \\ 7 & 19 & 31 \end{bmatrix}$ **15.** Not possible

17. $\begin{bmatrix} 3 & 2 & 3 \\ 0 & 1 & -1 \\ 5 & 1 & 0 \end{bmatrix}$ **19.** Not possible

21. $\begin{bmatrix} 7 & 16 \\ 4 & 7 \\ 2 & 1 \end{bmatrix}$ **23.** $\begin{bmatrix} 29 & 16 \\ 42 & 18 \\ 0 & 0 \end{bmatrix}$

25. $\begin{bmatrix} 1 & 3 & 5 \\ 2 & 4 & 6 \\ 0 & 0 & 0 \end{bmatrix}$ **27.** Not possible

Answers to Odd-Numbered Problems

29. Not possible

31. $\begin{bmatrix} 34 & 24 & 29 \\ 3 & 9 & 16 \end{bmatrix}$

33. Not possible

35. Not possible

37. Not possible

39. $\begin{bmatrix} 42 & 3 \\ 6 & -1 \\ 41 & -6 \end{bmatrix}$

41. $\begin{bmatrix} 34 & 24 & 29 \\ 3 & 9 & 16 \end{bmatrix}$

43. No

45. Matrix multiplication *may* distribute over addition

47. $\begin{bmatrix} 2 & 3 \\ -1 & 4 \end{bmatrix} \cdot \begin{bmatrix} x \\ y \end{bmatrix} = \begin{bmatrix} 4 \\ -2 \end{bmatrix}$

49. $\begin{bmatrix} 150 & 50 & 200 \\ 300 & 100 & 100 \\ 80 & 150 & 500 \end{bmatrix}$

51. $\begin{bmatrix} 3950 & 1250 \\ 3400 & 1300 \\ 8870 & 2610 \end{bmatrix}$

35. $\begin{bmatrix} \frac{3}{2} & -\frac{1}{2} & 1 & -1 \\ \frac{3}{2} & -\frac{1}{2} & 0 & 0 \\ 1 & -\frac{1}{2} & \frac{1}{2} & 0 \\ -\frac{7}{2} & \frac{3}{2} & -1 & 1 \end{bmatrix}$

37. $\begin{bmatrix} 0 & 1 \\ -1 & 0 \end{bmatrix}$

39. $\begin{bmatrix} 0 & 0 & -1 \\ -\frac{1}{10} & -\frac{2}{5} & 0 \\ -\frac{1}{5} & \frac{1}{5} & 0 \end{bmatrix}$

41. No

43. No

45. $I, -I, \begin{bmatrix} 0 & 1 \\ 1 & 0 \end{bmatrix}, \begin{bmatrix} 0 & -1 \\ -1 & 0 \end{bmatrix}$

47. $\begin{bmatrix} \frac{5}{2} & -\frac{1}{2} \\ -4 & 2 \end{bmatrix}$

49. $\begin{bmatrix} -\frac{20}{13} & \frac{11}{13} & \frac{4}{13} \\ -\frac{33}{13} & \frac{11}{13} & \frac{17}{13} \\ -\frac{17}{13} & \frac{10}{13} & \frac{6}{13} \end{bmatrix}$

PROBLEM SET 12.4

1. $\begin{bmatrix} 1 & 0 & 0 & 0 \\ 0 & 1 & 0 & 0 \\ 0 & 0 & 1 & 0 \\ 0 & 0 & 0 & 1 \end{bmatrix}$

9. $\begin{bmatrix} 1 & 0 \\ 0 & 1 \end{bmatrix}$

11. $\begin{bmatrix} 1 & 0 & 0 \\ 0 & 1 & 0 \\ 0 & 0 & 1 \end{bmatrix}$

13. No

15. Yes

17. Yes

19. $\begin{bmatrix} \frac{1}{6} \end{bmatrix}$

21. $\begin{bmatrix} 1 & 0 \\ 0 & \frac{1}{4} \end{bmatrix}$

23. Not possible

25. $\begin{bmatrix} \frac{1}{5} & 0 \\ 0 & \frac{1}{5} \end{bmatrix}$

27. $\begin{bmatrix} -2 & 1 \\ \frac{3}{2} & -\frac{1}{2} \end{bmatrix}$

29. $\begin{bmatrix} 0 & 0 & -\frac{1}{2} \\ 0 & -\frac{1}{2} & 0 \\ -\frac{1}{2} & 0 & 0 \end{bmatrix}$

31. $\begin{bmatrix} \frac{3}{13} & -\frac{1}{13} & \frac{2}{13} \\ -\frac{15}{13} & \frac{5}{13} & \frac{3}{13} \\ \frac{10}{13} & \frac{1}{13} & -\frac{2}{13} \end{bmatrix}$

33. Not possible

PROBLEM SET 12.5

1. 12   **3.** 28   **5.** -2 **7.** -13
9. 7   **11.** 4   **13.** -11 **15.** -11
17. -1 **19.** 12   **21.** 1   **23.** 0
25. 0   **27.** 15   **29.** 0   **31.** 1
33. 0   **35.** 0   **37.** 15   **39.** 0

41. $\begin{bmatrix} 1 & 0 \\ -\frac{1}{2} & \frac{1}{4} \end{bmatrix}$

43. $\begin{bmatrix} \frac{1}{11} & \frac{2}{11} \\ \frac{4}{11} & -\frac{3}{11} \end{bmatrix}$

45. $\begin{bmatrix} \frac{1}{11} & \frac{4}{11} \\ \frac{2}{11} & -\frac{3}{11} \end{bmatrix}$

47. $\begin{bmatrix} 0 & 1 \\ 1 & 0 \end{bmatrix}$

49. $\begin{bmatrix} -\frac{1}{2} & \frac{2}{3} \\ -\frac{1}{4} & \frac{1}{6} \end{bmatrix}$

51. $\begin{bmatrix} 1 & 0 & 0 \\ 0 & 1 & 0 \\ 0 & 0 & 1 \end{bmatrix}$

53. No inverse

55. No inverse

57. $\begin{bmatrix} \frac{1}{3} & -\frac{1}{3} & \frac{2}{3} \\ -\frac{1}{15} & \frac{7}{15} & -\frac{11}{15} \\ -\frac{2}{15} & -\frac{1}{15} & \frac{8}{15} \end{bmatrix}$

59. No inverse

61. Yes

63. $D = (A * B) - (C * D)$

PROBLEM SET 12.6

1. $(2, 0)$ **3.** $(-1, 1)$
5. Infinite number of solutions
11. $(4, 3)$ **13.** $(\frac{8}{5}, -\frac{4}{5})$
15. $(2, 2)$ **17.** No solution
19. Infinite number of solutions

21. (3, 2, 0)   23. (1, 2, 3)
25. (−1, 0, 4)   27. No solution
29. Infinite number of solutions
31. (1, 0, 2, −3)   33. 6 and 2
35. 90¢ per can of orange juice, 60¢ per can of grapefruit juice
37. $\begin{bmatrix} 2 & 1 & -1 \\ 1 & -2 & 1 \\ 3 & -1 & 2 \end{bmatrix}$ 39. $\begin{bmatrix} 1 \\ 0 \\ 7 \end{bmatrix}$

41. $\begin{bmatrix} \frac{3}{10} & \frac{1}{10} & \frac{1}{10} \\ -\frac{1}{10} & -\frac{7}{10} & \frac{3}{10} \\ -\frac{1}{2} & -\frac{1}{2} & \frac{1}{2} \end{bmatrix}$

PROBLEM SET 12.7

7. (2, 3)   9. $(\frac{22}{7}, -\frac{5}{7})$ 11. (2, 2)
13. Infinite number of solutions
15. $(\frac{24}{41}, -\frac{43}{41})$ 17. (3, 2, −3)   19. (2, 6, −1)
21. (−4, 7, 2)   23. No solution   25. (−3, 2, 6)
27. (0, 0, 1, 0)   29. 18 by 24
31. 20 22¢ stamps, 13 14¢ stamps

REVIEW PROBLEMS, CHAPTER 12

1. 3   3. 3   5. 0
7. [3 3 0 5]   9. $\begin{bmatrix} 6 & 3 & 4 \\ 3 & 0 & 5 \\ 3 & 11 & -3 \end{bmatrix}$

11. $\begin{bmatrix} -3 & 2 \\ 2 & -4 \end{bmatrix}$ 13. $\begin{bmatrix} 8 & 12 & 20 \\ 4 & -4 & 8 \\ 16 & 24 & -12 \end{bmatrix}$

15. $\begin{bmatrix} 1 & 3 \\ -2 & 7 \end{bmatrix}$ 17. $\begin{bmatrix} 4 & 6 & 23 \\ 17 & 23 & 3 \\ 3 & -8 & 5 \end{bmatrix}$

19. $\begin{bmatrix} -10 & 8 \\ 3 & -19 \end{bmatrix}$ 21. $\begin{bmatrix} \frac{7}{13} & \frac{2}{13} \\ -\frac{3}{13} & \frac{1}{13} \end{bmatrix}$

23. $\begin{bmatrix} -\frac{9}{65} & \frac{3}{5} & \frac{11}{65} \\ \frac{11}{65} & -\frac{2}{5} & \frac{1}{65} \\ \frac{2}{13} & 0 & -\frac{1}{13} \end{bmatrix}$ 25. 13

27. 65   29. (2, 0)
31. No solution   33. (2, −1)
35. (−2, 3, −3)   37. $\frac{3}{2}$ and $\frac{29}{2}$
39. (a) X = 1
    (b) DO-WHILE (X ≤ M)
        (i) Y = 1
        (ii) DO-WHILE (Y ≤ N)
            (a) S(X,Y) = A(X,Y) + B(X,Y)
            (b) Y = Y + 1
        (iii) END DO-WHILE
        (iv) X = X + 1
    (c) END DO-WHILE

PROBLEM SET 13.1

1. Yes   3. No   5. No
7. Yes   9. No
11. Minimize: $z = x_1 + x_2$
    Constraints: $x_1 \geq 0, x_2 \geq 0, 2x_1 + 3x_2 \leq 12$
13. Minimize: $z = x_1 - 2x_2$
    Constraints: $x_1 \leq 0, \ x_2 \leq 0, \ x_1 \geq -10, \ x_2 \geq -10, x_1 + x_2 \geq -15$
15. Maximize: $z = 5x_1 + 7x_2$
    Constraints: $x_1 \geq 0, x_2 \geq 0, x_1 \leq 200, x_2 \leq 100, x_1 + x_2 \leq 150$
17. Maximize: $z = 100x_1 + 150x_2$
    Constraints: $x_1 \geq 0, \ x_2 \geq 0, \ x_1 \leq 7, \ x_2 \leq 6, \ x_1 + x_2 \leq 9, 40x_1 + 50x_2 \leq 400$
19. Minimize: $z = 1000x_1 + 500x_2 + 400x_3$
    Constraints: $x_1 \geq 0, \ x_2 \geq 0, \ x_3 \geq 0, \ x_1 + x_2 + x_3 \leq 12, \ 100x_1 + 100x_2 \geq 1000, 200x_1 + 50x_2 + 150x_3 \geq 1000, 300x_1 + 100x_3 \geq 1000$

PROBLEM SET 13.2

1.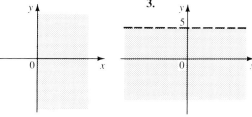

3.

5.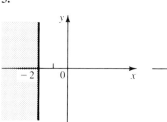

7.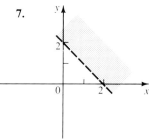

Answers to Odd-Numbered Problems

9.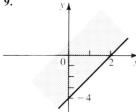
11.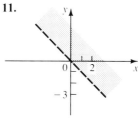
25.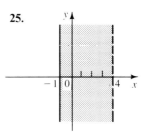
27.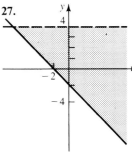

13.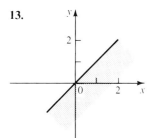
15.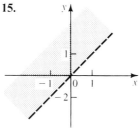
29.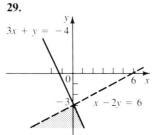
31.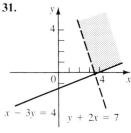

17.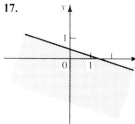
19.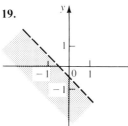

33. $-x + y \leq 1$ **35.** $x < y$
37. $y \leq 2x + 4$ **39.** $x < 3$
41. $x < 3$ and $3x + 2y \leq 6$
43. $x > y$ (answers may vary); $x < y$

PROBLEM SET 13.3

3. $z = 60, x_1 = 3, x_2 = 3$
5. $z = 20, x_1 = 0, x_2 = 4$
7. $z = -20, x_1 = -1, x_2 = 0$
9. $z = 17, x_1 = 2, x_2 = 1$
11. $z = 20, x_1 = 2, x_2 = 0$
13. $z = 32, x_1 = 8, x_2 = 0$
15. $z = \frac{200}{3}, x_1 = \frac{4}{3}, x_2 = \frac{4}{3}$
17. $z = \frac{950}{7}, x_1 = \frac{10}{7}, x_2 = \frac{10}{7}$
19. $z = 4, x_1 = 1, x_2 = 0$
21. $z = 320, x_1 = 0, x_2 = 4$
23. $z = -10, x_1 = -10, x_2 = 0$
25. $z = 950, x_1 = 50, x_2 = 100$
27. $z = 1100, x_1 = 5, x_2 = 1$
29. Minimize: $z = -x_1 - x_2$ (answers may vary)
Constraints: $x_1 \geq 0, x_2 \geq 0, 2x_1 + x_2 \leq 4$
Optimal solution: $z = -4, x_1 = 0, x_2 = 4$

21.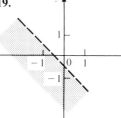
23.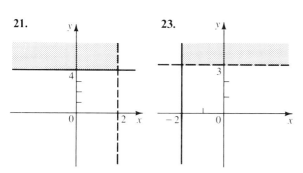

PROBLEM SET 13.4

1. Slack variables are added to the smaller side of the inequality to attain equality, so they can't be negative

5.

	x_1	x_2	x_3	B
x_3	1	3	1	6
	-6	-2	0	0

7.

	x_1	x_2	x_3	B
x_3	2	1	1	2
	-7	-3	0	0

9.

	x_1	x_2	x_3	x_4	B
x_3	2	5	1	0	11
x_4	1	2	0	1	5
	-10	-13	0	0	0

11.

	x_1	x_2	x_3	x_4	B
x_3	1	4	1	0	10
x_4	3	3	0	1	7
	-3	-4	0	0	0

13.

	x_1	x_2	x_3	x_4	x_5	x_6	B
x_4	1	2	0	1	0	0	4
x_5	0	1	4	0	1	0	6
x_6	3	-1	-1	0	0	1	3
	-5	3	-1	0	0	0	0

15.

	x_1	x_2	x_3	B
x_3	2	3	1	12
	-1	-1	0	0

17.

	x_1	x_2	x_3	x_4	x_5	B
x_3	1	0	1	0	0	200
x_4	0	1	0	1	0	100
x_5	1	1	0	0	1	150
	-5	-7	0	0	0	0

19. $z = 36, x_1 = 6, x_2 = 0$
21. $z = 7, x_1 = 1, x_2 = 0$
23. $z = 50, x_1 = 5, x_2 = 0$
25. $z = \frac{28}{3}, x_1 = 0, x_2 = \frac{7}{3}$
27. $z = 5, x_1 = 1, x_2 = 0, x_3 = 0$
29. $z = 6, x_1 = 6, x_2 = 0$
31. $z = 950, x_1 = 50, x_2 = 100$

Answers to Odd-Numbered Problems

PROBLEM SET 13.5

1. $x_1 + x_2 + x_3 = 4$; x_3: slack variable
3. $2x_1 - 7x_2 + x_3 = 2$; x_3: artificial variable
5. $-3x_1 - x_2 + 3x_3 + x_4 = 6$; x_4: slack variable
7. $10x_1 - 3x_2 + 2x_3 + x_4 - x_5 = 4$; x_4: artificial variable; x_5: surplus variable
9. $-8x_1 + 3x_2 - x_3 + x_4 = 4$; x_4: artificial variable
11. $-2x_1 + 9x_2 + x_3 - x_4 = 11$; x_3: artificial variable; x_4: surplus variable

13.

	x_1	x_2	x_3	x_4	B
x_3	2	-1	1	0	3
x_4	1	2	0	1	2
	-2	-6	0	0	0

15.

		x_1	x_2	x_3	x_4	B
x_3	0	1	1	1	0	4
x_4	$-M$	2	1	0	1	8
		$-4-2M$	$2-M$	0	0	$-8M$

17.

		x_1	x_2	x_3	x_4	B
x_3	M	1	1	1	0	5
x_4	0	-2	1	0	1	3
		$1-M$	$3-M$	0	0	$-5M$

19.

		x_1	x_2	x_3	x_4	x_5	B
x_4	$-M$	1	11	6	1	0	5
x_5	0	1	5	-1	0	1	4
		$-7-M$	$-9-11M$	$-1-6M$	0	0	$-5M$

21.

		x_4	x_5	x_6	x_7	x_8	x_9	x_{10}	x_{11}	x_{12}	B
x_{10}	$-M$	-3	3	-7	7	-4	4	1	0	0	3
x_{12}	$-M$	-1	1	-1	1	-3	3	0	-1	1	5
		-4 $+4M$	-4 $-4M$	-12 $+8M$	-12 $-8M$	-1 $+7M$	-1 $-7M$	0	M	0	$-8M$

23.

		x_3	x_4	x_5	x_6	x_7	x_8	x_9	x_{10}	x_{11}	B
x_8	M	-3	3	7	-7	-1	1	0	0	0	3
x_{10}	M	2	-2	5	-5	0	0	-1	1	0	8
x_{11}	M	1	-1	-3	3	0	0	0	0	1	11
		14	-14	23 $-9M$	-23 $+9M$	M	0	M	0	0	$-22M$

25. $z = 6$, $x_1 = 0$, $x_2 = 1$
27. $z = 16$, $x_1 = 4$, $x_2 = 0$

29. Artificial variables are added to *equations*. So that equality is preserved, they must be zero.

31. Yes; $x_1 + x_2 \geq 4$, $x_1 + x_2 - x_3 + x_4 = 4$; surplus variable: x_3; artificial variable: x_4 (answers may vary)

33. No

REVIEW PROBLEMS, CHAPTER 13

1. **3.**

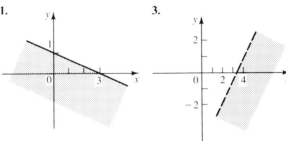

5. **7.**

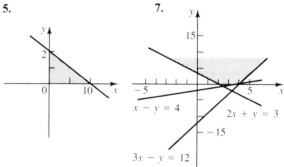

9. $z = 21, x_1 = 1, x_2 = 10$
11. $z = -5, x_1 = 0, x_2 = 1$
13. $z = 4, x_1 = 0, x_2 = 4$
15. No maximum
17. $z = 150, x_1 = 10, x_2 = 10$
19. $z = 0, x_1 = 0, x_2 = 0, x_3 = 0$

21.

		x_1	x_2	x_3	x_4	x_5	x_6	x_7	x_8	x_9	x_{10}	B
x_6	M	20	0	50	30	-1	1	0	0	0	0	200
x_8	M	15	30	60	0	0	0	-1	1	0	0	700
x_{10}	M	5	10	15	0	0	0	0	0	-1	1	100
		2000	1500	4000	1000	M	0	M	0	M	0	$-1000M$
		$-40M$	$-40M$	$-125M$	$-30M$							

Answers to Odd-Numbered Problems

PROBLEM SET 14.1

1. 720
3. 3,628,800
5. 840
7. 1680
9. 28
11. 7
13. 5!
15. $\dfrac{7!}{3!}$
17. $\dfrac{n!}{(n-3)!}$
19. 42
21. 2520
23. 49
25. $7^5 = 16,807$
27. 175,760
29. 84
31. 210
33. 50!
35. 1
37. 625
39.

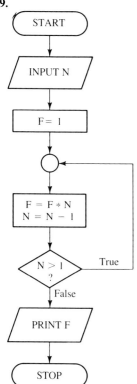

41. 5

PROBLEM SET 14.2

1. $\{5, 6, 7\}$
3. {winter, spring, summer, fall}
5. {AEI, AEO, AEU, EIO, EIU, IOU}
7. $\{2, 3, 4, 5, 6, 7, 8, 9, 10, 11, 12\}$
9. $\{x \mid x$ is a card in a normal deck of 52 cards$\}$
11. *George* and Sam; *George* and Janet; *George* and Carol; Sam and Janet; Sam and Carol; Janet and Carol
13. 4545, 6570, 1649, 7192
15. 0.25
17. 0.50
19. 0.25
21. $\frac{1}{36}$
23. $\frac{1}{6}$
25. $\frac{25}{36}$
27. $\frac{5}{21}$
29. $\frac{16}{21}$
31. $\frac{13}{21}$
33. 0
35. 0.5625
37. 0.25
39. 0.5
41. 0.3
43. 0.2
45. 0.2
47. $\frac{1}{3}$
49. $\frac{1}{3}$
51. 0.5125
53. 0.52
55. (a) T = 0
    (b) H = 0
    (c) C = 1
    (d) DO-WHILE (C ≤ 500)
        (i) R = INT (RND * 2)
        (ii) IF R = 1 THEN H = H + 1
             ELSE T = T + 1
        (iii) C = C + 1
    (e) END DO-WHILE
    (f) PRINT H, T
57. (a) N = 0
    (b) F = 0
    (c) REPEAT
        (i) D = INT (RND * 6) + 1
        (ii) E = INT (RND * 6) + 1
        (iii) S = D + E
        (iv) IF S = 7 THEN F = 1
        (v) N = N + 1
    (d) UNTIL (F = 1)
    (e) PRINT N

PROBLEM SET 14.3

1. $\sum\limits_{i=1}^{10} X_i$
3. $\sum\limits_{i=1}^{15} (10 + 2i)$
5. $\sum\limits_{i=1}^{N} (2X_i - 3)$
7. $(\sum\limits_{i=3}^{N} X_i)^2$
9. $\sum\limits_{i=1}^{N} (3X_i^2 - 2X_i + 6)$
11. 15
13. 84
15. 24
17. 14
19. 2106
21. (a) I = 1
    (b) S = 0
    (c) DO-WHILE (I ≤ 100)
        (i) S = S + (4I–2)

(ii) $I = I + 1$
(d) END DO-WHILE
(e) PRINT S
23. (a) $I = 50$
(b) $S = 0$
(c) DO-WHILE ($I \leq 250$)
   (i) $S = S + [(2I - 11)^2]$
   (ii) $I = I + 1$
(d) END DO-WHILE
(e) PRINT S

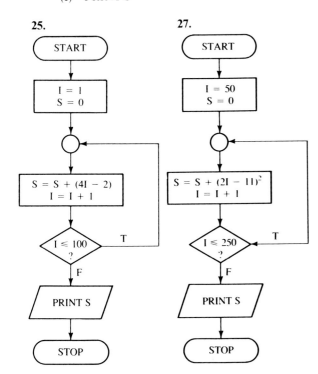

29. (a) $I = 1$
(b) $S = 0$
(c) DO-WHILE ($I \leq 30$)
   (i) INPUT T
   (ii) $S = S + T$
   (iii) $I = I + 1$
(d) END DO-WHILE
(e) $A = S^2$
(f) PRINT A
31. (a) $N = 2$
(b) $S = 0$
(c) DO-WHILE ($N \leq 100$)
   (i) $S = S + N$
   (ii) $N = N + 2$
(d) END DO-WHILE
(e) PRINT S
33. 25
35. 345
37. (a) INPUT N
(b) $S = 0$
(c) $C = 1$
(d) DO-WHILE ($C \leq N$)
   (i) $S = S + C$
   (ii) $C = C + 1$
(e) END DO-WHILE
(f) PRINT S

PROBLEM SET 14.4

1. ≈ 39.2 3. 76   5. 42   7. 76
9. 25.5   11. 76   13. 10   15. 76
17. 6 different modes   19. 76
21. 0 and 93   23. 76
25. 90
27. 65, 65, 65, 70, 70, 75, 80
29. 40, 60, 60, 80, 82, 83, 85
31. 0, 0, 0, 0, 100, 300, 300
33. 67, 68, 69, 76, 78, 78, 173
35. 20, 25, 50, 60, 65, 100, 100
37. Median   39. Mean   41. Mode
43. 0, 0, 100, 200, 300, 400, 400; mode = 0 and 400; mean = 200; median = 200 (answers may vary)
45. 10, 10, 10, 20, 20, 20; mode = 10 and 20; median = 15
47. 4054.4   49. 58.4
51. (a) $I = 1$
(b) $T = 0$
(c) DO-WHILE ($I \leq N$)
   (i) $S = A(I)^2$
   (ii) $T = T + S$
   (iii) $I = I + 1$
(d) END NO-WHILE
(e) $M = T/N$
53. No   55. $\dfrac{\left(\sum_{i=1}^{N} X_i\right)}{N^2}$

PROBLEM SET 14.5

1. 68   3. 0   5. 93

7.	54	9.	≈ 761	11.	0	
13.	≈ 1850	15.	329.5	17.	≈ 27.6	
19.	0	21.	≈ 43	23.	≈ 18	
25.	≈ 76	27.	≈ 103	29.	1	

31. Any score between 57 and 93
33. When all values are identical
35. ≈ -1.4 37. 15
39. $\sigma = \sqrt{\dfrac{\sum_{i=1}^{N}(X_i - \mu)^2}{N}}$ 41. ≈ 761
43. Wilma

REVIEW PROBLEMS, CHAPTER 14

1.	40,320	3.	36	5.	3024
7.	59,049	9.	6760	11.	0.375
13.	0.24	15.	0.5		

17. (a) C = 1
    (b) T = 0
    (c) DO-WHILE (C $\leq$ 600)
        (i) A = INT (RND * 6) + 1
        (ii) B = INT (RND * 6) + 1
        (iii) IF A = B THEN T = T + 1
        (iv) C = C + 1
    (d) END DO-WHILE
    (e) PRINT T

19. $\sum_{i=7}^{56}(3x_i - 7)$ 21. 115

23. Mean ≈ 61; median = 69.5; mode = 6 different numbers
25. $31,421
27. 40, 40, 50, 65, 70, 75, 75, 75, 95 (answers may vary)
29. range: 86; variance: 861.5; standard deviation: 29.4
31. 861.5   33. -5
35. All numbers are the same

INDEX

Accuracy, 175
Algorithm, 274
Alphanumeric expression, 194
APL, 221
Arguments, 237–240
Arithmetic IF, 282
Arrays:
 dimensions of, 300
 equal, 300
 one-dimensional, 298
 three-dimensional, 300
 two-dimensional, 299
ASCII code, 194–196
Assignment statement, 22, 43, 191–192, 196
Associative property:
 of Boolean algebra, 258, 268
 of logic, 234, 242
 of real numbers, 6, 31
 of sets, 213–214, 217
Axis of symmetry, 100, 116

BASIC, 9, 12, 29, 42, 105, 110, 221, 275–276, 373
BCD code, 186–187
Binary number system, 122–124, 142
 addition in, 146–148
 Bit, 122
 Byte, 122
 converting from decimal, 131–136
 converting from hexadecimal, 140–141
 converting from octal, 140–141
 converting to decimal, 128–130
 converting to hexadecimal, 137–139
 converting to octal, 137–139
 division in, 167–168
 expanded form, 123
 face value, 122
 multiplication in, 164–165
 place value, 122
 subtraction in, 158–160
Boole, George, 246
Boolean algebra, 246–269
 operations in, 250–251
 properties of, 258–260
Built-in function, 42

Cartesian coordinate system, 38
Closure property:
 of real numbers, 7, 31
 of sets, 214, 217
COBOL, 221
Coefficient, 17–18
Cofactor, 317
Combination, 366
Common logarithm. See logarithm
Commutative property:
 of Boolean algebra, 258, 268
 of logic, 234, 242
 of real numbers, 6, 31
 of sets, 213, 217
Complement addition, 154–163
Complement property:
 of Boolean algebra, 258–259, 269
 of sets, 216, 218
Completing the square, 92, 116
Compound decision, 283
Compound statement, 221
Conclusion, 237
Conjunction, 221
Connective, 221
Constant, 15
Contradiction, 229
Contrapositive, 230–231
Converse, 230–231
Conversion error, 185
Coordinate, 37
Counter, 287
Cramer's rule, 326–329
Cube, 10
Cube root, 25

Decimal number system, 120–121, 142
 addition in, 146
 converting from binary, 128–130
 converting from hexadecimal, 128–130
 converting from octal, 128–130
 converting to binary, 131–136
 converting to hexadecimal, 131–136
 converting to octal, 131–136
 expanded form, 121
 face value, 120
 place value, 120

 subtraction in, 154–157
Decision. See logical decision
Decision structure, 280–285, 295
DEF FN statement, 42–43
DeMorgan property:
 of Boolean algebra, 258, 269
 of logic, 235, 242
 of sets, 215, 217
Descartes, René, 38
Determinant, 316–320
DIM statement, 301
Discriminant, 96
Disjunction, 221
Distributive property:
 of Boolean algebra, 258, 269
 of logic, 235, 242
 of real numbers, 8, 31
 of sets, 215, 217
Domain, 37, 58
DO-WHILE, 287, 295
Double negation, 235, 242
DYNAMO, 9

e, 105
EBCDIC code, 194–196
Elementary event, 370
Equality, rules of, 21, 31, 69
equivalent statements, 231
Equation, 20
Event, 370
Exclusive or, 228
EXP, 105
Explicit formula, 112
Exponent, 10
Exponential function, 103, 116
Exponentiation, rules of, 10, 31
Expression, 15
Extraction of roots, 90

Factor, 26
Factorial, 365
Factoring, 90
FIX, 382
Flowcharts, 277
Flowchart symbols:
 calculation, 277
 connector, 280
 decision, 280–281
 input/output, 277

438

INDEX

terminal, 277
FORTRAN, 282
function, 40, 58
function notation, 41

Gaussian elimination, 321–322
Googol, 179
Googolplex, 179

Half-adder circuit, 267
Hexadecimal number system, 126–127, 142
   addition in, 151–152
   converting from binary, 137–139
   converting from decimal, 131–136
   converting from octal, 141
   converting to binary, 140
   converting to decimal, 128–130
   converting to octal, 141
   division in, 169
   expanded form, 127
   face value, 127
   multiplication in, 165–166
   place value, 127
   subtraction in, 161–163
hypothesis, 237

Idempotent property:
   of Boolean algebra, 258, 269
   of logic, 235, 242
   of sets, 216, 217
Identity property:
   of Boolean algebra, 258, 269
   of logic, 234–235, 242
   of matrices, 310
   of real numbers, 7, 31
   of sets, 214, 217
Index, 25, 299
Inequality, 22
Inequality, rules of, 23, 32
Input, 275
INT, 373
Integers, set of, 3, 31
Intersection. See sets, intersection of
Inverse, 230
Inverse property:
   of matrices, 311
   of real numbers, 7, 31
   of sets, 214
Inverter, 265

Linear equation, 44, 58

   horizontal, 50, 58
   parallel, 51, 58
   perpendicular, 51, 58
   point-slope form, 54, 58
   slope-intercept form, 55, 58
   standard form, 53, 58
   system of. See system of linear equations
   vertical, 50, 58
Linear inequality:
   solution of, 23–24
   graphing, 337–340
Linear programming, 333–360
   constraint, 334
   graphical solution of, 344–348
   objective function, 334
   optimal solution, 334
   simplex method, 349–359
LOG, 110
Logarithm, 107
   common, 109
   natural, 110
Logarithmic function, 107, 109, 116
Logic circuits:
   parallel, 250–251
   series, 251
Logic gates:
   AND, 264
   NOT, 265
   OR, 264–265
Logical decision, 24
Loop, 287, 295

Matrices, 299
   addition of, 302
   augmented, 322
   difference of, 303
   identity, 310
   inverse of, 310–311, 319
   multiplication of, 307–308
   of cofactors, 317–319
   row transformations of, 311
   scalar multiplication of, 304
   singular, 314
   square, 310
   transpose of, 304
Mean, 380, 390
Mean square, 384–385, 392
Measures of central tendency, 379–383
Measures of dispersion, 385–389
Median, 380–381, 390

Memory map, 141
Minor, 316
Mode, 382–383, 390

Natural logarithm. See logarithm
Natural numbers, set of, 2, 31
Negation, 221
Network, 258–263
Normalized notation, 181

Octal number system, 124–125, 142
   addition in, 150
   converting from binary, 137–139
   converting from decimal, 131–136
   converting from hexadecimal, 141
   converting to binary, 140
   converting to decimal, 128–130
   converting to hexadecimal, 141
   expanded form, 125
   face value, 125
   division in, 168
   multiplication in, 165
   place value, 125
   subtraction in, 161–163
Order of operations:
   in mathematics, 15
   in computers, 188–189
Ordered pair, 36
Output, 275
Overflow error, 185

Parabola, 99
Pascal, 110, 216, 221
Permutation, 364
Pi, 4
Pivot, 351
Polynomial, 16
Precision, 175
Probability, 364, 369–373
Pseudocode, 281

Quadrant, 38
Quadratic equation, 90
Quadratic formula, 95, 116
Quadratic function, 98, 116

Radical, 25
Radical expressions, rules for, 27, 32
Radical sign, 25
Radicand, 25
Random function, 372–373

Random numbers, 372–373
Range:
  of relation, 37, 58
  of set of numbers, 385, 390
Rational numbers, set of, 3–4, 31
Real number line, 4–5
Real numbers, set of, 4, 31
Reciprocal, 7, 12
Recursive formula, 112
Relation, 37, 58
REPEAT-UNTIL, 287–288, 295
Repetition structure, 287–292, 295
RND, 373
Rounding, 185

Sample space, 370
Scalar, 304
Scientific notation, 176–179
Sequence, infinite, 111, 116
  alternating, 112
  arithmetic, 113, 116
  Fibonacci, 112
  geometric, 113, 116
Sequence structure, 278, 295
Set-builder notation, 3–4, 200
Sets:
  complement of, 207
  difference of, 209
  disjoint, 207
  empty, 203
  equal, 201
  equivalent, 201–202
  intersection of, 207
  proper subset, 204
  properties of, 213–216
  subset, 203–205
  union of, 206
  universal, 205
  Venn diagrams, 209–212, 238–240
Significant digits, 174–175
Similar terms, 18
SIMSCRIPT, 392
Slope, 49, 58
Solution set, 37
Spreadsheet, 298
SQR, 29, 42
Square, 10
Square root, 4, 25
Standard deviation, 387–388, 391
Statement, 220
Statistics, 364, 379–389
Subscript, 298–299
Summation notation, 375–378
System of linear equations, 62
  consistent system, 74
  independent system, 76
  solution by Cramer's rule, 326–329
  solution by graphing, 62–64, 86
  solution by matrices, 321–325
  solution by multiplication–addition, 69–72, 86
  solution by substitution, 66–68, 86

Tautology, 229
Truncation, 184, 373, 382
Truth tables:
  biconditional, 231
  conditional, 230
  conjunction, 224
  disjunction, 225
  exclusive or, 228
  NAND, 227
  negation, 225
  NOR, 228

Union. *See* sets, union of
User-defined function, 42–43

Variable, 15
  artificial, 356
  dependent, 41
  independent, 41
  slack, 350
  surplus, 356
Variance, 386, 390
Vector, 298, 344
Venn diagrams, 209–212, 238–240
Venn, John, 209
Vertex, 100, 116, 344
Voyager 2, 124

x axis, 38
x intercept, 47

y axis, 38
y intercept, 47